T0235090

Communications
in Computer and Information Science   786

Commenced Publication in 2007
Founding and Former Series Editors:
Alfredo Cuzzocrea, Xiaoyong Du, Orhun Kara, Ting Liu, Dominik Ślęzak,
and Xiaokang Yang

More information about this series at http://www.springer.com/series/7899

Przemysław Różewski · Christoph Lange (Eds.)

Knowledge Engineering and Semantic Web

8th International Conference, KESW 2017
Szczecin, Poland, November 8–10, 2017
Proceedings

 Springer

Editors

Przemysław Różewski
West Pomeranian University of Technology
  in Szczecin
Szczecin
Poland

Christoph Lange ⓘ
University of Bonn
Bonn
Germany

ISSN 1865-0929   ISSN 1865-0937 (electronic)
Communications in Computer and Information Science
ISBN 978-3-319-69547-1   ISBN 978-3-319-69548-8   (eBook)
https://doi.org/10.1007/978-3-319-69548-8

Library of Congress Control Number: 2017956727

Printed on acid-free paper

This Springer imprint is published by Springer Nature
The registered company is Springer International Publishing AG
The registered company address is: Gewerbestrasse 11, 6330 Cham, Switzerland

Preface

These proceedings contain the papers accepted for oral presentation at the 8th International Conference on Knowledge Engineering and Semantic Web (KESW 2017). The conference was held in Szczecin, Poland, during November 8–10, 2017.

The principal mission of the KESW conference series is to provide a discussion forum for the community of researchers currently underrepresented at the major International Semantic Web Conference (ISWC) and Extended/European Semantic Web Conference (ESWC). This mostly includes researchers from Eastern and Northern Europe, Russia, and former Soviet republics. This year, the conference was held in Szczecin to catalyze discussions between the traditional KESW community and the European research community.

As in previous years, KESW 2017 aimed at helping the community to get used to the common international standards for academic conferences in computer science. To this end, KESW features a peer-reviewing process in which every paper was reviewed in a rigorous but constructive way by three members of the Program Committee (PC), supported by EasyChair. As before, the PC was international, representing countries ranging from Russia to Germany and the USA.

We received a total of 58 submissions. The strict reviewing policies resulted in the acceptance of 16 full research papers. This translates into an acceptance rate of 28 %. An additional eight papers (14 %) were accepted for short presentation and have also been given space in these proceedings. The authors represent mainly various parts of Russia, as well as EU countries, including Poland and Germany.

KESW 2017 continued the tradition of inviting established researchers for keynote presentations. We are grateful to the keynote speakers for their insightful talks. The program also included posters and position paper presentations to help attendees, especially younger researchers, to discuss preliminary ideas and promising PhD topics.

We thank Dmitry Mouromtsev and Pavel Klinov, who helped us immensely during the conference preparation. Next, we would like to thank the organizing institutions, the West Pomeranian University of Technology in Szczecin and ITMO University, for their support, in particular the local organization team composed of Magdalena Kieruzel, Wojciech Sałabun, and Tomasz Lipczynski. We would also like to express our thanks to this year's sponsors and partners, namely, Polskie Stowarzyszenie Zarządzania Wiedzą and MDPI – without their support the event would not have been possible. We would also like to thank the hard-working PC as well as our publicity chair, Maxim Kolchin, for their reliable and quick work.

September 2017
Przemysław Różewski
Christoph Lange

Organization

Organizing Committee

General Chair

Przemysław Różewski   West Pomeranian University of Technology, Poland

Program Chair

Christoph Lange   University of Bonn and Fraunhofer IAIS, Germany

Publicity Chair

Maxim Kolchin   NRU ITMO, Russia

Program Committee

Alessandro Adamou   Knowledge Media Institute, The Open University, UK
Sören Auer   TIB Leibniz Information Center Science
                          and Technology and Leibniz Universität Hannover,
                          Germany
Ina Blümel   TIB Leibniz Information Center Science, Germany
Long Cheng   Eindhoven University of Technology, The Netherlands
Jeremy Debattista   ADAPT Centre, School of Computer Science
                          and Statistics, Trinity College, Dublin, Ireland
Elena Demidova   L3S Research Center, Germany
Ivan Ermilov   Universität Leipzig, Germany
Ralph Ewerth   Leibniz Universität Hannover, Germany
André Freitas   University of Passau, Germany
Irini Fundulaki   ICS-FORTH, Greece
Kleanthi Georgala   University of Leipzig, Germany
Peter Haase   metaphacts, Germany
Lavdim Halilaj   University of Bonn and Fraunhofer IAIS, Germany
Ali Hasnain   Insight Centre for Data Analytics, Ireland
Konrad Höffner   University of Leipzig, Germany
Dmitry Ignatov   National Research University Higher School
                          of Economics, Russia
Vladimir Ivanov   Innopolis University, Russia
Ernesto Jimenez-Ruiz   University of Oslo, Norway
Natalya Keberle   Zaporizhzhya National University, Ukraine
Evgeny Kharlamov   University of Oxford, UK
Alexander Kirillovich   Kazan Federal University, Russia
Pavel Klinov   Stardog Union, Germany
Jakub Klímek   Charles University, Czech Republic

Andrea Kohlhase	University of Applied Sciences Neu-Ulm, Germany
Boris Konev	University of Liverpool, UK
Roman Kontchakov	Birkbeck, University of London, UK
Liubov Kovriguina	NRU ITMO, Russia
Petr Kremen	Czech Technical University in Prague, Czech Republic
Dmitry Kudryavtsev	St. Petersburg State Polytechnical University, Russia
Oliver Kutz	KRDB Research Centre for Knowledge and Data, Free University of Bozen-Bolzano, Italy
Martin Ledvinka	Czech Technical University in Prague, Czech Republic
Steffen Lohmann	Fraunhofer IAIS, Germany
Till Mossakowski	University of Magdeburg, Germany
Dmitry Mouromtsev	NRU ITMO, Russia
Adam Naumowicz	Institute of Informatics, University of Bialystok, Poland
Axel-Cyrille Ngonga Ngomo	University of Paderborn, Germany
Rafael Peñaloza	Free University of Bozen-Bolzano, Italy
Denis Ponomaryov	A.P. Ershov Institute of Informatics Systems, Novosibirsk State University, Russia
Julia Rayz	Purdue University, USA
Mariano Rodríguez Muro	IBM Research, USA
Yuliya Rubtsova	A.P. Ershov Institute of Informatics Systems, Russia
Wojciech Salabun	West Pomeranian University of Technology, Poland
Muhammad Saleem	AKSW, University of Leipzig, Germany
Tzanina Saveta	ICS-FORTH, Greece
Marvin Schiller	Ulm University, Germany
Kuldeep Singh	Fraunhofer IAIS, Germany
Wojtek Sylwestrzak	ICM University of Warsaw, Poland
Ioan Toma	STI Innsbruck, Austria
Jörg Unbehauen	University of Leipzig, Germany
Josef Urban	Czech Technical University in Prague, Czech Republic
Dmitry Ustalov	Ural Federal University/Lappeenranta University of Technology, Finland
Sahar Vahdati	University of Bonn, Germany
Jarosław Wątróbski	West Pomeranian University of Technology, Poland
Oleg Zaikin	Warsaw School of Computer Science, Poland
Amrapali Zaveri	Maastricht University, The Netherlands

Additional Reviewers

Elias, Mirette	Hedblom, Maria
Galkin, Mikhail	Hoppe, Anett
Galliani, Pietro	Khiat, Abderrahmane
Glauer, Martin	Maldonado, Alfredo
Gottschalk, Simon	Moodley, Kody
Grangel, Irlan	Neuhaus, Fabian

Otto, Christian
Pandit, Harshvardhan Jitendra
Pei, Yulong
Petersen, Niklas
Rouleaux, Nadine

Saeeda, Lama
Tempelmeier, Nicolas
Vyskocil, Jiri
Yang, Zhengyu

Organizers

West Pomeranian University of Technology

http://www.zut.edu.pl/

ITMO University

http://www.ifmo.ru/

Sponsors

Silver Sponsors

AKAENE

http://www.akaene.com/

VISmart

http://www.vismart.biz/

SlideWiki.eu

http://www.slidewiki.eu/

Journal Special Issue Sponsor

MDPI

http://www.mdpi.com/

Partner

Polskie Stowarzyszenie Zarządzania Wiedzą

http://www.pszw.edu.pl/

Contents

Ontologies and Controlled Vocabularies

Scalable Data Access and Storage Solutions

Semantic Web and Education

Linked Data

Semantic Technologies in Manufacturing and Business

Natural Language Processing

Reducing the Degradation of Sentiment Analysis for Text Collections Spread over a Period of Time

Yuliya Rubtsova[(✉)]

A.P. Ershov Institute of Informatics Systems, Novosibirsk State Universiry,
Novosibirsk, Russia
yu.rubtsova@gmail.com

Abstract. This paper presents approaches to improve sentiment classification in dynamically updated text collections in natural language. As
social networks are constantly updated by users there is essential to take
into account new jargons, vital discussed topics while solving classification task. Therefore two fundamentally different methods for solution
this problem are suggested. Supervised machine learning method and
unsupervised machine learning method are used for sentiment analysis.
The methods are compared and it is shown which method is most applicable in certain cases. Experiments comparing the methods on sufficiently
representative text collections are described.

Keywords: Natural language processing · Sentiment analysis · Sentiment classification · Machine learning

1 Introduction

Automatic sentiment classification is a rather topical subject. The huge amount
of information contained in the network is represented as text in a natural language. It is requires computational linguistics methods to proceed this information. Over the past ten years, a lot of scientists and researchers worldwide were
involved in the task of automatically extracting and analyzing the comments
and opinions of social media. Moreover as one of the main tasks was considered
the problem of sentiment classification of texts in natural language.

The first works were dedicated to sentiment classification of a whole document [1,2]. Classification at the level of short phrases, rather than entire documents or paragraphs, has been carried out by Wilson, Wiebe and Hoffmann [3].
In their paper, the authors showed that it is important to determine the sentiment (positive or negative) of a single sentence, not the whole text in its entirety.
In a long document, the author's opinion about a matter can change from positive to negative and vice versa. In addition, the author may speak negatively
about minor shortcomings, but overall retain a positive attitude regarding the
subject described in the text. In other words, a long document or review cannot
always be clearly classified as positive or negative in sentiment.

© Springer International Publishing AG 2017
P. Różewski and C. Lange (Eds.): KESW 2017, CCIS 786, pp. 3–13, 2017.
https://doi.org/10.1007/978-3-319-69548-8_1

Microblog posts do not exceed 140 characters, which allows us to consider that their classification takes place at the phrase or sentence level. Despite the fact that microblogging is quite a young phenomenon, researchers are actively involved in analysing the sentiment of blog posts, in general, and tweets, in particular [4–7]. Microblog posts are short enough to describe all the different aspects of a product or service and, at the same time, are full of opinions and emotional assessments, so short-text sentiment classification is dealt with not only on the phrase and sentence level, but also relative to the stated subject [8,9].

One of the challenges in developing and using a sentiment analysis system is that their performance constantly deteriorates over time. This occurs mainly due to the fact that the vocabulary used in texts changes. Also the focus of the discussion shifted from one topic to another. This paper proposes approaches to solving this problem.

The study is organised as follows: the second chapter identifies and sub-stantiates the problem of quality degradation in the sentiment classification of text collections identical in composition and characteristics, but staggered over time. In this regard, the collections on which experiments were conducted are described, as are the measures for assessing the quality of results. The results of experiments regarding the classifier's performance for text collections collected 6–18 months apart are given. The third chapter proposes approaches to solving this problem. The final section consists of conclusion.

2   Reduce in Quality of Sentiment Classification Due to Changes in Emotional Vocabulary

Users of social networks are among the first to use new terms in everyday life. For instance the 40 new words added to the Oxford Dictionary in 2013 included terms from social networks, such as "srsly" and "selfie". The active vocabulary is constantly updated, therefore automatic classifiers must take this into account in their models. When it comes to machine learning, the training collection of texts must be expanded. In the context of rules and dictionaries, it is necessary to take into account the slang that social networks are saturated with in order to improve the quality of classifiers.

2.1   Short Text Collections

Operations and experiments with automatic text classification show that the results of classification generally depend on the training text set and the subject are that the training collection corresponds to. Today, many projects centre on feature engineering and the involvement of additional data, such as external text collections (that do not overlap with the training collection) or sentiment lexicons. Additional information can reduce the reliance on the training collection and improve classification results.

In order to successfully classify texts by sentiment, it is necessary to have tagged by sentiment text collections. Moreover, in order to improve sentiment classification in dynamically updated collections, it is necessary to have several text collections, compiled in different periods of time.

First of all we prepare three text collections which are uniform on structure but collected in a different time. The first corpus was collected between December 2013 and February 2014. For the sake of brevity, we shall call it the 2013 collection. Using the distinct supervision method firstly proposed by [10] and filtration [11], a training collection was formed from the 2013 texts.

Next, it is necessary to collect and prepare test text collections. The second corpus, which consists of about 10 million short texts, was collected in July–August 2014. The third corpus, consisting of about 20 million texts, was compiled in July and November 2015.

Two test collections were formed from the 2014 and 2015 texts. The distribution of texts in the collections by sentiment class is presented in Table 1. All three collections are domain-independent, i.e. they do not belong to any pre-defined subject area.

Table 1. Distribution of texts in the collections by sentiment class

	POSITIVE MESSAGES	NEGATIVE MESSAGES	NEUTRAL MESSAGES
2013	114 911	111 922	107 990
2014	5 000	5 000	4 293
2015	10 000	10 000	9 595

The compiled text collections formed the basis for training and test collections of Twitter posts used to assess the sentiments of tweets towards a given subject at classifier competition at SentiRuEval [9,12,13] in 2015 and 2016.

It was shown that the collections are complete and sufficiently representative [4].

2.2 Classifier Quality Measures

The quality of sentiment classification system is done by comparing the results obtained from the automatic classification system and reference marked results. Based on the difference in the values of the reference collections and collections automatically marked by the evaluates algorithm the following commonly accepted metrics are used: accuracy, formula 1; precision, formula 2; recall, formula 3; and F-measure formula 4 [14].

$$Accuracy = \frac{(TP + TN)}{(TP + FP + TN + FN)} \tag{1}$$

$$Precision = \frac{TP}{(TP + FP)} \tag{2}$$

$$Recall = \frac{TP}{(TP + FN)} \tag{3}$$

$$F - measure = 2 \times \frac{Precision Recall}{Precision + Recall} \tag{4}$$

Where TP is true positive decision, the number of texts correctly assigned to the class P by authomatic classificator; FP - false positive decision, the number of texts that are not correctly assigned to the class P; FN - false negative decision, the number of texts that are not correctly assigned to the Class N; TN - true negative decision, the number of texts correctly assigned to the class N.

2.3 The Problem of Reducing Quality in Sentiment Classification Due to Changes in Emotional Texts

To simulate a real situation, when language or the topics discussed on social media may change over time, a second and third collection of short texts were prepared. The first and second collections were compiled about six months apart, the first and third – a year and a half. At first glance, it would seem that vocabulary cannot change so quickly, however the topics of tweets, which affect the overall mood in general and reputation in particular, are significantly dependent on positive or negative events that occur involving the target subject; usually, such events cannot be predicted in advance. For example, in January and February 2014, about 12% of all tweets were about the Olympics, whereas in August 2014 mentions of the Olympic Games did not exceed 0.5% of all posts.

Consider two methods for determining the weight of unigram. In order to show decrease in results of sentiment classification for the texts collections separated in time, the Boolean model representation of texts in vector and feature vector weighted by a measure TF-IDF were used.

1. Boolean model or there attendance of a term in the text. Pang et al. [1] were the first who used the presence or absence of the word form as a feature for the sentiment text classification task. The idea of the Boolean model based on the fact that the word form receives weight 1 if it is present in the text. Repetition of the word form in the text is not considered, i.e. we are interested in the presence or absence of word form in the text, but not the number of repetitions of the particular word form. As a result, applying boolean model to the above-described collection of tweets, got feature space, the dimension of 219 280.
2. Regarding the second model each unigramma weighed using TF-IDF measure (formula 5). The authors of [28] have shown that it is possible to improve the classification results if you use the TF-IDF measure for weighing the features for the classification method of support vector machine.

$$tfidf = tf \times log\frac{T}{T(t_i)} \tag{5}$$

where tf – is the frequency of occurrence of the term in the text, T - total number of messages in the training collection, $T(t_i)$ - the number of messages in the positive and negative collections, containing term.

Than, it is necessary to show the decrease in classification quality for collections staggered over time. To do this, we trained the classifier model on the 2013 collection and apply it to the 2014 and 2015 collections. The lexicons men_3, men_5 and BOW (bag-of-words) were selected to build a feature space. The men_N prefix indicates that a term is found no less than N times in one of the collections that corresponds to a sentiment class (positive, negative or neutral). The total quantity of terms in the training collection is designated as BOW.

The experiment results that show the reduction in quality of text classification are presented in Table 2. Table 2 shows that over a year and a half the classification quality of microblog texts can fall up to 15–20% according to F-measure, depending on the selected set of features.

Table 2. Quality measurements for the classification of microblog posts by sentiment for collections staggered over time

BOW				Men_3_tfidf				Men_5_tfidf			
Acc	P	R	F	Acc	P	R	F	Acc	P	R	F
2013 text collection											
0,7459	0,7595	0,7471	0,7505	0,6457	0,6591	0,6471	0,6506	0,6189	0,6542	0,6184	0,6223
2014 text collection											
0,6964	0,6984	0,7062	0,6933	0,5086	0,5829	0,5040	0,5026	0,5745	0,5823	0,5795	0,5808
2015 text collection											
0,6118	0,6317	0,6156	0,5996	0,4651	0,5218	0,4638	0,4549	0,5343	0,5337	0,5360	0,5344

3   Ways to Reduce Deterioration of Classification Results for Text Collections Staggered over Time

The SVM (support vector machine) method and LIBLINEAR library [15] were used as a classifier. The LIBLINEAR library is an implementation of the SVM algorithm with a linear kernel. Experiments show that the LIBLINEAR library significantly surpasses its counterparts in speed while training a model, therefore it was used for this paper.

3.1   Using External Lexicons of Emotional Words and Expressions

The first hypothesis is that the use of external lexicons with emotive and/or evaluative vocabulary will improve the quality of text classification by sentiment, as well as reducing the classifier's dependency on the training collection. The terms in the lexicon can be used as features in machine learning [16] or as part of approaches based on dictionaries and rules [17]. There have been studies describing the derivation and configuration of sentiment lexicons on a certain pre-determined subject area [18,19]. Examples are given of terms that can describe positive features in one subject area, but neutral or even negative ones in another. However, according to [16,20], combining training data from different domains improves the quality of sentiment classification in each of the selected subject areas. Consequently, there are many evaluative words with strongly pronounced tonal orientation that are suitable for different subject areas.

Two general-topic lexicons of emotional language, labeled by experts, were used as additional external dictionaries for this paper: RuSentiLeks and Linis-crowd.

RuSentiLeks [21] is a lexicon compiled from several sources: evaluative words from Russian thesaurus RuTez, slang words from Twitter and words with pos-itive or negative associations (connotations) from a news corpus. The lexicon contains more than ten thousand words and phrases. It includes emotional terms automatically extracted from text and checked by experts.

Another dictionary used in this paper is Linis-crowd [22]. Despite the fact that the authors used socio-politically themed texts to form the lexicon, it is noted that the dictionary contains vocabulary that is not specific to this subject area, but conveys an emotional assessment, which is why the authors of the dictionary decided to include it in the Linis-crowd prototype. The dictionary contains 9539 terms. Each one is weighted from -2 (strongly negative) to $+2$ (strongly positive).

Activation of sentiment lexicons. For a sentiment classifier based on machine learning methods, lexicon features were added in addition to features generated on the basis of training data. For each term w in the lexicon with polarity p, a value (w, p) is determined:

$$(w, p) = \begin{cases} > 0, & w - positive \\ < 0, & w - negative \\ = 0, & w - neutral \end{cases}$$

The following are added as features:

- The total number of terms (w, p) in the text of the tweet
- The sum of all polarity values of words in the lexicon: $\sum_{w \in tweet}(w, p)$;
- The maximum polarity value: $\max_{w \in tweet}(w, p)$.

Each of the lexicons was activated separately, comparison of their perfor-mance can be seen in Table 3. As can be seen from the table, both lexicons show quite similar results when used on the training and test collections.

Consequently, external lexicons make it possible to stop the loss in quality when classifying collections staggered over time. Since the main features are generated by the training collection, the trend towards degradation nevertheless persists. However, it is reduced from 15% when using the bag-of-words Table 2 to 5.6% when activating emotional vocabulary lexicons.

Table 3. Classifier results when activating lexicons RuSentiLeks and Linis-Crowd

	RuSentiLex				Linis-Crowd			
	Acc	P	R	F	Acc	P	R	F
2013	0,7273	0,74	0,7284	0,7318	0,7272	0,7398	0,7283	0,7316
2014	0,7245	0,7387	0,7259	0,7295	0,7244	0,7386	0,7258	0,7294
2015	0,6724	0,6802	0,6733	0,6759	0,6725	0,6803	0,6733	0,6760

Clearly, it makes sense to use this method when external sentiment lexicons are available, as it inhibits the reduction in quality when sentiment classifying collections staggered over time.

3.2   Using Distributed Word Representations as Features

In the previous method, the feature space for training the classifier was based on the training collection and is therefore highly dependent on the quality and completeness of this collection. Despite the good results of the models described above, there are no semantic relationships between the terms, and the continuous addition of new terms leads to an increase in the dimention of feature vector space. Another way to overcome the obsolescence of a lexicon is the use of the distributed word representations as features to train the classifier.

Distributed Word Representations. Distributed word representation (word embedding) is a k-dimensional feature vector $w = (w_1, \ldots, w_k)$, where $w_i \in R$ is the vector coordinate [23]. When compared with the boolean or other weighted vector models, the number of coordinates k of such a vector is much smaller. Usually, this number does not exceed several hundred, whereas in the boolean model it is measured in tens of thousands, depending on the original size of the lexicon.

In addition to reducing feature vector length, distributed word representation takes into account the meaning of a word in context. In other words, it allow us to extend "fast car", for example, into "speedy automobile", which is absent in the training sample, thereby reducing dependence on the latter.

Unsupervised Machine learning models are used to obtain distributed word representations, for instance, CBOW, Skip-Gram, AdaGram [24] and Glove. Recent studies show [25] that the neural language model Skip-gram is superior to others in the quality of obtained vector representations. Therefore, the Skip-Gram model is used in this paper.

Using the Skip-Gram Model to Reduce Dependence on the Training Collection. The Skip-Gram model was proposed by Thomas Mikolov et al. in 2013 [26]. An unlabeled corpus of texts is input into the model and the number of occurrences in the corpus is calculated for each word.

In [27] it is shown that neural networks using vector representations of words obtained by means of the word2vec algorithm [24] can effectively solve the problem of natural language text processing in general, and can be applicable to sentiment text classification in particular. This algorithm has shown the best results on the selected text collections in comparison to others.

In order to train the Skip-Gram model, 5 million texts were arbitrarily selected from the original 2013 collection, which were not split into sentiment classes. The 2014 and 2015 collections did not take part in training, since it is assumed that the trained model should be transferrable to later collections.

Word2Vec [24] was used as a software implementation of the Skip-gram model with the following parameters:

- size 300 – every word is represented as a vector of this length;
- window 5 – how many words of context the training algorithm should take into account;
- negative 10 – the number of negative examples for negative sampling;
- sample 1e-4 – sub-sampling (the usage of sub-sampling improves performance). The recommended parameter for sub-sampling is from 1e-3 to 1e-5;
- threads 10 – the number of threads to use;
- min-count 3 – limits the size of the lexicon to significant words. Words that appear in the text less than this specified number of times are ignored. The default value is 5;
- iter 15 – the amount of training iterations.

Emoticons were filtered out of the text, as they designate that a text belongs to a particular sentiment class as emoticons used for the distinct supervision.

Each text is represented as an averaged vector of its constituent words formula 6:

$$d = \frac{\sum w_i}{n} \tag{6}$$

where w_i is the vector representation of the its word in the studied text, $i = (1,..,n)$; n is the number of unique words from the lexicon found in the analysed text.

The classifier was trained on the 2013 collection, and then the trained classifier model was used to test the 2014 and 2015 collections. The classifier results are shown in Table 4; the quality measurements for the 2013 collection are given for clarity.

Table 4. The results of sentiment classification with the aid of vectors obtained by using Word2Vec as features

	Acc	Precision	Recall	F-measure
2013	0,7206	0,7250	0,7221	0,7226
2014	0,7756	0,7763	0,7836	0,7787
2015	0,7289	0,7250	0,7317	0,7252

Table 4 clearly demonstrates that the quality of classification into three classes is not reduced for collections complied 6–18 months apart and remains at the level of the best values obtained using the bag-of-words model when cross-checking on a collection from one year (Table 2). Nevertheless, the number of coordinates in the word vector is exactly 300 (according to the settings), rather than exceeding 200,000, like in the boolean or vector models.

This method is well suited for use in the case that we have an external, fairly representative collection of texts, which is similar in vocabulary to the training and test collections, meaning that here, as for other neural networks methods, a large training sample of texts is required. This method makes it possible to obtain stable and consistent results for text classification by sentiment.

4 Conclusion

This paper suggests two fundamentally different models to overcome the deterioration of sentiment classification results for collections staggered over time. In Table 2, it was shown that the quality of text classification by sentiment can be reduced to 15% according to F-measure over 18 months. Therefore, the aim of the approaches proposed in this paper is to minimise the decrease according to F-measure when classifying text collections that are staggered over time.

1. The first approach is based on adding lexicons of emotional vocabulary: RuSentiLeks and Linis-Crowd. The use of external dictionaries makes it possible to reduce the gap in classification quality between the 2013 and 2015 collections to 5.6%, according to F-measure. The difference between the classification results of the 2013 and 2014 is less than 1% – only 0.2%. At the same time, the quality of the classifier remains at the 0.68-0.73 level, which is comparable with the best results. Therefore, the generation of features based on external lexicons does not entail a large increase in the feature space and makes it possible to achieve good classification results. Despite this, since the feature space is still dependent on a training collection, there is a negligible reduction in classification quality for later collections.
2. The foundation of the second approach is the concept of a distributed word representation and the Skip-gram neural language model. As in the second approach, external resources were used here. The distributed word representation space was built on an untagged collection of tweets that was many times larger than the automatically tagged training collection. The averaged word vectors from one tweet were used as features for the classificator. Thus, the length of the vector space was only 300 – this is the first advantage of the approach. A second advantage of the approach is the classification results: the difference between the 2013 and 2015 F-measure is 0.26%, with the classification results for the 2015 collection being higher. The classification results of the 2013 and 2014 collections are similar: the 2014 ones exceed the 2013 values by 5.6%, according to F-measure. This can be explained by the fact that a cross-validation method was used on the 2013 collection, i.e. the collection was divided into training and test ones at a ratio of 4:5, whereas the

full 2013 collection was used to train the classifier for testing on the 2014 and 2015 collections.

In summary, all proposed approaches can reduce the deterioration of sentiment classification results for collections staggered over time.

References

1. Pang, B., Lillian, L., Vaithyanathan, S.: Thumbs up?: sentiment classification using machine learning techniques. In: Proceedings of the ACL-02 Conference on Empirical Methods in Natural Language Processing, vol. 10. Association for Computational Linguistics (2002)
2. Turney, P.D.: Thumbs up or thumbs down?: semantic orientation applied to unsupervised classification of reviews. In: Proceedings of the 40th Annual Meeting on Association for Computational Linguistics, pp. 417–424. Association for Computational Linguistics (2002)
3. Wilson, T., Wiebe, J., Hoffmann, P.: Recognizing contextual polarity in phrase-level sentiment analysis. In: Proceedings of the Conference on Human Language Technology and Empirical Methods in Natural Language Processing, pp. 347–354. Association for Computational Linguistics, October 2005
4. Rubtsova, Y.V.: Research and development of domain independent sentiment classifier. Trudy SPIIRAN **36**, 59–77 (2014)
5. Agarwal, A., Xie, B., Vovsha, I., Rambow, O., Passonneau, R.: Sentiment analysis of twitter data. In: Proceedings of the Workshop on Languages in Social Media, pp. 30–38. Association for Computational Linguistics, June 2011
6. Kouloumpis, E., Wilson, T., Moore, J.D.: Twitter sentiment analysis: the good the bad and the OMG!. ICWSM **11**(538–541), 164 (2011)
7. Pak, A., Paroubek, P.: Twitter as a corpus for sentiment analysis and opinion mining. In: LREc, vol. 10, no. 2010 (2010)
8. Lek, H.H., Poo, D.C.: Aspect-based Twitter sentiment classification. In: 2013 IEEE 25th International Conference on Tools with Artificial Intelligence (ICTAI), pp. 366–373. IEEE (2013)
9. Loukachevitch, N., Rubtsova, Y.: Entity-oriented sentiment analysis of tweets: results and problems. In: Král, P., Matoušek, V. (eds.) TSD 2015. LNCS, vol. 9302, pp. 551–559. Springer, Cham (2015). doi:10.1007/978-3-319-24033-6_62
10. Read, J.: Using emoticons to reduce dependency in machine learning techniques for sentiment classification. In: Proceedings of the ACL Student Research Workshop, pp. 43–48. Association for Computational Linguistics (2005)
11. Rubtsova, Y.V.: A method for development and analysis of short text corpus for the review classification task. In: Digital Libraries: Advanced Methods and Technologies, RCDL 2013, pp. 269–275 (2013)
12. Loukachevitch, N., Rubtsova, Y.: Entity-oriented sentiment analysis of tweets: results and problems. In: Král, P., Matoušek, V. (eds.) TSD 2015. LNCS, vol. 9302, pp. 551–559. Springer, Cham (2015). doi:10.1007/978-3-319-24033-6_62
13. Loukachevitch, N., Rubtsova, Y.: SentiRuEval-2016: overcoming time gap and data sparsity in tweet sentiment analysis. In: Proceedings of International Conference on Computational Linguistics and Intellectual Technologies Dialog 2016, pp. 375–384 (2016)

14. Manning, C.D., Schütze, H.: Foundations of Statistical Natural Language Processing, vol. 999. MIT Press, Cambridge (1999)
15. Fan, R.-E., et al.: LIBLINEAR: a library for large linear classification. J. Mach. Learn. Res. **9**, 1871–1874 (2008)
16. Mohammad, S.M., Kiritchenko, S., Zhu, X.: NRC-Canada: building the state-of-the-art in sentiment analysis of tweets. arXiv preprint arXiv:1308.6242 (2013)
17. Taboada, M., Brooke, J., Tofiloski, M., Voll, K., Stede, M.: Lexicon-based methods for sentiment analysis. Comput. Linguist. **37**(2), 267–307 (2011)
18. Klekovkina, M.V., Kotelnikov, E.V.: The automatic sentiment text classification method based on emotional vocabulary. In: Digital libraries: Advanced Methods and Technologies, Digital Collections (RCDL-2012), pp. 118–123 (2012)
19. Chetviorkin, I., Loukachevitch, N.: Extraction of domain-specific opinion words for similar domains. In: Proceedings of the Workshop on Information Extraction and Knowledge Acquisition Held in Conjunction with RANLP 2011, pp. 7–12, September 2011
20. Mansour, R., Refaei, N., Gamon, M., Abdul-Hamid, A., Sami, K.: Revisiting the old kitchen sink: do we need sentiment domain adaptation? In: RANLP, pp. 420–427 (2013)
21. Loukachevitch, N.V., Levchik, A.V.: Creating a general Russian sentiment lexicon. In: Proceedings of the Tenth International Conference on Language Resources and Evaluation (LREC 2016), European Language Resources Association (ELRA) (2016)
22. Alexeeva, S., Koltsov, S., Koltsova, O.: Linis-crowd. org: a lexical resource for Russian sentiment analysis of social media. Comput. Linguist. Comput. Ontology, 25–34 (2015). Proceedings of the XVIII joint Conference on Internet and modern society (IMS) 2015
23. Titov, I., McDonald, R.: Modeling online reviews with multi-grain topic models. In: Proceedings of the 17th International Conference on World Wide Web, pp. 111–120. ACM, April 2008
24. Mikolov, T., Sutskever, I., Chen, K., Corrado, G. S., Dean, J.: Distributed representations of words and phrases and their compositionality. In: Advances in Neural Information Processing Systems, pp. 3111–3119 (2013)
25. Levy, O., Goldberg, Y., Dagan, I.: Improving distributional similarity with lessons learned from word embeddings. Trans. Assoc. Comput. Linguist. **3**, 211–225 (2015)
26. Mikolov, T., et al.: Efficient estimation of word representations in vector space. arXiv preprint arXiv:1301.3781 (2013)
27. Kim, Y.: Convolutional neural networks for sentence classification. arXiv preprint arXiv:1408.5882 (2014)
28. O'Keefe, T., Koprinska, I.: Feature selection and weighting methods in sentiment analysis. In: Proceedings of the 14th Australasian Document Computing Symposium, Sydney, pp. 67–74, December 2009

Searching for the Most Negative Opinions

Sattam Almatarneh[(✉)] and Pablo Gamallo

Centro Singular de Investigación en Tecnoloxías da Información (CITIUS),
Universidad de Santiago de Compostela, Rua de Jenaro de la Fuente Domínguez,
15782 Santiago de Compostela, Spain
{sattam.almatarneh,pablo.gamallo}@usc.es

Abstract. Studies in sentiment analysis and opinion mining have been focused on several aspects of opinions, such as their automatic extraction, identification of their polarity (positive, negative or neutral), the entities or facets involved, and so on. However, to the best of our knowledge, no sentiment analysis approach has considered the automatic identification and extraction of the most negative opinions, in spite of their significant impact in many fields such as industry, trade, political and socials issues.

In this article, we will use diversified linguistic features and supervised machine learning algorithms so as to examine their effectiveness in the process of searching for the most negative opinions.

Keywords: Sentiment analysis · Opinion mining · Linguistic features · Classification · Most negative opinion

1 Introduction

A fundamental task in opinion mining is polarity classification. Polarity classification occurs when a piece of text stating an opinion is classified into a predefined set of polarity categories (e.g., positive, neutral, negative). Categorizing reviews into classes such as "thumbs up" versus "thumbs down," or "like" versus "dislike" are examples of two-class polarity classification [8,9,13,16,17,24–26].

A still not usual way of performing sentiment analysis is to detect and classify the most negative opinions about a topic, object or individual. The most negative opinion is the worst judgment, or appraisal formed in mind about a particular matter. These opinions only constitute a small portion of all opinions found in Social Media. According to [16], only about 5% of all opinions are in the most negative level of the opinion scale, which makes their automatic search a challenge. There is a need for systematic studies that attempt to understand how to mine the vast amounts of unorganized text data and extract the most negative comments.

The objective of the article is to investigate the effectiveness of linguistic features and supervised machine learning classification to search for the most negative opinions. The rest of the paper is organized as follows. In Sect. 2 we discuss the related work. Then, Sect. 3 describes the method. The Experiments are introduced in Sect. 4, where we also describe the evaluation and discuss the results. We draw conclusions in Sect. 5.

© Springer International Publishing AG 2017
P. Różewski and C. Lange (Eds.): KESW 2017, CCIS 786, pp. 14–22, 2017.
https://doi.org/10.1007/978-3-319-69548-8_2

2   Related Work

There are two main approaches to find the sentiment polarity at a document level. First, machine learning techniques based on training corpora annotated with polarity information and, second, strategies based on polarity lexicons.

In machine learning techniques there are are two methods, supervised learning, where the most existing techniques for document-level classification use, although there are also unsupervised methods. The success of both mainly depends on the choice and extraction of the proper set of features used to identify sentiments. The current reviews and books in sentiment analysis [1,3,4,14,15,22] included all issues in this field. For instance, the most important linguistic features that used in sentiment classification are listed in Chap. 3 of [15] book. [5] presented a systematic study of different sentence features for two tasks in sentiment classification in (polarity classification and subjectivity classification) our study.

On the other hand, Sentiment words are the core component in opinion mining and have been used in many studies [2,7,11,12,21,23] they relied on lexicons as a source for determining the polarity of documents.

In this study, we focused on searching for the most negative opinions by use linguistic features, because of the vast importance of these views. Previous works analyzed this importance, such as the experiments reported in [6], which found that one-star reviews hurt book sales on Amazon.com. The impact of 1-star reviews that represent the most negative views is higher than the impact of 5-star reviews. [18] also stated that the negative reviews have more impact than positive reviews.

3   The Method

Sentiment analysis typically works at three levels of granularity, namely, document level, sentence level, and aspect level.

Document-level works with whole documents as the basic information unit. Analogously, at the sentence level, sentiment classification is applied to individual sentences in a document. But concerning aspect level, the system performs at a finer-grained level of analysis. Instead of looking at language constructs such as documents, paragraphs, sentences, clauses or phrases, a system working at the aspect level directly looks at the opinion itself. It is based on the idea that an opinion consists of, at least, a sentiment (positive, negative or neutral) and a target, namely the aspect of an entity receiving that opinion.

In this paper, however, we are involved with document-level classification issues, more precisely with the identification of most negative opinion *vs.* other opinions at the document level. This binary categorization can be achieved by the use of classifiers built from training data. Converting a portion of text into a feature vector is the essential and basic step in any data-driven approach to Sentiment Analysis. Selection of features is a requirement to make the learning task efficient and accurate. In our experiments, we studied different strategies and examined the following sets of features.

3.1   Unigram Features

First, all stop words are removed from the document collection. Then, the vocabulary is cleaned up by eliminating those terms appearing in less than 12 documents so as to eliminate terms that are too infrequent. Finally, we assign a weight to all terms by using Term Frequency - Inverse Document Frequency (TF-IDF), which is computed in Eq. 1.

$$tf/idf_{t,d} = (1 + log(tf_{t,d})) \times log(\frac{N}{df_t}). \tag{1}$$

where $tf_{t,d}$ in the term frequency of the term t in the document d, N is the number of documents in the collection and df_t is the number of documents in the collection containing t.

3.2   Part of Speech Features

A part of speech (PoS) is a category classifying words with similar grammatical properties. PoS tag information is usually used in sentiment analysis and opinion mining. Several researchers [5,8,25] used PoS tags, especially adjectives, as features to classify opinions, such they are a good indicator of sentiment. We processed the document collection using the Natural Language Toolkit (NLTK)[1], which provides words with Penn Treebank PoS tags (see Table 1). Then we counted the occurrences of each tag in the document.

Table 1. Penn Treebank Part-Of-Speech (POS) tags.

CC	conjunction, coordinating	PRP\$	pronoun, possessive
CD	cardinal number	RB	adverb
DT	determiner	RBR	adverb, comparative
EX	existential there	RBS	adverb, superlative
FW	foreign word	RP	adverb, particle
IN	conjunction, subordinating or preposition	SYM	symbol
JJ	adjective	TO	infinitival to
JJR	adjective, comparative	UH	interjection
JJS	adjective, superlative	VB	verb, base form
LS	list item marker	VBZ	verb, 3rd person singular present
MD	verb, modal auxillary	VBP	verb, non-3rd person singular present
NN	noun, singular or mass	VBD	verb, past tense
NNS	noun, plural	VBN	verb, past participle
NNP	noun, proper singular	VBG	verb, gerund or present participle
NNPS	noun, proper plural	WDT	wh-determiner
PDT	predeterminer	WP	wh-pronoun, personal
POS	possessive ending	WP\$	wh-pronoun, possessive
PRP	pronoun, personal	WRB	wh-adverb

---

[1] http://www.nltk.org/.

3.3 Syntactic Patterns

We used in this study the patterns defined by Turney [25]. More precisely, he used five patterns of PoS tags to extract opinions from reviews, as the example depicted in Table 2. We define two types of features based on PoS patterns: counting patterns frequency and considering presence or absence of patterns in each document.

Table 2. Pattern of POS by Turney [25]

First word	Second word	Third word
JJ	NN or NNS	Anything
RB, RBR, or RBS	JJ	not NN nor NNS
JJ	JJ	not NN nor NNS
NN or NNS	JJ	not NN nor NNS
RB, RBR, or RBS	VB, VBD, VBN, or VBG	Anything

3.4 Sentiment Lexicons

In our approach, we have experimented with some lexicons: the Opinion Lexicon or (Sentiment Lexicon), Linguistic Inquiry and Word Count (LIWC) and VADER Lexicon.

- Opinion Lexicon (or Sentiment Lexicon): This is a list of negative and positive sentiment words for English: 5,789 words, 2,006 are positive words and 3,783 are negative. This list has been compiled for many years and its construction was reported in [11]. It includes mis-spellings, morphological variants, slang, and social-media mark-up. The features based on this lexicon are defined by considering the number of negative and positive terms in the document, as well as the proportion of negative and positive terms.
- Linguistic Inquiry and Word Count (LIWC): [21] LIWC dictionary consists of 290 words and word-stems. Each word or word-stem defines one or more word categories or sub-dictionaries. We believe that the use of features derived from the LIWC dictionary (Linguistic Inquiry and Word Count) would be helpful in the search for the most negative opinions since negative opinions can also be associated with psychological factors. We obtained 65 features based on the lexical categories defined in LIWC.
- Valence Aware Dictionary and Sentiment Reasoner (VADER): This is a lexicon and rule-based sentiment analysis tool that is specifically attuned to sentiments expressed in social media and works well on texts from other domains [12]. We obtained over 7,500 lexical features with validated valence scores indicating both sentiment polarity (negative/positive) and sentiment intensity on a scale from -4 to $+4$. Intensity was classified as follows.

Words were split into four groups according to valence scores: -4 to -2 most negative, -1.9 to -0.1 negative, $+0.1$ to $+1.9$ positive and $+2$ to $+4$ most positive. Then the number and proportion of each group of words were considered to define the intensity-based features. Also, we included additional features: namely, the total scores for all the words that appear in the documents and the total scores of words that are only provided with negative scores in the documents.

4   Experiments

4.1   Data Collection

In order to extract the most negative opinions, we require to analyze document collections with scaled opinion levels (e.g. rating) and extract those documents associated with the lowest scale. So, we have adopted (Pang & Lee Sentiment scale dataset)[2], which was described in [16]. This dataset contains four corpora of movie reviews, where each corpus includes documents written by the same author. The total number of documents in all corpus are 5,006.

4.2   Training Set

Since we are facing a text classification problem, any existing supervised learning method can be applied. Support vector machines (SVMs) have been shown to be highly effective at traditional text categorization [17]. We decided to utilize *scikit*[3] which is an open source machine learning library for the Python programming language [20]. This library implements several classifiers, including regression and clustering algorithms. We chose SVMs as our classifier for all experiments, hence, in this study we will only summarize and discuss results for this learning model. More specifically, we utilized the sklearn.svm.LinearSVC module[4]. Our collection has 5,006 reviews and our method handles a large number of features for each example. To do classification, we need two samples of documents: training and testing. The training sample will be used to learn various characteristics of the documents and the testing sample was used to predict and next verify the efficiency of our classifier in the prediction. So we divided the dataset into two stratified samples: we have allocated 25% of the collection for the testing sample and 75% of the collection for the training sample.

There are only 615 most negative reviews out of 5,006 in our dataset and 4,394 labeled as a negative class (not most negative), which results in an unbalanced two-class classification problem. To deal with this problem there are many frameworks and approaches such as undersampling and oversampling, even if undersampling gives rise to loss of information. As recommended in [10,19], we examined the performance by giving more importance to the positive class.

---

[2] http://www.cs.cornell.edu/people/pabo/movie-review-data/.

[3] http://scikit-learn.org/stable/.

[4] http://scikit-learn.org/stable/modules/generated/sklearn.svm.LinearSVC.html.

We found that performance was insensitive to the SVM cost parameter (C) but very sensitive to the weights that modify the relative cost of misclassifying positive and negative samples.

In our analysis, we employed 5-fold cross-validation and the effort was put on optimizing F1 which is computed with respect to the most negative opinions (which is the target class):

$$F1 = 2 * \frac{P * R}{P + R} \tag{2}$$

where P and R are defined as follows:

$$P = \frac{TP}{TP + FP} \tag{3}$$

$$R = \frac{TP}{TP + FN} \tag{4}$$

where TP stands for true positive, FP is false positive, and FN is false negative.

To optimize F1, we tried out a grid search approach with exponentially growing sequences of the value of the parameter *class_weight*. More precisely, we tested class_weight with different values: $2^{-5}, 2^{-4}, 2^{-3}, 2^{-2}, ..., 2^{10}$. After finding the best value of class_weight within that sequence, we conducted a finer grid search on that better district (e.g. if the optimal value of class_weight is 8, then we test all the neighbors in this region: e.g. 3, 4, 5, 6, 7, 9, 10, 11, 12, 13, 14, 15 and 16).

The class_weight was finally set to the value returning the highest F1 across all these experiments (see Table 3).

Table 3. The best (F1) performance with varying class weights

Features	Class-weight
Unigram (TF-IDF)	893
Unigram (Pres)	2
POS	4
Pattern (Freq)	8
Pattern (Presence)	8
Opinion Lexicon	6
LIWC	6
VADER	6
ALL	4

Figure 1 shows the average of F1 performance across the variation of class_weight for each set of features.

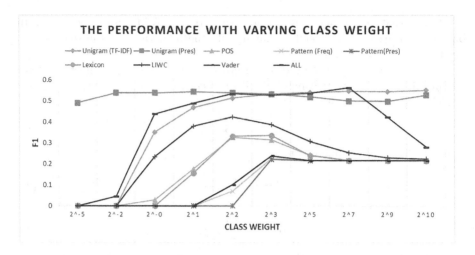

Fig. 1. The average performance of F1 with across different values of class_weight

4.3   The Results

In the test collection, there are 1,252 reviews and 157 of them belong to the target class (the most negative opinions). The proportion of positive examples in the training and test collections are similar (around 12%); consequently, both datasets are similarly unbalanced. The results depicted in Table 4 reveal that all combined features give the best performance in terms of precision and F1, even though just unigrams work reasonably well.

In order to select the best and most influential singular features for finding most negative opinions, we need to perform further fine-grained experiments with different groups of feature combinations.

Table 4. The best results for the collection, in terms of precision, recall, and F1 scores

Features	Precision	Recall	F1
Unigram (TF-IDF)	0.60	0.54	0.57
Unigram (Pres)	0.63	0.47	0.54
POS	0.25	0.33	0.29
Pattern (Freq)	0.14	0.73	0.24
Pattern (Presence)	0.13	0.71	0.22
Opinion Lexicon	0.25	0.61	0.36
LIWC	0.30	0.62	0.40
VADER	0.18	0.29	0.22
ALL	0.69	0.54	0.61

5   Conclusions

In this article, we have studied different linguistic features for a particular task in Sentiment Analysis. More precisely, we examined the performance of these features within supervised learning methods (using Support Vector Machine (SVM)), to identify the most negative documents on movie review datasets.

The experiments reported in our work shows that the evaluation values for identifying the most negative class are low. This can be partially explained by the difficulty of the task, since the difference between very negative and not very negative is a subjective continuum without clearly defined edges. The border-line between very negative and not very negative is still more difficult to find than that discriminating between positive and negative opinions, since there are a quite clear space of neutral/objective sentiments between the two opinions. However, there is not such an intermediate space between *very* and *not very*.

In future work, there is much room for improvement. First, use more data sets such as products reviews as well as movie reviews. Second, we will provide the classifiers with a set of features which would be sensitive to the concept of *most negative*. Third, it would be useful to make experiments with unsupervised learning approaches and lexicon-based methods to improve the performance for this difficult task.

References

1. Agarwal, B., Mittal, N.: Prominent Feature Extraction for Sentiment Analysis. Socio-Affective Computing. Springer, Cham (2016). doi:10.1007/978-3-319-25343-5
2. Almatarneh, S., Gamallo, P.: Automatic construction of domain-specific sentiment lexicons for polarity classification. In: De la Prieta, F., Vale, Z., Antunes, L., Pinto, T., Campbell, A.T., Julián, V., Neves, A.J.R., Moreno, M.N. (eds.) PAAMS 2017. AISC, vol. 619, pp. 175–182. Springer, Cham (2018). doi:10.1007/978-3-319-61578-3_17
3. Benamara, F., Taboada, M., Mathieu, Y.: Evaluative language beyond bags of words: linguistic insights and computational applications. Comput. Linguist. **43**, 201–264 (2017)
4. Cambria, E., Schuller, B., Xia, Y., Havasi, C.: New avenues in opinion mining and sentiment analysis. IEEE Intell. Syst. **28**(2), 15–21 (2013)
5. Chenlo, J.M., Losada, D.E.: An empirical study of sentence features for subjectivity and polarity classification. Inf. Sci. **280**, 275–288 (2014)
6. Chevalier, J.A., Mayzlin, D.: The effect of word of mouth on sales: online book reviews. J. Mark. Res. **43**(3), 345–354 (2006)
7. Ding, X., Liu, B., Yu, P.S.: A holistic lexicon-based approach to opinion mining. In: Proceedings of the 2008 International Conference on Web Search and Data Mining, pp. 231–240. ACM (2008)
8. Hatzivassiloglou, V., McKeown, K.R.: Predicting the semantic orientation of adjectives. In: Proceedings of the Eighth Conference on European Chapter of the Association for Computational Linguistics, pp. 174–181. Association for Computational Linguistics (1997)

9. Heerschop, B., Goossen, F., Hogenboom, A., Frasincar, F., Kaymak, U., de Jong, F.: Polarity analysis of texts using discourse structure. In: Proceedings of the 20th ACM International Conference on Information and Knowledge Management, pp. 1061–1070. ACM (2011)

10. Hsu, C.W., Chang, C.C., Lin, C.J., et al.: A practical guide to support vector classification (2003)

11. Hu, M., Liu, B.: Mining and summarizing customer reviews. In: Proceedings of the Tenth ACM SIGKDD International Conference on Knowledge Discovery and Data Mining, pp. 168–177. ACM (2004)

12. Hutto, C.J., Gilbert, E.: Vader: a parsimonious rule-based model for sentiment analysis of social media text. In: Eighth International AAAI Conference on Weblogs and Social Media (2014)

13. Kamps, J., Marx, M., Mokken, R.J., De Rijke, M., et al.: Using wordnet to measure semantic orientations of adjectives. In: LREC, vol. 4, pp. 1115–1118 (2004)

14. Liu, B.: Sentiment analysis and opinion mining. Synth. Lect. Hum. Lang. Technol. **5**(1), 1–167 (2012)

15. Liu, B.: Sentiment Analysis: Mining Opinions, Sentiments, and Emotions. Cambridge University Press, Cambridge (2015)

16. Pang, B., Lee, L.: Seeing stars: exploiting class relationships for sentiment categorization with respect to rating scales. In: Proceedings of the 43rd Annual Meeting on Association for Computational Linguistics, pp. 115–124. Association for Computational Linguistics (2005)

17. Pang, B., Lee, L., Vaithyanathan, S.: Thumbs up?: Sentiment classification using machine learning techniques. In: Proceedings of the ACL 2002 Conference on Empirical Methods in Natural Language Processing, vol. 10, pp. 79–86. Association for Computational Linguistics (2002)

18. Papathanassis, A., Knolle, F.: Exploring the adoption and processing of online holiday reviews: a grounded theory approach. Tourism Manag. **32**(2), 215–224 (2011)

19. Parapar, J., Losada, D.E., Barreiro, A.: A learning-based approach for the identification of sexual predators in chat logs. In: CLEF (Online Working Notes/Labs/Workshop) (2012)

20. Pedregosa, F., Varoquaux, G., Gramfort, A., Michel, V., Thirion, B., Grisel, O., Blondel, M., Prettenhofer, P., Weiss, R., Dubourg, V., et al.: Scikit-learn: machine learning in python. J. Mach. Learn. Res. **12**, 2825–2830 (2011)

21. Pennebaker, J.W., Boyd, R.L., Jordan, K., Blackburn, K.: The development and psychometric properties of LIWC2015. Technical report (2015)

22. Serrano-Guerrero, J., Olivas, J.A., Romero, F.P., Herrera-Viedma, E.: Sentiment analysis: a review and comparative analysis of web services. Inf. Sci. **311**, 18–38 (2015)

23. Taboada, M., Brooke, J., Tofiloski, M., Voll, K., Stede, M.: Lexicon-based methods for sentiment analysis. Comput. Linguist. **37**(2), 267–307 (2011)

24. Takamura, H., Inui, T., Okumura, M.: Extracting semantic orientations of phrases from dictionary. In: HLT-NAACL, vol. 2007, pp. 292–299 (2007)

25. Turney, P.D.: Thumbs up or thumbs down?: Semantic orientation applied to unsupervised classification of reviews. In: Proceedings of the 40th Annual Meeting on Association for Computational Linguistics, pp. 417–424. Association for Computational Linguistics (2002)

26. Yu, H., Hatzivassiloglou, V.: Towards answering opinion questions: separating facts from opinions and identifying the polarity of opinion sentences. In: Proceedings of the 2003 Conference on Empirical Methods in Natural Language Processing, pp. 129–136. Association for Computational Linguistics (2003)

Diversified Semantic Query Reformulation

Rubén Manrique$^{(\boxtimes)}$ and Olga Mariño

Systems and Computing Engineering Department, School of Engineering,
Universidad de Los Andes, Bogotá, Colombia
{rf.manrique,olmarino}@uniandes.edu.co

Abstract. One main challenge for search engines is retrieving the user's intended results. Diversification techniques are employed to cover as many aspects of the query as possible through a tradeoff between the relevance of the results and the diversity in the result set. Most diversification techniques reorder the final result set. However, these diversification techniques could be inadequate for search scenarios with small candidate set sizes, or those for which response time is a critical issue. This paper presents a diversification technique for such scenarios. Instead of reordering the result set, the query is reformulated, thus taking advantage of the knowledge available in Linked Data Knowledge Bases. The query is annotated with semantic data and then expanded to related resources. An adapted Maximal Marginal Relevance technique is applied to select resources from this expanded set whose properties form the expanded query. Experiments conducted on federated and non-federated scenarios show that this method has superior diversification capacity and shorter response times than algorithms based on result set reordering.

1 Introduction

The main goal of Web-based Information Retrieval Systems (IRS) is to retrieve the best result set from an input query. To achieve this goal, the IRS must consider the different possible aspects of the input query and the possible intentions of the user [1,16,19]. Since query terms are often ambiguous, and users' intentions are seldom known, discovering the best result set entails finding a tradeoff between the relevance of a retrieved document and the novelty of this document [2,9,20,22]. To solve the tradeoff problem, diversification techniques can be implemented.

Result diversification is useful in certain settings. In exploratory searches where the search domain is partially known by the user, and the query has deviations and expresses ambiguous meanings, it is especially practical. Most information retrieval systems search in a centralized settings in only one corpus of documents [8,11] and then forward the query unchanged to the search engine. In such cases, diversification methods are applied in the final step to reorder the recovered set of documents and produce the final result set [20]. Therefore, these types of techniques are susceptible to the size of the candidate set of results. If the set is very small, the diversification ability is decreased since it is expected

© Springer International Publishing AG 2017
P. Różewski and C. Lange (Eds.): KESW 2017, CCIS 786, pp. 23–37, 2017.
https://doi.org/10.1007/978-3-319-69548-8_3

to have less novel documents. On the other hand, if the set is very large, there is an increase in the computational cost of the result rearrangement. This implies increasing response times [17].

One of the scenarios where the above problems could emerge is a federated search, and more specifically, non-Web federated search engines. Our proposed technique arose from the need to operate in a federated search engine where (i) there is a limited amount of independent collections, (ii) each collection has restrictions on the number of documents retrieved per query, and (iii) the system is uncooperative, that is, information about the collections is not shared. Although our main objective is to solve a diversification problem in federated search, our strategy could also be applied in centralized settings.

The proposed method does not seek to reorder the result set, but rather to reformulate the query before the search. To our knowledge, very few systems use this strategy [2,17]. The main characteristic of our approach is the analysis of Semantic Web data which guides query expansion with semantically related terms. The proposed query expansion process is based on the generation of a graph of Semantic Web resources and relations connected to the original query terms. A set of resources from the graph are then selected using a modified version of the Maximal Marginal Relevance (MMR) technique and similarity measures for RDF graphs. Finally, labeling properties of the resources are recovered and integrated in the expanded query. The experiments we conducted in centralized and federated scenarios show that this approach has a competitive time performance and better outcomes in terms of novelty.

To explain our findings, Sect. 2 first reviews work related to our proposal. Section 3 presents the architecture of our proposal and describes each step of the process. Section 4 explains the experiments conducted and their results, and the paper ends with conclusions and further work.

2   Related Work

In search result diversification, a broad spectrum of strategies has been proposed. Most strategies are based on a reordering of the final result set. They are classified based on how they express user intentions: implicitly or explicitly. The implicit approach assumes that dissimilar documents should cover diverse query aspects. It uses dissimilarity measures between documents as a main criteria in document selection for the final result set [20]. In the implicit approach, diversification involves three main components: a relevance measure, a dissimilarity measure and a diversification optimization problem (DOP) [13]. The first measure provides a relevance score for each result (e.g. standard IR techniques measure). The second evaluates the dissimilarity between the result items. Finally, the DOP formalizes the relationship between these two conflicting measures in an objective function to be maximized. Most objective functions state an NP-hard problem [9,22] and are solved using a greedy approximation [21]. A representative method of this approach is MMR [4].

The more recent explicit approach includes coverage, which is the incorporation of query aspects into the DOP formulation [20]. The most representative methods of this group are IA-Select [1], xQuAD [19] and PM-2 [7]. IA-select uses a common taxonomy for both queries and documents. Two documents are considered similar if they share common categories covered by the query. xQuAD uses the probability of relevance of a document to all aspects of the query to decide whether to incorporate it in the result set. Sub-queries, derived from query suggestions from Web search engines, represent the different aspects of the query. PM-2 addresses the problem of finding a diverse set of documents that preserves the proportionality between document presence and each aspect of the query. Like xQuAD, PM-2 uses suggestions provided by a commercial search engine as representations of aspects of the query. In all these methods, modeling aspects of a query and their relevance to a document is fundamental [10]. According to Santos *et al.*, explicit strategies outperform implicit approaches [19], yet they depend on the correct mining of query aspects from external engines. A comprehensive study on mining query aspects is presented by He *et al.* [10]. The aforementioned methods were developed mainly for diversification in a centralized setting, or those in which only one corpus of documents is available [8,11]. Little work on diversification in federated scenarios or distributed information retrieval has been done up to this point [8,11,18].

In a federated search (FS), queries are submitted to multiple independent collections. To begin with, the central piece of the federated search is the broker that first receives the user query. The broker submits the query to different collections, and finally merges the results returned by the search. In some federated search scenarios where many collections are searched, or there are constraints such as bandwidth limits, it is not feasible to search all collections. In such a setting, the broker is also in charge of selecting the collections in which it expects to find the most relevant and diverse answers to the given query.

Researchers examining diversification in federated searches have focused mainly on the selection of the collections [8,11]. However, in some non-Web federated systems, the process of selection is not necessary. There are scenarios in which information is distributed across a limited number of different independent collections and it is feasible to submit the query to all of them. In such federated systems, centralized algorithms described above could operate in the merging step.

While the use of centralized algorithms in federated searches could overcome the diversification problem, it might also be too expensive [17]. As an alternative, the reformulation of the query is a promising option. To reformulate the query, [17] use a Pseudo Relevance Feedback (PRF) technique in which the original query is submitted to a subset of the Wikipedia corpus to find related terms to expand the query. PRF shows diversification levels comparable to the IA-select algorithm which means that reformulation techniques could be applied to the diversification problem.

Our work keeps with query reformulation though we do not rely on a PRF strategy to search related terms. We use the rich relational knowledge available in Linked Open Data (LOD) Knowledge Bases (KB) (e.g. DBpedia) to reformulate the query. In order to select the most appropriate terms, we adapt the MMR technique [4] to select resources whose labeling properties are used to expand the original query. To select terms in ConceptNET for query expansion, [2,3] used a similar adaptation. However, to the best of our knowledge, this work is the first one to exploit LOD-KBs and its graph-base structure. The similarities measures employed were designed to operate under RDF graphs and exploit the rich background knowledge available.

3   Query Expansion Process

This section presents our method for expanding the query. Query expansion considers semantic information of the query terms recovered from the Semantic Web. The process is performed in four steps (Fig. 1). In the first step, called the semantic query annotator, the user query is received. Mentions of Linked Open Data resources in the query terms (i.e. annotations) are identified. Then, the set of annotations is sent to the Query Knowledge Graph Builder so that it can build a graph of related resources. This builder enriches the graph with resources in the neighborhood of the query annotations by considering selected properties. The resulting query knowledge graph (QKG) is the backbone of the system, and it represents the pertinent knowledge related to the query. The resource selection step chooses the most adequate resources from the QKG. How suitable a resource is for diversifying the query is based on a reformulation of the MMR [4]. Finally, the query expansion is performed with the label properties of the selected resources; this step is called query reformulation. In the following sub-sections, each step in the query expansion process is described in detail.

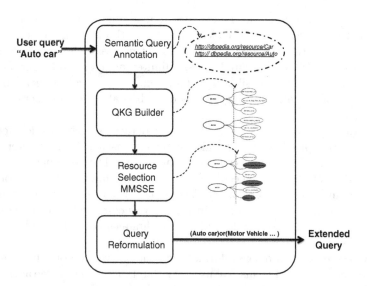

Fig. 1. System query reformulation process

To implement the first two steps of the expansion, we use DBpedia as a Linked Open dataset. Its comprehensive vocabularies and its extensive relationships between entities, combined with continuous updating, enable cross domain modeling capabilities. DBpedia consists of a set of resources R and literals L interrelated through a set of properties P which are used to denote specific relationships. The data consist of a set of statements involving the resources $E \subset R \times P \times (R/L)$ under an RDF model. Each $e \in E$ is a triplet composed of a subject, a predicate and an object (resource or literal). Recent experiments conducted by [14] show the potential of this knowledge source for the efficient calculation of similarity between entities.

3.1 Semantic Query Annotation

As was mentioned above, the first step of the process is the semantic annotation of the query. The goal of this step is to obtain Semantic Web resources from the initial query terms. Using DBpedia Spotlight[1] to find annotated mentions of DBpedia resources, first the query is processed. Then, a free text search in disambiguation resources in DBpedia for all words in the query is performed (see Sect. 3.2). If no resource is identified, the system sends the query without reformulation. Otherwise, the set of DBpedia annotations identified in the query becomes the input parameter for the next step.

3.2 Query Knowledge Representation Builder

The second step, which is the backbone of the process, is the generation of a Query Knowledge Graph. To do so, the builder receives the set S of annotations discovered in the preceding step, and it returns a QKG. This QKG is built from DBpedia resources and their relations.

Our query representation is basically a set of connections between the input set $S \subset R$ and an expanded set of resources $S_{expand} \subset R$ disjoint from S (i.e. $S \cap S_{expand} = \emptyset$). Each vertex represents a resource $r \in R$, and each edge represents the existence of a specific relationship between one node in the input set S and one node in S_{expand}. To build the set S_{expand}, we search in DBpedia for resources from selected relationships, i.e. candidate resources connected to resources in S through the properties $p \in P_{select}$, where $P_{select} \subset P$. We choose dbo:wikiPageDisambiguates and rdfs:seeAlso as P_{select}. These properties allow us to add resources to our graph with associated labels/literals that are useful in the following query expansion step. In DBpedia, disambiguation nodes are a way to resolve conflicts when a term is associated with more than one topic, so they are useful for diversification purposes. These ambiguous resources are associated with dbo:wikiPageDisambiguates property, which makes them easy to identify. Similarly, rdfs:seeAlso relates one resource to another one that might provide additional information about the subject resource[2]. In order to control the size of

---

[1] https://github.com/dbpedia-spotlight/dbpedia-spotlight.
[2] https://www.w3.org/wiki/UsingSeeAlso.

S_{expand}, we use a parameter f. This parameter represents the maximum number of resources that could be added to S_{expand}.

Figure 2 shows a fragment of the graph built for the input set $S=\{$dbr:Auto, dbr:Car$\}$. This input set represents the annotations identified in the user's query: Auto car. Right side nodes are resources from S_{expand} set, while left side nodes are the input set nodes. Orange and violet edges represent resources are added through dbo:wikiPageDisambiguates and rdfs:seeAlso respectively. The nodes dbr:Autonomous_car, dbr:List_of_Mega_Man_characters and dbr:Auto _(play) are added through the disambiguation node dbr:Auto. Once the graph is formed, it is enriched with additional information related to the graph nodes such as the hierarchical structure of Wikipedia categories (NHSW), neighborhood resources (LNR), and a specific set of chosen labels (NL).

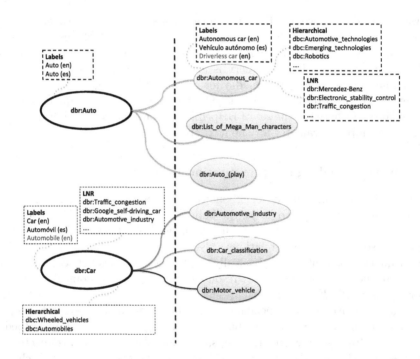

Fig. 2. Fragment of the QKG for the input set S=\{dbr:Car,dbr:Auto\}

Node hierarchical structure (NHSW): The hierarchical structure of a node is drawn from categories in a Wikipedia categorical system. Categories are extracted through dct:subject predicate and the classification attributes discovered are used to calculate the hierarchical similarity between two nodes. Only categories in the hierarchical structure processed by [12] were used to avoid disconnected categories and cycles.

List of neighboring resources (LNR): The LNR is the set of resources recovered by following the path that starts with the resource node and moves through the properties: $\{p : (p \in P_{DbOntology}) \wedge (p \notin P_{select})\}$ where $P_{DbOntology}$ is the set of non-categorical properties that are described in the DBpedia ontology. The maximum path length is defined by the configurable parameter $l_{maxpath}$. We limit $l_{maxpath}$ to a maximum of three following the recommendations presented in [14]. We also avoid adding resources already discovered in S_{expand}, or those that are in the initial S set. It is important to note that we do not store whole paths; instead, we add the discovered neighborhood as a list of neighboring resources (n_r) with an associated damping factor $\beta^{l_{nr}}$ $(0 < \beta \leq 1)$ that represents a penalization parameter depending on the path length (l_{nr}). So, the final LNR of the node r_i is expressed as $r_{i \to LNR} = \left\{ (n_{r_{i,1}}, \beta^{l_{nr_{i,1}}}), (n_{r_{i,2}}, \beta^{l_{nr_{i,2}}}), \ldots, (n_{r_{i,m}}, \beta^{l_{nr_{i,m}}}) \right\}$. Since we traverse the neighborhoods with a breadth-first strategy, the path found is the shortest one between the two resources. This process follows similar approaches that also penalize the path length we set as $\beta = 0.5$ [14]. The main use of LNR is the calculation of the similarity between two nodes. However, in cases when there is a limited amount of resources in S_{expand} (i.e. $|S_{expand}| < f$), we take resources from this list (starting from $l_{nr} = 1$) and incorporate them as nodes in the graph. In this case we only take into account LNR of nodes belonging to the input set. The black node in Fig. 2 is an example of resources added from LNR.

Node labeling (NL): Node labels are extracted from rdfs:label property. The two most frequent terms of the pairCount dataset[3] available through DBpedia Spotlight [6] are also selected. This data set associates terms with resources and gives each term a frequency.

3.3   Resource Selection

The third step of the process selects a subset of resources from the QKG for the query reformulation. The selection of resources is based on the MMR principle. As described earlier, MMR is a method that seeks result diversification. In MMR, the documents are selected according to their dissimilarity with previously selected documents [4]:

$$MMR\,(d_i) = \lambda\,sim\,(d_i,\,Q) - (1 - \lambda)\,max\,sim_{d_j\,\in\,C}\,(d_i,\,d_j) \qquad (1)$$

where sim could be any similarity metric used in document retrieval and relevance ranking. The candidate document d_i with the highest MMR score is selected in each step. C is the set of documents already selected and λ is a parameter that controls the tradeoff between the relevance of the document and its novelty. In our approach, instead of selecting documents, MMR is applied to select a set of resources in the QKG. The strategy, which we call MMSSE (Maximal Marginal Semantic Similarity Expansion), is calculated as follows:

$$MMSSE\,(r_i) = \alpha\,sim\,(r_i,\,S) - (1 - \alpha)\,max\,sim_{r_j\,\in\,C_r}\,(r_i,\,r_j) \qquad (2)$$

---

[3] http://spotlight.sztaki.hu/downloads/latest_data/.

where S is input set of the QKG, $r_i \in S_{expand}$ is a candidate resource and C_r is the set of already selected resources. In order to compute MMSSE, direct similarity between two resources $sim(r_i, r_j)$ and a joint similarity between a resource and a set of resources $sim(r_i, S)$ must be calculated. We use α instead of λ to denote that in our proposal the relevance/novelty tradeoff is performed between resources; however, both parameters are used to explicitly specify the desired degree of diversity.

Direct similarity measure: To calculate the similarity between two nodes, we calculate a hierarchical (HS) and non-hierarchical similarity (NHS). The total similarity measure is expressed as the sum of HS and NHS:

$$sim\,(r_i, r_j) = HS\,(r_i, r_j) + NHS(r_i, r_j) \tag{3}$$

The taxonomic similarity proposed by [15] was used to calculate HS. If A is the set of categories of the resource r_i and B is the set of categories of the resource r_j, we compute HS as:

$$HS\,(r_i, r_j) = \max_{c_i \in A, c_j \in B} taxsim(c_i, c_j) \tag{4}$$

$$\text{where } taxsim(c_i, c_j) = \frac{\delta(root, c_{lca})}{\delta\,(c_i, c_{lca}) + \delta\,(c_j, c_{lca}) + \delta(root, c_{lca})}$$

where $\delta(a, b)$ is the number of edges on the shortest path between a and b, and c_{lca} is the lowest common ancestor (LCA) of c_i and c_j. Given the taxonomic categories T, the LCA of two nodes c_i and c_j is the vertex of greatest depth in T that is the common ancestor of both c_i and c_j. We use the nodes in LNR to calculate the NHS. Given two nodes r_i, r_j and a set of common neighborhood resources $M_{shared} = \{r_{i \to LNR} \cap r_{j \to LNR}\}$ the NHS is calculated as:

$$NHS\,(r_i, r_j) = \sum_{nr \,\in\, M_{shared}} \beta^{l_{nr_i}} * \beta^{l_{nr_j}} = \sum_{nr \,\in\, M_{shared}} \beta^{l_{nr_i} + l_{nr_j}}$$

$$NHS_{norm}\,(r_i, r_j) = \frac{NHS\,(r_i, r_j)}{\min(NHS\,(r_i, r_i), NHS\,(r_j, r_j))} \tag{5}$$

Joint similarity measure: We compute the joint similarity as the average of the individual direct similarities between the resource involved and resources in the set S:

$$sim\,(r_i, S) = \frac{\sum_{r_j \,\in\, S} sim\,(r_i, r_j)}{|S|} \tag{6}$$

Using the formulas above, we construct the set C_r by sequentially incorporating the resource that maximizes Eq. 2.

3.4   Query Reformulation

In the final step of the process, we extract node-labeling attributes from the resulting resources in C_r to reformulate the query. Because we use labels instead of terms in the reformulation, we cannot anticipate the number of words that will be added to the query. We can summarize the reformulation subprocess as follows: for each resource in C_r we take two different labels and add them to the original query as a disjunction: (Original Query Terms) OR (TermsLabel1 OR ... OR Term Label2). If the specific search engine does not support query operators, we merely append the labels to the query at the expense of reduced diversification of results [17].

4   Evaluation

To evaluate our approach, we conducted two different experiments with different diversity measures. The first one, called categorical diversity, was performed on a small centralized collection of learning resources crawled from MERLOT[4]. In this experiment, we took advantage of MERLOT's taxonomic categories to measure novelty in terms of a hierarchical distance between the results categories. In the second experiment, called federated diversity, we simulated a federated scenario from three datasets: (i) DBLP data set[5], (ii) MERLOT crawled dataset and a (iii) set of news articles crawled from different sources[6]. In this scenario, we compared our approach (MMSSE) with two algorithms commonly used as baselines for implicit diversification strategies [2,9,22]: MMR and maximum sum dispersion (MSD).

The set of documents returned by the search engine mentioned above will be referred to here as the candidate set. In the case of the federated scenario, the candidate set is the set of documents produced after the merging step. Result ordering algorithms operate on this candidate set to obtain the result set L. For all experiments, we set $N = 10$ (number of resources to be selected), and a result set size $k = |L| = 20$. We selected 50 queries in the English language from navigation logs of a user seeking learning materials on the Web. The queries chosen differ in the number of annotations of DBpedia resources (e.g. 10 queries with one annotation, 20 queries with 2 annotations and 20 with 3 annotations). All of our experiments in DBpedia were performed in a Virtuoso installation with a memory configuration of 16 GB.

4.1   Categorical Diversity

We used 20,000 unique crawled documents from the MERLOT portal; in this corpus, each document belonged to one or more categories. We added duplicates

---

[4] https://www.merlot.org/merlot/index.htm.

[5] http://dblp.uni-trier.de/db/conf/dbpl/.

[6] Datasets can be downloaded from https://github.com/Ruframapi/Diversified-Semantic-Query-Reformulation.

of each document according to the number of categories that it belonged to so that the final data set consisted of more than 70,000 document-category pairs. As a diversification measure, we computed the number of categories in the result set L for the query q according to the formula [9]:

$$CN_q(L) = \frac{2}{k(k-1)} \sum_{u,v \in L} I(lca(u,v) \notin \{u,v\}) \tag{7}$$

where the indicator function I assigns 1 if the expression is true, and 0 otherwise. $lca(u,v)$ is the lowest common ancestor between two categories in the result set. In this formula, categories that are not descendant of one another are more representative in the result set. To simplify the analysis of the results, we normalized the categorical novelty with the novelty value of the result set produced by the original query (q without expansion). To perform this, we used Solr[7] as a retrieval engine without any special configuration. In the first round of experiments using this engine, we were interested in evaluating the performance of MMSSE as we changed the different parameters of our approach. Table 1 presents the tested ranges and their default values.

Table 1. Parameters tested in the categorical diversity experiments

Parameter	Value (**default**)
α	0.1,**0.3**,0.5,0.7,0.9
$l_{maxpath}$	1,**2**,3
f	10,**20**,50,100,500,1000

Figure 3 shows the average CN_{norm} for queries with one (MMSSE-q1), two (MMSSE-q2) and three annotations (MMSSE-q3). For each query, we vary one specific parameter. Note that with the increase of α, there is a natural decrease in the proportion of the number of categories covered in the result set. In general, as annotations increase, diversification measures increase because more precise similarity and dissimilarity measures can be calculated. The analysis of our query logs shows that queries with more than three annotations are rare, which is consistent with the literature stating that queries are often short [5]. Hence, we did not perform experiments with more annotations.

Concerning the size of the set S_{expand}, we noticed an improvement in category diversity as the size of S_{expand} increased. However, a considerable number of similarity measures had to be calculated in order to select the resources. So for further iterations of the experiment, we fixed $f = 20$ as the max size as it has a competitive computational cost. Unlike the behavior of α and f, there was no significant change with the increase $l_{maxpath}$.

Finally, we compared MMSSE with MMR and MSD algorithms. We selected a high value for the tradeoff parameter of all algorithms. We used $\alpha = 0.1$ for

---

[7] http://lucene.apache.org/solr/.

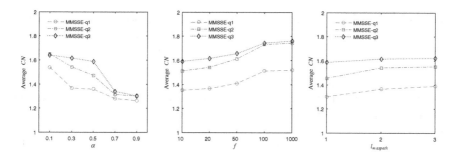

Fig. 3. Centralized search test: average CN vs α, f and $l_{maxpath}$

MMSSE and $\lambda = 0.1$ for MMR and MSD. For MMR and MSD we varied the candidate set size from 100 to 1000 and we used the tf-idf *cosine* as measure of similarity. In the case of MSD, we employed $1 - cosine$ as the diversification measure. For the construction of tf-idf dictionaries and the calculation similarities, we used Gensim and Stanford Core NPL.

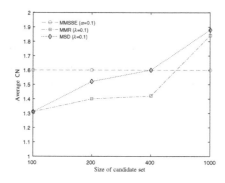

Fig. 4. Centralized search test, algorithms comparison

In Fig. 4, we show the superiority of our approach in terms of novelty when the size of the candidate set is less than 400 documents. It is important to notice that we could improve MMSSE results if we performed an additional reordering on the result set; however, we are interested in evaluating our approach with no intervention on the candidate result set. With this constraint in place, MMSSE is independent of the result set. In the following experiments, we explore not only novelty, but also relevance in a federated scenario.

4.2   Federated Diversity

In the second experiment, we worked with five different collections used as independent sources. Three of them came from the DBLP dataset in which random selection was performed. Each subset had more than one million documents related to publications in computer science. We used three different instances of Terrier[8] to support the search across these collections and the MERLOT dataset on an instance of Solr. Finally, we use a dataset of crawled news with 4,300 articles. This dataset operated on an instance of Elasticsearch[9]. No special configurations were implemented in the selected search engines. The same set of queries used in the categorical novelty evaluation was employed in this set of experiments.

We also employed a low-cost merging strategy by taking the most sequentially relevant result from each collection. Unlike previous experiments, we not only considered a measure of novelty, but also a relevance measure. Following the evaluation approach of [22], given a query q, we computed the relevance and novelty of the result set as:

$$Relevance\,(q, L) = \sum_{i=1, l_i \in L}^{k} \delta_{sim}\,(q, l_i) \tag{8}$$

$$Novelty(L) = \sum_{i=1, s_i \in L}^{k-1} \sum_{j=i+1, s_j \in L}^{k} \delta_{div}\,(s_i, s_j) \tag{9}$$

where δ_{sim} is a similarity measure and δ_{div} is a measure of diversity. We normalized both measures dividing them by the values of the result set with the highest relevance L_{max} (i.e. k most similar documents against the query) in the candidate set obtained by q (not expanded version).

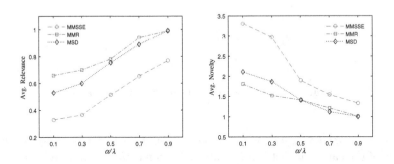

Fig. 5. Federated search evaluation: search engines limited to a maximum of 100 retrieval documents per query

Given the absence of common categories associated with all documents, we rely on *cosine* similarity for δ_{sim} and δ_{div} ($1 - cosine$). We used the defaults in Table 1 for our approach except for α values. In our first experiment, we fixed the

---

[8] http://terrier.org/.
[9] https://www.elastic.co/.

number of results returned from each search engine to a maximum of 100 while
different values of α and λ were employed. Figure 5 shows the results obtained.

As was expected, MMR and MSD achieved the most relevant results with
values of λ close to 0.9. Our approach does not have the capacity to return
the most relevant documents in the evaluated result set of size $k = 20$. Despite
this shortcoming, however, we noticed that in most of the experiments, the
complete set of documents in L_{max} were presented in the merged list. Conversely,
our approach shows a superior novelty capacity for all values of the tradeoff
parameter.

In the following experiment, we limited the number of results from each
collection to a maximum of 20. The same results were achieved for MMSSE
while MMR and MSD show few differences in the degree of novelty achieved for
different alpha values (Fig. 6). In general, our approach showed its full potential
in the scenario where reordering techniques failed due to the reduced number of
results.

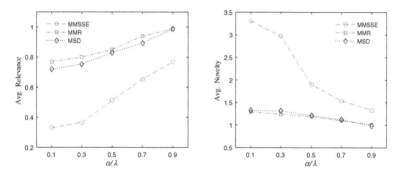

Fig. 6. Federated search evaluation: search engines limited to a maximum of 20 retrieval
documents per query.

Finally, Table 2 shows the average execution time for the process in the federated scenario (without taking into account the time spent by the different search
engines). For these experiments we calculated the average time with default
MMSSE parameters. In the MMS and MSD columns, time spent in the construction of the dictionaries and frequencies is considered. Although we tried
different tf-idf implementations, we found similar (and longer) processing times
for this construction. Since different strategies could be followed in order to
reduce the time, (e.g. precalculated inverse frequency values over a representative set), we show the processing time for only the calculations of similarly in
each approach (columns MMS_reduce and MSD_reduce). Even in a setting with
the aforementioned advantages, MMSSE has a constant competitive computational cost.

Table 2. Average processing times in milliseconds

Candidate size	MMSSE (ms)	MMS (ms)	MSD (ms)	MMS_reduce (ms)	MSD_reduce (ms)
50	1131	8115	8485	650	953
100	1131	19254	19450	1211	2450
400	1131	112845	133140	5120	8452
1000	1131	131254	171020	12123	22410

Note: Average processing times for each step of MMSSE: Semantic Query Annotation (250.11 ms), QKG Builder (520.88 ms), Resource Selection MMSSE (330 ms), Query reformulation (30.07 ms).

5   Conclusion

In this paper, we presented a new approach for result diversification based on semantic representation and expansion of the query. We employed recent advances in similarity calculations over knowledge graphs in order to perform a fast selection of resources for query reformulation. The experiments conducted in centralized and federated scenarios show a superior capacity in terms of novelty in comparison with reordering techniques when operating in comparably small candidate sets. We also found a competitive processing time that made our approach suitable for systems where the response time is critical. Future work on the reformulation process will explore the use of knowledge databases different from DBpedia. Additionally, we would like to explore the automatic adjustment of the parameter α based on an estimated degree of ambiguity obtained analyzing the QKG.

Acknowledgment. This work was partially supported by COLCIENCIAS PhD scholarship (Call 647-2014).

References

1. Agrawal, R., Gollapudi, S., Halverson, A., Ieong, S.: Diversifying search results. In: Proceedings of the Second ACM International Conference on Web Search and Data Mining, pp. 5–14, WSDM 2009. ACM, New York (2009)
2. Bouchoucha, A., He, J., Nie, J.Y.: Diversified query expansion using conceptnet. In: Proceedings of the 22nd ACM International Conference on Information Knowledge Management, CIKM 2013, pp. 1861–1864. ACM, New York (2013)
3. Bouchoucha, A., Liu, X., Nie, J.Y.: Integrating Multiple Resources for Diversified Query Expansion, pp. 437–442. Springer, Cham (2014)
4. Carbonell, J., Goldstein, J.: The use of MMR, diversity-based reranking for reordering documents and producing summaries. In: Proceedings of the 21st Annual International ACM SIGIR Conference on Research and Development in Information Retrieval, SIGIR 1998, pp. 335–336. ACM, New York (1998)
5. Carpineto, C., Romano, G.: A survey of automatic query expansion in information retrieval. ACM Comput. Surv. **44**(1), 1–50 (2012)
6. Daiber, J., Jakob, M., Hokamp, C., Mendes, P.N.: Improving efficiency and accuracy in multilingual entity extraction. In: Proceedings of the 9th International Conference on Semantic Systems, pp. 121–124 (2013)

7. Dang, V., Croft, W.B.: Diversity by proportionality: an election-based approach to search result diversification. In: Proceedings of the 35th International ACM SIGIR Conference on Research and Development in Information Retrieval - SIGIR 2012, p. 65 (2012)
8. Ghansah, B., Wu, S.: A mean-variance analysis based approach for search result diversification in federated search. Int. J. Uncertain. Fuzziness Knowl. Based Syst. **24**(02), 195–211 (2016)
9. Gollapudi, S., Sharma, A.: An axiomatic approach for result diversification. In: Proceedings of the 18th International Conference on World Wide Web, pp. 381–390, WWW 2009 (2009)
10. He, J., Hollink, V., de Vries, A.: Combining implicit and explicit topic representations for result diversification. In: Proceedings of the 35th International ACM SIGIR Conference on Research and Development in Information Retrieval, p. 851 (2012)
11. Hong, D., Si, L.: Search result diversification in resource selection for federated search. In: Proceedings of the 36th International ACM SIGIR Conference on Research and Development in Information Retrieval, SIGIR 2013, pp. 613–622 (2013)
12. Kapanipathi, P., Jain, P., Venkataramani, C.: Hierarchical interest graph. Technical report (2015)
13. Minack, E., Demartini, G., Nejdl, W.: Current approaches to search result diversification. In: Proceedings of 1st International Workshop on Living Web (2009)
14. Paul, C., Rettinger, A., Mogadala, A., Knoblock, C.A., Szekely, P.: Efficient graph-based document similarity. In: Sack, H., Blomqvist, E., d'Aquin, M., Ghidini, C., Ponzetto, S.P., Lange, C. (eds.) ESWC 2016. LNCS, vol. 9678, pp. 334–349. Springer, Cham (2016). doi:10.1007/978-3-319-34129-3_21
15. Pekar, V., Staab, S.: Taxonomy learning: factoring the structure of a taxonomy into a semantic classification decision. In: Proceedings of the 19th International Conference on Computational Linguistics, vol. 1, pp. 1–7 (2002)
16. Rafiei, D., Bharat, K., Shukla, A.: Diversifying web search results. In: Proceedings of the 19th International Conference on World Wide Web, WWW 2010, p. 781 (2010)
17. Rubien, R., Ziak, H., Kern, R.: Efficient search result diversification via query expansion using knowledge bases. In: Proceedings of 12th International Workshop on Text-based Information Retrieval (TIR), p. 5 (2015)
18. Santos, R.L.T., Macdonald, C., Ounis, I.: Aggregated search result diversification. In: Amati, G., Crestani, F. (eds.) ICTIR 2011. LNCS, vol. 6931, pp. 250–261. Springer, Heidelberg (2011). doi:10.1007/978-3-642-23318-0_23
19. Santos, R.L.T., Macdonald, C., Ounis, I.: On the role of novelty for search result diversification. Inf. Retrieval **15**(5), 478–502 (2012)
20. Vargas, S., Castells, P., Vallet, D.: Explicit relevance models in intent-oriented information retrieval diversification. In: Proceedings of the 35th International ACM SIGIR Conference on Research and Development in Information Retrieval, SIGIR 2012, p. 75 (2012)
21. Vee, E., Srivastava, U., Shanmugasundaram, J., Bhat, P., Yahia, S.A.: Efficient computation of diverse query results. In: Proceedings - International Conference on Data Engineering, pp. 228–236 (2008)
22. Vieira, M.R., Razente, H.L., Barioni, M.C.N., Hadjieleftheriou, M., Srivastava, D., Traina, C., Tsotras, V.J.: On query result diversification. In: Proceedings of the 2011 IEEE 27th International Conference on Data Engineering, pp. 1163–1174, ICDE 2011. IEEE Computer Society, Washington, DC (2011)

RuThes Cloud: Towards a Multilevel Linguistic Linked Open Data Resource for Russian

Alexander Kirillovich[1(✉)], Olga Nevzorova[2], Emil Gimadiev[1], and Natalia Loukachevitch[3]

[1] Kazan Federal University, Kazan, Russia
alik.kirillovich@gmail.com, gimpwl@gmail.com
[2] Institute of Applied Semiotics, Tatarstan Academy of Sciences, Kazan, Russia
onevzoro@gmail.com
[3] Lomonosov Moscow State University, Moscow, Russia
louk_nat@mail.ru

Abstract. In this paper we present a new multi-level Linguistic Linked Open Data resource for Russian. It covers four linguistic levels: semantic, lexical, morphological and syntactic. The resource has been constructed on base of the well-known RuThes thesaurus and the original hitherto unpublished Extended Zaliznyak grammatical dictionary. The resource is represented in terms of SKOS, Lemon, and LexInfo ontologies and a new custom ontology. Building the resource, we automatically completed the following tasks: merging source resources upon common lexical entries, decomposing complex lexical entries, and publishing constructed resource as LLOD-compatible dataset. We demonstrate the use case in which the developed resource is exploited in IR task. We hope that our work can serve as a crystallization point of the LLOD cloud in Russian.

Keywords: Linguistic Linked Open Data · Linked data · Language resources · Ontology · Thesaurus · Lexicon · Grammatical dictionary · Dependency grammar · RuThes · Russian language

1 Introduction

Publishing linguistic data according to the principles of Linked Open Data is a recently established trend, that has led to the emergence of the fast-growing Linguistic Linked Open Data (LLOD) cloud [1, 2]. It has the following advantages:

- Structural and conceptual interoperability.
- Integration of linguistic resources (for example, of a corpus and a thesaurus linked to it) to solve a common problem.
- Integration of linguistic and extralinguistic resources from the Linked Open Data cloud. This integration is exploited in such tasks as: (1) Ontology-based information extraction from text. (2) Semantic annotation and search for documents on the LOD cloud. (3) Generation of SPARQL queries for user-provided natural language questions. (4) Production of human understandable descriptions of LOD resources.

© Springer International Publishing AG 2017
P. Różewski and C. Lange (Eds.): KESW 2017, CCIS 786, pp. 38–52, 2017.
https://doi.org/10.1007/978-3-319-69548-8_4

– Exploitation of the powerful Semantic Web infrastructure and ecosystem, that include triplestores, logical reasoners, ontologies, etc.

Linguistic resources for many major languages have already been published on the LLOD cloud. Examples of such resources are the WordNet (W3C's WordNet [3], Lemon WordNet [18], lemonUby [4] and official WordNet RDF [5]), English and German Wiktionary (lemonUby [4] and DBnary [6]), FrameNet [4], VerbNet [4], BROWN corpus, and dozens of subject-related thesauri, such as EuroVoc [7], Agro-Voc [8, 9], TheSoz [10] and the Library of Congress Subject Headings [11].

However, Russian linguistic resources are underrepresented in the LLOD cloud. There are several projects in this area, but its results are quite limited. (1) The first and the most valuable is RTLOD [12], the project of converting three Russian general-purpose thesauri into RDF dataset in terms of LLOD-related ontologies. However, obtained resources do not fully comply with the LLOD standards: first, they do not contain resolvable URIs, and second, they are not interlinked with external resources. (2) The second is the set of small multilingual resources for narrow subject areas, such as the agricultural thesaurus AgroVoc, the thesaurus for social sciences TheSoz, and the ontology of mathematical knowledge OntoMath^{PRO} [13, 14]. Most of these resources are fully comply with the LLOD standards, but they cover only restricted domains. (3) The third is BabelNet [15, 16], the large-scale multilingual thesaurus, automatically constructed on base of WordNet and Wikipedia. This resource covers general domain and is fully comply with the LLOD standards, but its Russian lexical entries were translated automatically and not manually checked. Moreover, none covers more than the semantical and the lexical layers.

To overcome these limitations, we are launching a large project to create RuThes Cloud—Linguistic Linked Open Data resource built on base of the conceptual network of the RuThes thesaurus. We intend that this resource will become the core of the LLOD cloud in Russian and the languages of the peoples of Russia. According to the project, RuThes Cloud will include:

1. A hierarchy of language-independent concepts.
2. Lexical entries in Russian, English, Tatar, and, later, in other languages of the peoples of Russia.
3. Forms of lexical entries (in different cases, numbers, tenses, etc.)
4. Semantic and syntactic frames. The semantic frame contains semantic roles associated with a concept (for example, for the concept of *Gift*, the roles are *giver*, *recipient*, *gift*, *time*, etc.). The syntactic frame contains a subcategorization model for a particular lexical entry and its mapping to semantic roles (for example, for the verb 'подарить' ('to give') dependent word in the dative case expresses a *recipient* of the *Gift*).
5. Syntactic trees of multi-word lexical entries, where these entries are decomposed to their constituting words.
6. Other relations between lexical entries (antonymy, derivation, abbreviation, etc.)
7. Links to external resources, such as WordNet, BabelNet, DBpedia, etc.

In this paper we present the first results of the project and describe the current state of RuThes Cloud. The resource has been constructed on base of the well-known

RuThes thesaurus and the original hitherto unpublished Extended Zaliznyak grammatical dictionary. It contains (1) a hierarchy of concepts, (2) lexical entries in Russian referring to concepts, (3) forms of lexical entries associated with their grammatical tags and (4) syntactic trees of multi-word lexical entries. Thus, the resource covers four major language levels: semantic, lexical, morphological and partially syntactic.

The main contributions of this paper are: (1) the Extended Zaliznyak grammatical dictionary, we have been developing for decade; (2) merging this dictionary with the RuThes-Lite thesaurus; (3) decomposition of multi-word lexical entries; (4) representation of the constructed resource as LLOD-compatible dataset and its publishing on the LLOD cloud.

The rest of the paper is organized as follows. In Sect. 2 we give an overview of the models and the ontologies for the LLOD resources. In Sect. 3 we describe two source resources, the RuThes thesaurus and the Extended Zaliznyak grammatical dictionary. In Sect. 4, we describe the result of our work, the current version of RuThes Cloud. Section 5 presents examples of queries to the resource. In conclusion, we describe the directions of future work.

2   The Linguistic Linked Open Data Cloud: Models and Ontologies

In this section we briefly describe the main models and ontologies of the Linguistic Linked Open Data cloud.

To represent the simplest lexical resources, the SKOS ontology is used [17]. This ontology allows us to describe a set of concepts related by hierarchical and association relationships. Each concept can have a representation in different languages.

Lemon [18–20] ontology is used to represent complex lexical resources. This ontology is based on the international standard ISO 24613:2008 "Language resource management - Lexical markup framework (LMF)" [21]. The basic elements of this ontology are: lexicons, lexical entries, forms of a lexical entry, senses of a lexical entry, and concepts from ontologies of subject domains. Lemon makes it possible to describe: (1) word forms with grammatical tags (gender, number, case, tense, etc.); (2) decomposition of a complex lexical entry into simple ones; (3) syntactic frame of a lexical entry; (4) binding a lexical entry to the concept referred by it from an external ontology; (5) relations between different lexical entries/lexical forms (for example, etymological relations, relation between the full form of a word and its abbreviation, etc.); (6) semantic relations between senses (antonym, translation, meronym, hyponym, hypernym, etc.); (7) morphological patterns for lexical entries.

To describe language categories (gender, number, case, tense, direct object, indirect object, synonym, antonym, etc.) in Lemon, the external ontology LexInfo is used.

To describe language categories in LLOD, different ontologies are used that are linked to the ISOcat data category registry [22–24], which is the implementation of the international standard ISO 12620: 2009 "Terminology and other languages and content resources - Specification of data categories and management of the Data Category" [25]. Examples of such ontologies are LexInfo [26] and ubyCat [4].

OLiA [27, 28] is a family of ontologies designed to map different systems of language categories and annotation schemes to each other using an intermediate ontology.

NLP Interchange Format (NIF) [29] and Web Annotation [30] ontologies are used for corpora annotation.

3 Source Resources

In this section we briefly describe the two linguistic resources used to construct RuThes Cloud: the Extended Zaliznyak grammatical dictionary and the RuThes-Lite thesaurus.

3.1 Extended Zaliznyak Grammatical Dictionary

We use the grammatical dictionary of our "OntoIntegrator" system [31] when constructing our resource. This dictionary contains about 140 thousand word paradigms of the Russian language (more than 3 million wordforms). The dictionary was based on the grammatical dictionary by A.A. Zaliznyak, and then we essentially supplemented the basic vocabulary with modern terminology from different areas.

Each word paradigm has a unique identifier. Paradigms of lexical homonyms have different identifiers, but their paradigms coincide. Grammatical (related to different parts of speech) homonyms have unified identifiers and a different composition of paradigms.

3.2 RuThes-Lite Thesaurus

The semantic core of the multilevel open linked resource is the RuThes-Lite thesaurus [32].

RuThes thesaurus is a linguistic ontology, i.e., an ontology, the concept of which is based on the meanings of words and expressions. Each concept of this ontology has a unique and unambiguous name and is associated with a set of words and expressions that can be expressed in the text (so-called text entries). The set of text entries includes synonyms, derivatives (words of different parts of speech), word combinations and other types of expressions.

Figure 1 contains the description of the concept *Computer virus* in RuThes thesaurus.

The meanings of ambiguous words refer to different concepts. There are four types of relationships in the RuThes thesaurus: hierarchical relation, part-whole relation, relation of the asymmetric association (the relation of ontological dependence) and relation of symmetric association.

The published version of the RuThes-Lite thesaurus (about 115 thousand lexical units of the Russian language) includes a set of xml files containing:

- a list of concepts with an indication of the domain (general lexicon, socio-political area, geography);
- a list of relations between concepts;

```
COMPUTER VIRUS
BT: MALWARE
WHOLE: VIRUS ATTACK
NT: BOOT VIRUS
NT: MACRO VIRUS
NT: MAIL VIRUS
PART: VIRUS LENGTH
PART: VIRUS CODE
ASC2: ANTIVIRUS PROTECTION
ASC2: VIRUS ACTIVATION
ASC2: VIRUS WRITING
```

Fig. 1. The description of the concept *Computer virus* (from [33])

- a list of text entries of concepts, where the text entry description contains the syntactic type (NG, VG, etc.) and the main word (head) of the noun group.
- a list of correspondences of text entries to the concepts of thesaurus.

4   RuThes Cloud Description

In this section we describe the current version of RuThes Cloud. It incorporates the following components:

1. The hierarchy of language-independent concepts;
2. Lexical entries in Russian, related to the concepts they refer to through lexical senses;
3. Forms of lexical entries associated with grammatical tags;
4. Syntactic decomposition trees of multi-word lexical entries.

RuThes Cloud was constructed on base of RuThes-Lite thesaurus and the Extended Zaliznyak grammatical dictionary. The hierarchy of concepts is derived from the thesaurus, the forms of lexical entries are derived from the dictionary, the lexical entries are derived from the both sources, and the syntactic trees are generated automatically.

The resource is expressed in terms of SKOS [17], Lemon [18, 19], LexInfo [26] and PROV [34] ontologies.

Also, we have developed a new custom ontology that defines classes and properties which cannot be expressed by other ontologies. However, our ontology has been developed as an extension of these pre-existing ontologies, whereby our resource retains compatibility with applications that are based on these ontologies. Wherever possible, we linked the entities of our ontology to the ISOcat data category registry (see [24]).

4.1   The Hierarchy of Concepts

The concept hierarchy of our RuThes Cloud reflects the concept hierarchy of RuThes. Concepts are considered as "units of thought" and are not language-dependent.

The description of a concept contains its unambiguous name and glosses. Currently, they are only in Russian, but we are going to add names in other languages (first of all, in English and Tatar). Also, the description contains a link to the resource, from which this concept has been obtained (now it is RuThes only for all concepts).

There are seven types of sematic relations between concepts. The first two are the hypernymic and hyponymic relations. Hypernymy is the union of the traditional ontological instanceOf and isA relations. Examples of such relations are the relation between *Poland* and *European state* (instanceOf relation) as well as the relation between the *European state* and *State* concepts (isA relation).

The third and fourth relations are holonymy and meronymy (whole and part) relations.

The fifth and sixth relations are two directed association relations: association1 and association2. These relations express the relation of the external ontological dependence between two concepts [35]. The association1 relation is established between two concepts C_1 and C_2 when C_1 is ontologically dependent on C_2 and C_1 is not a part of C_2. An example of this relation is the relation between *Auto racing* and *Car* concepts. The association2 relation is the inverse of the association1.

The seventh relation is the undirected association between two concepts.

Concepts of our resource are represented as instances of skos:Concept class. The name of the concept is expressed by the rdfs:label property, and the gloss is expressed by skos:definition property.

The relationship between the concept and the resource from which it was obtained is expressed by prov:wasDerivedFrom property of PROV ontology.

Usually, in the thesauri the skos:broader and skos:narrower properties are used to express the hypernym-hyponym relationship. However, semantics of these properties is not clear and they can also express a whole-part relationship. In this regard, we have defined two new properties: ruthes-ontology:hypernym and ruthes-ontology: hyponym, which are defined as subproperties of skos:broader and skos:narrower.

In BabelNet [16], as well as in the RTLOD project [12], holonymy and meronymy relations between concepts are expressed by lexinfo:meronymTerm and lexinfo:holonymTerm properties. But it is not correct, because these properties are subproperties of lemon:senseRelation property, and their domain and range are lexical senses, not concepts. In the WordNet RDF [5] as well as the W3C's WordNet [3] projects, holonymy and meronymy relations were defined from scratch. The disadvantage of this approach is the problem of interoperability. In our resource, we have chosen a middle way and defined new ruthes-ontology:holonym and ruthes-ontology:meronym properties, which, however, were defined as subproperties of the standard skos:broader and skos:narrower properties.

Representation of ruthes-ontology:hyponym and ruthes-ontology:meronym as subproperties of the common skos:narrower property is convenient in the query expansion task, when all subordinated concepts of a given concept are to be selected.

For the two directed and one undirected association relations we have defined the new ruthes-ontology:association1, ruthes-ontology:association2 and ruthes-ontology: association properties respectively, which are subproperties of the standard skos:related property.

4.2 Lexical Entries, Lexical Forms and Senses

Lexical entries are of two types: single words and multi-word expressions.

The description of the lexical entry contains its name and part of the speech tag, as well as a link to the resource from which this lexical entry was obtained (RuThes, grammatical dictionary or both).

Lexical entries are also related to their lexical forms. The description of a form contains its written representation, part of speech tag, as well as other grammatical tags.

Lexical entries refer to concepts. The same lexical entry may refer to more than one concepts, and the same concept may be referred to by several lexical entries. Lexical entries are related with concepts through lexical senses.

Single words and multi-word expressions are represented as instance of lemon:Word and lemon:Phrase classes respectively. The name of the lexical entry is expressed as the rdfs:label property, part of speech is expressed by the lexinfo:partOfSpeech property.

A lexical entry is related to its main form by the lemon:canonicalForm property, and to other forms by the lemon:otherForm property.

Forms of the lexical entry are represented as instances of the class lemon:Form. The text representation of a lexical entry is expressed by the lemon:writtenRep property, the part of speech is expressed by the lexinfo:partOfSpeech property, the case, the number, the gender, the tense, the person and the mood are expressed by the lexinfo:case, lexinfo:number, lexinfo:gender, lexinfo:tense, lexinfo:person, and lexinfo:verbFormMood properties respectively.

The sense of the lexical entry is represented as instance of the lemon:LexicalSense class. The lexical entry is related to the sense by the lemon:sense property and the sense is related to the concept by the lemon:reference property (Fig. 2).

4.3 Decomposition of Multi-word Lexical Entries

Multi-word lexical entries are decomposed into its constituting single words according to the following syntactic models.

The common syntactic models of a Russian multi-word terminological phrase are the Noun Phrase (NP) and Prepositional Phrase (PN). A Noun Phrase is a syntactic unit, resulting from expanding a lexical noun (a head) by the addition of a variety of arguments and modifiers. The structure of NP is determined by the set of syntactic relations between the head and dependent words. We consider three relations between components of NP (coordination, subordination, and cumulation). In subordination we have the head and the dependent, and the dependent is subordinated to the head. Such syntactic relations are found in all headed phrases, e.g. 'beautiful girl', 'country doctor'. The examples of cumulation one can found in non-headed dependent phrases, e.g. 'my old (friend)', 'big red (ball)'.

However, we distinguish the two cases in Russian: (1) NP with Verbal Noun; and (2) NP with Noun. In the first case, we apply the attribute Sub to mark the relation between the head and dependent words. In the second case, we apply the attribute Atr.

We use the dependency relation of dependency grammars for representing the NP structure. The head directly dominates its dependent in the dependency tree.

The basic element of decomposition is a constituent component, which represents a particular occurrence of the word in the multi-word expression. The component description contains a link to the corresponding word and grammatical tags of its occurrence.

The decomposition result is presented in the following structures. The first is an ordered list of constituent components. The order of the components in this list corresponds to the word order in the multi-word lexical entry.

The second structure is a dependency tree, where each node contains a link to a corresponding constituent component, and dependency relations to subordinated nodes.

Constituent components are represented as instances of the lemon:Component class. A link to a corresponding word is defined by the lemon:element property. Grammatical tags of the components are defined in the same way as for lexical forms.

The list of components is expressed as an ordered RDF list, and the relation between the multi-word lexical entry and the list of its components is defined by the lemon:decomposition property.

```
@prefix rdfs: <http://www.w3.org/2000/01/rdf-schema#>.
@prefix skos: <http://www.w3.org/2004/02/skos/core#>.
@prefix lemon: <http://www.lemon-model.net/lemon#>.
@prefix lexinfo: <http://www.lexinfo.net/ontology/2.0/lexinfo#>.
@prefix prov: <http://www.w3.org/ns/prov#>.
@prefix ruthes-ontology: <http://lod.ruthes.org/ontology#>.

<http://lod.ruthes.org/resource/concept/154>
   a skos:Concept;
   rdfs:label "АВТОМОБИЛЬ"@ru;
   skos:definition
     "самоходное автономное безрельсовое колёсное транспортное средство"@ru;
ruthes-ontology:hypernym
   <http://lod.ruthes.org/resource/concept/4029>;
ruthes-ontology:hyponym
   <http://lod.ruthes.org/resource/concept/156>,
   <http://lod.ruthes.org/resource/concept/162>, ...;
ruthes-ontology:meronym
   <http://lod.ruthes.org/resource/concept/2653>,
   <http://lod.ruthes.org/resource/concept/108798>, ...;
ruthes-ontology:association2
   <http://lod.ruthes.org/resource/concept/160>,
   <http://lod.ruthes.org/resource/concept/166>, ...;
lemon:isReferenceOf
   <http://lod.ruthes.org/resource/sense/154-RU-автомашина-n>,
   <http://lod.ruthes.org/resource/sense/154-RU-автомобиль-n>, ...;
prov:wasDerivedFrom <http://lod.ruthes.org/resource/ruthes-lite>;
skos:inScheme <http://lod.ruthes.org>.
```

Fig. 2. Code fragment with representation of *Автомобиль* (*Automobile*) concept and related lexical entries, senses and forms

```
<http://lod.ruthes.org/resource/sense/154-RU-автомобиль-n>
  a lemon:LexicalSense;
  lemon:reference
    <http://lod.ruthes.org/resource/concept/154>;
  lemon:isSenseOf
    <http://lod.ruthes.org/resource/entry/RU-автомобиль-n>.

<http://lod.ruthes.org/resource/entry/RU-автомобиль-n>
  a lemon:Word;
  rdfs:label "АВТОМОБИЛЬ";
  lexinfo:partOfSpeech lexinfo:noun;
  lemon:sense
    <http://lod.ruthes.org/resource/sense/154-RU-автомобиль-n>;
  lemon:canonicalForm
    <http://lod.ruthes.org/resource/form/RU-автомобиль-n-masc-sg-nom>;
  lemon:otherForm
    <http://lod.ruthes.org/resource/form/RU-автомобиля-n-masc-sg-gen>,
    <http://lod.ruthes.org/resource/form/RU-автомобилей-n-masc-pl-gen>,
    ...;
  prov:wasDerivedFrom
    <http://lod.ruthes.org/resource/ruthes-lite>,
    <http://lod.ruthes.org/resource/zaliznyak-dictionary>.

<http://lod.ruthes.org/resource/form/RU-автомобиль-n-masc-sg-nom>
  a lemon:Form;
  rdfs:label "автомобиль"@ru;
  lemon:writtenRep "АВТОМОБИЛЬ"@ru;
  lexinfo:partOfSpeech lexinfo:noun;
  lexinfo:gender lexinfo:masculine;
  lexinfo:case lexinfo:nominative;
  lexinfo:number lexinfo:singular.

<http://lod.ruthes.org/resource/form/RU-автомобилей-n-masc-pl-gen>
  a lemon:Form;
  rdfs:label "автомобилей"@ru;
  lemon:writtenRep "АВТОМОБИЛЕЙ"@ru;
  lexinfo:partOfSpeech lexinfo:noun;
  lexinfo:gender lexinfo:masculine;
  lexinfo:case lexinfo:genitive;
  lexinfo:number lexinfo:plural.
```

Fig. 2. (*continued*)

The decomposition tree and its nodes are represented by instances of the lemon:Node class. The lexical entry is linked to its tree by the lemon:phraseRoot property. The tree node is linked to a corresponding decomposition component by the lemon:leaf property. For representing dependency relations between nodes we defined four new properties as subproperties of lemon:edge property, namely ruthes-ontology:head, ruthes-ontology:attr, ruthes-ontology:prep and ruthes-ontology:sub.

Figure 3 represents decomposition of 'Объект культурного наследия' ('Cultural heritage object') multi-word lexical entry.

```
@base <http://lod.ruthes.org/resource/>.
@prefix rdfs: <http://www.w3.org/2000/01/rdf-schema#>.
@prefix lemon: <http://www.lemon-model.net/lemon#>.
@prefix lexinfo: <http://www.lexinfo.net/ontology/2.0/lexinfo#>.
@prefix ruthes-ontology: <http://lod.ruthes.org/ontology#>.

</entry/объект_культурного_наследия-ng#comp-объект>
    a lemon:Component;
    rdfs:label "объект";
    lexinfo:partOfSpeech lexinfo:noun;
    lexinfo:gender lexinfo:masculine;
    lexinfo:case lexinfo:nominative;
    lexinfo:number lexinfo:singular;
    lemon:element </entry/объект-n>.

</entry/объект_культурного_наследия-ng#comp-наследия>
    a lemon:Component;
    rdfs:label "наследия";
    lexinfo:partOfSpeech lexinfo:noun;
    lexinfo:gender lexinfo:neuter;
    lexinfo:case lexinfo:genetive;
    lexinfo:number lexinfo:singular;
    lemon:element </entry/наследие-n>.

</entry/объект_культурного_наследия-ng#comp-культурного>
    a lemon:Component;
    rdfs:label "культурного";
    lexinfo:partOfSpeech lexinfo:adjective;
    lexinfo:gender lexinfo:neuter;
    lexinfo:case lexinfo:accusative;
    lexinfo:number lexinfo:singular;
    lemon:element </entry/культурный-adj>.
```

Fig. 3. Decomposition of 'Объект культурного наследия' ('Cultural heritage object') multi-word lexical entry

```
</entry/объект_культурного_наследия-ng>
  lemon:decomposition
    (
    </entry/объект_культурного_наследия-ng#comp-объект>
    </entry/объект_культурного_наследия-ng#comp-культурного>
    </entry/объект_культурного_наследия-ng#comp-наследия>
    ).

</entry/объект_культурного_наследия-ng>
  lemon:phraseRoot
    [
    a lemon:Node;
    rdfs:label "объект культурного наследия";
    ruthes-ontology:head
      [
      a lemon:Node;
      rdfs:label "объект";
      lemon:leaf </entry/объект_культурного_наследия-ng#comp-объект>;
      ruthes-ontology:sub
        [
        a lemon:Node;
        rdfs:label "наследия";
        lemon:leaf </entry/объект_культурного_наследия-ng#comp-наследия>;
        ruthes-ontology:attr
          [
          a lemon:Node;
          rdfs:label "культурного";
          lemon:leaf </entry/объект_культурного_наследия-ng#comp-культурного>
          ]
        ]
      ]
    ].
```

Fig. 3. (*continued*)

4.4   The Schema of URIs

The entities of the resource are given URI in accordance with the following schema:

– Concepts: http://lod.ruthes.org/resource/concept/ namespace, followed by the concept numeric identifier. Example: http://lod.ruthes.org/resource/concept/154.
– Lexical entries: http://lod.ruthes.org/resource/entry/ namespace, followed by the lexical entry identifier, consisting of a language code, a lemma and a part of speech tag, separated by hyphens. Example: http://lod.ruthes.org/resource/entry/RU-машина-n.

– Lexical senses: http://lod.ruthes.org/resource/sense/ namespace, followed by the sense identifier, consisting of the concept and the lexical entry identifiers, separated by hyphens. Example: http://lod.ruthes.org/resource/sense/154-RU-машина.
– Forms: http://lod.ruthes.org/resource/form/ namespace, followed by the form identifier, consisting of the language code, the written representation, part of speech and grammatical tags, separated by hyphens. Example: http://lod.ruthes.org/resource/form/RU-машинами-n-fem-pl-instr.

4.5 Statistics

Statistics of RuThes Cloud is represented at Table 1.

Table 1. RuThes Cloud statistics

Entity type	Count
Concepts	31,538
Semantic relations	65,611
hypernym / hyponym	41,860
holonym / meronym	10,393
association1 / association2	11,000
association	2,358
Lexical entries	110,378
Words	58,484
Phrases	51,894
Senses	130,558
Forms	1,729,489
Dependency trees	36,666
Glosses	10,626

4.6 Publishing on the Web

The resource has been published on the Web and is available via:

– dereferenceable URIs: http://lod.ruthes.org;
– SPARQL endpoint: http://lod.ruthes.org/sparql;
– RDF dump: http://ruthes.org/download.

Access to the resource via dereferenceable URIs is supported by mechanisms of content negotiation. When a web browser requests a URI, it is redirected to a web page with an HTML view of the entity, but the Semantic Web agent request is redirected to the page with the RDF representation.

5 Query Example

In this section we present an example of the SPARQL 1.1 query to the RuThes Cloud, which exploits in combination its semantic, lexical and morphological levels. This query serves as a solution of a query expansion task, returning for a given query word

'Автомобиль' ('Automobile') all the lexical forms of all synonyms of this word, as well as of its hyponyms. The query is represented at Fig. 4.

```
SELECT ?result {
    ?form lemon:writtenRep "АВТОМОБИЛЬ"@ru.
    ?form
        ^lemon:canonicalForm /
        lemon:sense /
        lemon:reference /
        (ruthes-ontology:hyponym*) /
        lemon:isReferenceOf /
        lemon:isSenseOf /
        (lemon:canonicalForm|lemon:otherForm) /
        lemon:writtenRep ?result.
}
```

Fig. 4. SPARQL query example

The search result contains 1159 forms, such as 'Автомобиль' ('Automobile'), 'Машина' ('Car'), 'Машинами' ('By cars', i.e. 'Car' in plural number and instrumental case), 'Электромобиль' ('Electric car'), 'Лимузину' ('To a limousine'), 'Автобусы' ('Buses'), 'Бронетранспортер' ('Armoured personnel carrier'), 'БТР' ('APC', the abbreviation for 'Armoured personnel carrier'), 'Мусоровоз' ('Dustcart'), 'Фольксваген' ('Volkswagen'), etc.

6 Conclusion

In this paper we presented RuThes Cloud, a new multi-level Linguistic Linked Open Data resource for Russian. The current version of this resource contains (1) a hierarchy of concepts, (2) lexical entries in Russian referring to concepts, (3) forms of lexical entries associated with their grammatical tags and (4) syntactic trees of multi-word lexical entries.

According to the plan of subsequent work, we are going to extend this emerging resource with: (1) lexical entries in English and Tatar; (2) relations between lexical entries; (3) syntactic and semantic frames; (4) links to external resources.

We hope that our work can serve as a crystallization point of the Linguistic Linked Open Data cloud in Russian and in the languages of the peoples of Russia.

Acknowledgements. The main part of the reported work was funded by Russian Science Foundation according to the research project no. 16-18-02074. Developing the semantic publishing technological platform was funded by the subsidy allocated to Kazan Federal University for the state assignment in the sphere of scientific activities, grant agreement no. 1.2368.2017.

References

1. Chiarcos, C., McCrae, J., Cimiano, P., Fellbaum, C.: Towards open data for linguistics: Linguistic Linked Data. In: Oltramari, A., Vossen, P., Qin, L., Hovy, E. (eds.) New Trends of Research in Ontologies and Lexical Resources. Theory and Applications of Natural Language Processing, pp. 7–25. Springer, Heidelberg (2013). doi:10.1007/978-3-642-31782-8_2
2. McCrae, J.P., et al.: The open linguistics working group: developing the Linguistic Linked Open Data cloud. In: Calzolari, N., et al. (eds.) Proceedings of the 10th International Conference on Language Resources and Evaluation (LREC 2016), pp. 2435–2441 (2016)
3. van Assem, M., Gangemi, A., Schreiber, G.: Conversion of WordNet to a standard RDF/OWL representation. In: Calzolari, N., et al. (eds.) Proceedings of the 5th International Conference on Language Resources and Evaluation (LREC 2006), pp. 237–242 (2006)
4. Eckle-Kohler, J., McCrae, J.P., Chiarcos, C.: lemonUby - a large, interlinked, syntactically-rich lexical resource for ontologies. Semant. Web 6(4), 371–378 (2015). doi:10.3233/SW-140159
5. McCrae, J.P., Fellbaum, C., Cimiano, P.: Publishing and linking WordNet using Lemon and RDF. In: Chiarcos, C., et al. (eds.) Proceedings of the 3rd Workshop on Linked Data in Linguistics (LDL-2014) (2014)
6. Sérasset, G.: DBnary: Wiktionary as a Lemon-based multilingual lexical resource in RDF. Semant. Web 6(4), 355–361 (2015). doi:10.3233/SW-140147
7. Paredes, L.P., Álvarez Rodríguez, J.M., Azcona, E.R.: Promoting government controlled vocabularies for the Semantic Web: the EUROVOC thesaurus and the CPV product classification system. In: Kollias, S., Cousins, J. (eds.) Proceedings of the 1st International Workshop on Semantic Interoperability in the European Digital Library (SIEDL 2008), pp. 111–122 (2008)
8. Caracciolo, C., Stellato, A.: Thesaurus maintenance, alignment and publication as Linked Data: the AGROVOC use case. Int. J. Metadata Semant. Ontol. 7(1), 65–75 (2012). doi:10.1504/IJMSO.2012.048511
9. Caracciolo, C., Stellato, A., Morshed, A., Johannsen, G., Rajbhandari, S., Jaques, Y., Keizer, J.: The AGROVOC linked dataset. Semant. Web 4(3), 341–348 (2013). doi:10.3233/SW-130106
10. Zapilko, B., Schaible, J., Mayr, P., Mathiak, B.: TheSoz: a SKOS representation of the thesaurus for the social sciences. Semant. Web 4(3), 257–263 (2013). doi:10.3233/SW-2012-0081
11. Summers, E., Isaac, A., Redding, C., Krech, D.: LCSH, SKOS and Linked Data. In: Greenberg, J., Klas, W. (eds.) Proceedings of the 2008 International Conference on Dublin Core and Metadata Applications (DC 2008), pp. 25–33 (2008)
12. Ustalov, D.: Russian thesauri as Linked Open Data. In: Computational Linguistics and Intellectual Technologies: papers from the Annual conference "Dialogue", vol. 1, pp. 616–625. RGGU (2015)
13. Nevzorova, O., Zhiltsov, N., Kirillovich, A., Lipachev, E.: OntoMath^{Pro} ontology: a Linked Data hub for mathematics. In: Klinov, P., Mouromtsev, D. (eds.) KESW 2014. CCIS, vol 468, pp. 105–119. Springer, Cham (2014). doi:10.1007/978-3-319-11716-4_9
14. Elizarov, A.M., Kirillovich, A.V., Lipachev, E.K., Nevzorova, O.A., Solovyev, V.D., Zhiltsov, N.G.: Mathematical knowledge representation: semantic models and formalisms. Lobachevskii J. Math. 35(4), 348–354 (2014). doi:10.1134/S1995080214040143
15. Navigli, R., Ponzetto, S.P.: BabelNet: the automatic construction, evaluation and application of a wide-coverage multilingual semantic network. Artif. Intell. 193, 217–250 (2012). doi:10.1016/j.artint.2012.07.001

16. Ehrmann, M., Cecconi, F., Vannella, D., McCrae, J., Cimiano, P., Navigli, R.: Representing multilingual data as Linked Data: the case of BabelNet 2.0. In: Calzolari, N., et al. (eds.) Proceedings of the 9th International Conference on Language Resources and Evaluation (LREC 2014), pp. 401–408 (2014)

17. Baker, T., et al.: Key choices in the design of Simple Knowledge Organization System (SKOS). J. Web Semant. **20**, 35–49 (2013). doi:10.1016/j.websem.2013.05.001

18. McCrae, J., Spohr, D., Cimiano, P.: Linking lexical resources and ontologies on the Semantic Web with Lemon. In: Antoniou, G., et al. (eds.) ESWC 2011. Part I, LNCS, vol. 6643, pp. 245–259. Springer, Heidelberg (2011). doi:10.1007/978-3-642-21034-1_17

19. McCrae, J., et al.: The Lemon cookbook. http://lemon-model.net/lemon-cookbook.pdf

20. Cimiano, P., McCrae, J.P., Buitelaar, P.: Lexicon model for ontologies. Final community group report, 10 May 2016. https://www.w3.org/2016/05/ontolex/

21. ISO 24613:2008: Language resource management - Lexical markup framework (LMF)

22. Kemps-Snijders, M., Windhouwer, M., Wittenburg, P., Wright, S.E.: ISOcat: remodelling metadata for language resources. Int. J. Metadata Semant. Ontol. **4**(4), 261–276 (2009). doi:10.1504/IJMSO.2009.029230

23. Kemps-Snijders, M., Windhouwer, M., Wittenburg, P., Wright, S.E.: ISOcat: corralling data categories in the wild. In: Proceedings of the 6th International Conference on Language Resources and Evaluation (LREC 2008), pp. 887–891 (2008)

24. Windhouwer, M., Wright, S.E.: Linking to linguistic data categories in ISOcat. In: Chiarcos, C., Nordhoff, S., Hellmann, S. (eds.) Linked Data in Linguistics, pp. 99–107. Springer, Heidelberg (2012). doi:10.1007/978-3-642-28249-2_10

25. ISO 12620:2009: Terminology and other language and content resources—Specification of data categories and management of a Data Category Registry for language resources

26. LexInfo. http://www.lexinfo.net/

27. Chiarcos, C.: OLiA – Ontologies of Linguistic Annotation. Semant. Web **6**(4), 379–386 (2015). doi:10.3233/SW-140167

28. Chiarcos, C.: Ontologies of linguistic annotation: survey and perspectives. In: Calzolari, N., et al. (eds.) Proceedings of the 8th International Conference on Language Resources and Evaluation (LREC 2012), pp. 303–310 (2012)

29. Hellmann, S., Lehmann, J., Auer, S., Brümmer, M.: Integrating NLP using Linked Data. In: Alani, H., et al. (eds.) ISWC 2013, Part II. LNCS, vol 8219, pp. 98–113. Springer, Heidelberg (2013). doi:10.1007/978-3-642-41338-4_7

30. Sanderson, R., Ciccarese, P., Young, B.: Web annotation data model. W3C Recommendation, 23 February 2017. https://www.w3.org/TR/annotation-model/

31. Nevzorova, O., Nevzorov, V.: The Development Support System "OntoIntegrator" for Linguistic Applications. Information Science and Computing, vol. 13, Intelligent Information and Engineering Systems, vol. 3, pp. 78–84. ITHEA, Rzeszow-Sofia (2009)

32. Loukachevitch, N., Dobrov, B., Chetviorkin, I.: RuThes-Lite, a publicly available version of thesaurus of Russian language RuThes. In: Computational Linguistics and Intellectual Technologies: Papers from the Annual International Conference "Dialogue", pp. 340–349. RGGU (2014)

33. Loukachevitch, N., Dobrov, B.: Development of ontologies with minimal set of conceptual relations. In: Lino, M.T., et al. (eds.) Proceedings of the 4th International Conference on Language Resources and Evaluation (LREC 2004), pp. 1889–1892 (2004)

34. Gil, Y., Miles, S.: PROV Model Primer. W3C Working Group Note, 30 April 2013. https://www.w3.org/TR/prov-primer/

35. Guarino, N., Welty, C.A.: A Formal ontology of properties. In: Dieng, R., Corby, O. (eds.) EKAW 2000. LNCS, vol. 1937, pp. 97–112. Springer, Heidelberg (2000). doi:10.1007/3-540-39967-4_8

The Algorithm of Modelling and Analysis of Latent Semantic Relations: Linear Algebra vs. Probabilistic Topic Models

Nina Rizun[1(✉)], Yurii Taranenko[2], and Wojciech Waloszek[3]

[1] Department of Applied Informatics in Management,
Faculty of Management and Economics,
Gdansk University of Technology, Gdańsk, Poland
nina.rizun@zie.pg.gda.pl
[2] Department of Applied Linguistics and Methods of Teaching
Foreign Languages, Alfred Nobel University, Dnipro, Ukraine
taranen@rambler.ru
[3] Department of Software Engineering, Faculty of Electronics,
Telecommunications and Informatics,
Gdansk University of Technology, Gdańsk, Poland
wowal@eti.pg.gda.pl

Abstract. This paper presents the algorithm of modelling and analysis of Latent Semantic Relations inside the argumentative type of documents collection. The novelty of the algorithm consists in using a systematic approach: in the combination of the probabilistic Latent Dirichlet Allocation (LDA) and Linear Algebra based Latent Semantic Analysis (LSA) methods; in considering each document as a complex of topics, defined on the basis of separate analysis of the particular paragraphs. The algorithm contains the following stages: modelling and analysis of Latent Semantic Relations consistently on LDA- and LSA-based levels; rules-based adjustment of the results of the two levels of analysis. The verification of the proposed algorithm for subjectively positive and negative Polish-language film reviews corpuses was conducted. The level of the recall rate and precision indicator, as a result of case study, allowed to draw the conclusions about the effectiveness of the proposed algorithm.

Keywords: Latent semantic analysis · Latent dirichlet allocation · Rules of adjustment · Corpus · Linear algebra · Probability

1 Introduction

Modelling and Analysis of Latent Semantic Relations (LSR) – the approach of constructing a model of the corpus, reflecting the transition from a set of documents and set of words in the documents to a set of topics, describing the contents of documents. We can say that in the mathematical model of text collection, describing the words or documents is associated with a family of probability distributions on a variety of topics [4, 6, 13].

© Springer International Publishing AG 2017
P. Różewski and C. Lange (Eds.): KESW 2017, CCIS 786, pp. 53–68, 2017.
https://doi.org/10.1007/978-3-319-69548-8_5

Construction of the mathematical model can be considered as a problem of simultaneous clustering of documents and words for the same set of clusters, known as topics. In terms of the cluster analysis the topic is the result of bi-clustering, i.e. the simultaneous clustering of words and documents in accordance with their semantic closeness. Thus, compressed semantic description of words or of a document is a probability distribution on a variety of hidden variables (topics). The process of finding these distributions is called the topic model [18–20].

Those hidden variables (topics) allow presenting the document as a vector in the space of latent topics instead of submitting in the space of words. As a result, the document has a lower number of components, allowing faster and more efficient handling. Thus, the topic model is closely related to another class of problems known as a reduction of data dimension [14, 17–20].

The basic algorithms for modelling topics, on which we concentrate in this paper, are: determinant Latent Semantic Analysis (LSA), and probabilistic Latent Dirichlet Allocation (LDA). And although all of them share the fundamental assumption about latent semantic (topical) structure of the documents, they use different mathematical frameworks – Linear algebra (LSA) vs. Probabilistic Topic Modelling (LDA) [3, 4, 15].

With the aim of improving the quality of Topic Modelling Process (TMP), this paper focuses on:

- *analysing* the advantages and disadvantages of Latent Semantic Relations, revealing algorithms inside the textual collection, using two different mathematical frameworks;
- *developing* the complex Algorithm of Modelling and Analysis of Latent Semantic Relations, based on advantages of two different mathematical frameworks;
- demonstration of the *effectiveness* of proposed Algorithm implementation for specific, Argumentative, type of documents, via conducting a *case study* for the Polish-language Film Reviews Corpora.

The research results, presented in the paper, are supported by the Polish National Centre for Research and Development (NCBiR) under Grant No. PBS3/B3/35/2015, the project "Structuring and classification of Internet contents with the prediction of its dynamics".

2 Theoretical Background of the Research

2.1 Vector Space Models of the Semantic Relations Analysis

The aim of the LSR analysis is to extract "semantic structure" of the collection of information flow and automatically expands them into the underlying topic. Significant progress on the problem of presenting and analysing the data has been made by researchers in the field of information retrieval (IR) [1, 10, 11]. The basic methodology proposed by IR researchers for text collection reduces each document in the corpus to a vector of real numbers, each of which represents ratios of counts.

In the popular $TF \times IDF$ scheme [17–21], on the basis of vocabulary of "bag of words" the $A(m \times n)$ terms-document matrix is built, which contains as elements the counts of absolute frequency of words occurrence. After suitable normalization, this term frequency count is compared to an inverse document frequency count, which measures the number of occurrences of a word in the entire corpus:

$$F_{w_i} = TF \times IDF = tf(w, t) \cdot \log_2 \cdot \frac{D}{df} \tag{1}$$

where, $tf(w, t)$ – relative frequency of the wth word occurrence in document t:

$$tf(w, t) = \frac{k(w, t)}{df} \tag{2}$$

$k(w, L_t)$ – the number of wth word occurrences in the text t; df – the total number of words in the text of t; D – total number of documents in the collection.

Then, for solving the problem of finding the similarity of documents (terms) from the point of view of the relation to the same topic, the different metric can be applied. The most appropriate metric is cosine measure of the edge between the vectors [14, 20–22].

A further part of the algorithm is to divide the source data into groups corresponding to the events, as well as in determining whether a text document describes a set of any topic. The main idea of the solution is the use of clustering algorithms [12, 14, 17–21].

The *limitations* of this method are: the calculations measure the "surface" usage of words as patterns of letters; they can't distinguish such phenomena as polysemy and synonymy [10, 13, 16].

2.2   Latent Semantic Indexing

In 1988, Dumais et al. [7] proposed a method of Latent Semantic Indexing (LSI), most frequently referred to as LSA. Deerwester et al., 1990 [8], designed to improve the efficiency of IR algorithms and search engines by the projection of documents and terms in the space of lower dimension, which includes semantic concepts of the original set of documents.

LSA is a matrix algebra process. The most common version of LSA is based on the singular value decomposition (SVD) of a term-document matrix [10]. As a result of the SVD of the matrix A we have three matrices:

$$X_{t \times d} \approx X_{K_{t \times d}} = U_{K_{t \times d}} \Sigma_{K_{t \times d}} \left(V_{K_{t \times d}} \right)^T \tag{3}$$

$\Sigma_{K_{t \times d}} \left(V_{K_{t \times d}} \right)^T$ – represents terms in k-d latent space; $U_{K_{t \times d}} \Sigma_{K_{t \times d}}$ – represents documents in k-d latent space; $U_{K_{t \times d}}$, $V_{K_{t \times d}}$ – retain term–topic, document–topic relations for top k topics.

But, as [18, 19] proved, there are three *limitations* to apply LSA: documents having the same writing style (Lim#1); each document being centered on a single topic (Lim#2); a word having a high probability of belonging to one topic but low probability of belonging to other topics (Lim#3). The limitations of LSA are based on orthogonal characteristics of dimension factors as well as on the fact, that the probabilities for each topic and the document are distributed uniformly, which does not correspond to the actual characteristics of the collections of documents [7, 8, 23]. That is why, LSA tends to prevents multiple occurrences of a word in different topics and thus LSA cannot be used effectively to resolve polysemy issues (Lim#4).

2.3 Probabilistic Topic Models

In contrast to the so-called *discriminative* approaches (LSI, LSA), in a *probabilistic* approach the topics are given by the model, and then term-document matrix is used to estimate its hidden parameters, which can then be used to generate the simulated distributions [4, 6, 17, 25].

Latent Dirichlet Allocation
LDA – generative probabilistic graphical model proposed by David Blei [3, 4, 15]. LDA is a three-level hierarchical Bayesian model. The algorithm of the method is as follows: Each document is generated independently: randomly select its distribution for document on topics θ_d for each document's word; randomly select a topic from the distribution θ_d, obtained in the first step; randomly select a word from the distribution of words in the chosen topic φ_k(distribution of words in the topic k). In the classical model of LDA, the number of topics is initially fixed and specifies the explicit parameter k.

Methods of Evaluating the Quality of Results
The most common method of evaluating the quality of probabilistic topic models is the calculation of the *Perplexity* index on the test data set D_{test} [2–4]. In information theory, perplexity is a measurement of how well a probability model predicts a sample. A low perplexity indicates that the probability distribution is good at predicting the sample:

$$Perplexity(D_{test} = \exp\left\{ -\frac{\sum_{d=1}^{M} \log p(w_d)}{\sum_{d=1}^{M} N_d} \right\} \tag{4}$$

The *limitation* of LDA method is: it is possible to choose the optimum value of the k, but, even under condition of finding the optimal value of the k, the level of probability of a document belonging to a particular topic could be insignificant (Lim#5) [3, 4, 15].

3 Methodology

In this paper the following author's definitions will be used:

1. *Term* is a basic unit of discrete data.
2. *Latent Semantic/Probabilistic topic* (*topics*) is a basic unit of Latent Semantic Relations, received by *LSA/LDA* approach.
3. *Context Fragment* (CF) is indivisible, topically completed, sequence of terms, located within a document's paragraph.
4. *Document* is a set of CF.
5. *Corpus* (films reviews corpus, FRC) is a collection of the Documents.
6. *Semantic Cluster* (SC) is the set of CF that have hidden semantic closeness (HSC).

7. *Contextual Dictionary* (CD) is a set of terms that have HSC.
8. *Subjective Sentiment Corpus* (SSC) is a collection of Documents that have common sentiment closeness.

3.1 Novelty and Motivation

Motivation scenario of this research presupposes taking into account the *Specificity* of the Document Type (SDT) and concerns finding the ways to completely or partially eliminate the *Limitations* characterizing the Discriminant and Probabilistic approaches for Latent Semantic Relations revealing. In this regard the following scientific research questions were raised:

1. *Whether the taking into account of specific features of Argumentative type of document allows to affect Quality of the Topic Modelling Process Results.*
2. *Is it possible to increase the Level of Quality of the Topic Modelling Process Results via using the combination of the Discriminant and Probabilistic Methods?*

For finding the answers to these questions the following main heuristics and hypotheses were formulated:

Heuristic H1.1. Taking into account the specificity of chosen for this study Type of Documents and presence the nonofficial requirements of Film's Review structure and writing rules [22], assume that the writing style of each review is approximately the same (eliminating the Lim#1).

Hypothesis H1.2. Taking into account the chosen Document Type Specificity, assume, that each paragraph (CF) is centered on a single Topic and should be analyzed separately (eliminating the Lim#2).

Hypothesis H2.1. The combination of the Discriminant and Probabilistic methods have a synergistic effect to improve the recall rate and precision indicator of Topic Modelling Process realization. This effect is expected to be achieved via increasing:

– the quality of LDA-method of topics recognizing via increasing the level of probability of assigning the topic to particular CF by taking into account the hidden LSR phenomena (eliminating the Lim#5);
– the quality of LSA-method of LSR recognition via adjusting the consequences of influence the uniform distribution of the topics within the document by taking into account the probabilistic approaches (eliminating the Lim#3 and #4).

Basic version of *analysing* the part of proposed Algorithm of Two-Level Modelling and Analysis of the LSR includes 7 steps (Fig. 1). Each level additionally assumes a preliminary *modelling* stage (are not included to the Fig. 1).

As a sample for case study experiments the Polish-language film reviews from the filmweb.pl are used. For demonstration of the basic workability of the author's Algorithm, as a *preliminary case study* was used (the data set of only one, randomly chosen, Polish-language film review, which contains 7 CF). All words/terms of film reviews in this paper will be presented in Polish and English languages (separated by symbol "/"). The experimental part of all steps of author's Algorithm has been implemented in Python 3.4.1.

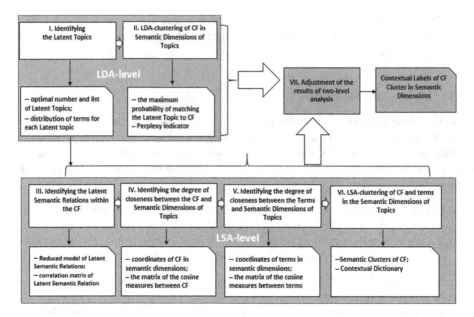

Fig. 1. The steps of the algorithm of two-level analysis of latent semantic relations *Source*: own research results

3.2 The Level of LDA-Based Modelling and Analysis of Latent Semantic Relations

LDA-based Modelling of LSR

LDA-based Modelling of LSR is the stage, which *aims* to ensure the implementation of the level of LDA-based Topic Analysis, presupposed the Forming the "Bag of Words" (preprocessing) step.

Taking into account the Specificity of chosen Document Type, as well as the case study language peculiarities (limited number of existing algorithms and software implementations for the analysis of texts in Polish) [22], in addition to standard procedures for text preprocessing, the authors have provided:

– text *adaptation* procedure, based on the specificity of the structure of reviews document layout (Fig. 2). This procedure is to implement the replacement of the Film's Titles, the Names/Surnames of Director/Actors/Characters into the corresponding position of the descriptive part of review (for example, the Title of the film is replaced by *"Film"*, Name and Surname of the actor – by *"Actor"* etc.);

– expanding by authors the list of *stop words* (near 400 Polish words) for improving the process of lemmatization (based on the dictionary *pyMorfologik* [13, 16, 22]);

– *part-of-speech* (adjective, nouns, verbs) morphological tagging and filtering procedures performing, allowed to increase the resolution of the of LSR analysis.

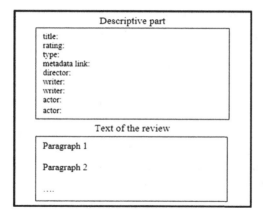

Fig. 2. Structure of text reviews layout *Source*: own research results

LDA-based Analysing of Latent Semantic Relations
Step I. Identifying the Topics

LDA-based Analysis is the stage, which *aims*: (1) to reveal the optimal number of latent probabilistic topics that describe the main content of the analyzed document; (2) to assign them to the CFs based on the probalistic LSR within the paragraphs. As a technical support, for the implementation this phase the LDA Gensim Python package (https://radimrehurek.com/gensim/models/ldamodel.html) was used.

Table 1 demonstrates the pretesting experiments results of preliminary case study (further – *PCS results*) of the main parameters of LDA model. The optimum value of the Perplexity index is achieved in the point, when further changes in the parameters do not lead to its significant decrease. In accordance with author's algorithm, obtained optimal number of latent probabilistic topics will be used as a recommended number of semantic clusters in the LSA-based level of SLR analysis.

Table 1. PCS results of the studying of the of LDA model parameters

Perplexity	Number of topics	Number of terms	Number of passes	Alpha parameter	Eta parameter	Max probability topic	Max probability of terms in the topics
3336	10	10	100	1.70	1.00	0.1025	0.057
633	7	7	100	1.50	1.00	0.6050	0.177
202	5	5	100	1.50	1.00	0.7134	0.167
64	3	5	100	1.50	1.00	0.8417	0.132
63	**3**	**7**	**100**	**1.50**	**1.00**	**0.8411**	**0.166**

The list of obtained latent probabilistic topics with information about most probable (significant) terms, described this topic, is presented in the Table 2.

Step II. LDA-clustering of CF in Semantic Dimensions of Corpus

Based on information about the maximum probability of matching the obtained Latent Probabilistic Topics to the CF, on this step the process of Semantic (topical) clustering of CF could be performed. The PCS results of this process are presented in Table 3.

Table 2. PCS results of the list of latent probabilistic topics with distribution of terms

Terms (Polish/English)	Probability	Terms (Polish/English)	Probability	Terms (Polish/English)	Probability
Topic #0		Topic #1		Topic #2	
fabuła/story	0.080	kino/cinema	0.109	bohater/character	0.166
akcja/action	0.062	twórca/creator	0.066	gra/playing	0.140
efekt/effect	0.050	kobieta/woman	0.062	dobry/good	0.130
bohater/character	0.047	obsada/cast	0.052	postać/character	0.090
ksiazka/book	0.046	scena/stage	0.051	rola/role	0.040
obraz/image	0.044	glowny/main	0.050	typowy/tipical	0.030
historia/history	0.042	reżyser/director	0.049	intryga/intrigue	0.029

Table 3. PCS results of the semantic clustering of CF

CF	CF_5	CF_0	CF_1	CF_4	CF_6	CF_2	CF_3
# topic (cluster)	0	1	1	1	1	2	2
Probability		0.8411	0.6228	0.8022	0.7039	0.4800	0.7957	0.6603

The values of the Perplexity in the Table 1 proves the validity of the assumptions about providing the analysis the Corpora by paragraphs (*Hypothesis H1.2*). But, on the other hand, we can note, that the level of probability of a CF belonging to a particular topic/cluster is not significant for all CF (for example, for CF_6 it is lower than 0.5).

3.3 The Level of LSA-Based Modelling and Analysis of Latent Semantic Relations

LSA-based Modelling of Latent Semantic Relations
LSA-based Modelling of LSR is the stage, which *aims* to ensure the implementation of the level of LSA-based Analysis of Latent Semantic Relations. As well as LDA-based level, this stage presupposed the preprocessing procedure, which contain additionally to forming the "Bag of Words", the Creating the Term-Document Matrix (TDM) step [20–22]. The fragment of the PCS results of LSA initial data building is presented in Table 4.

Table 4. The fragment of PCS results of the absolute frequency terms-cf matrix

Terms (Polish/English)	CF_0	CF_1	CF_2	CF_3	CF_4	CF_5	CF_6	Sum
bohater/character	1	1	4	5	2	2	1	16
akcja/action	0	1	0	2	1	3	2	9
kino/cinema	1	3	0	2	1	0	2	9
film/movie	0	2	1	0	0	1	1	5
główny/main	1	2	1	0	0	0	0	4
kobieta/woman	0	3	0	0	1	0	0	4

As for results of TF-IDF transformation of this matrix, we can state the following facts: differences in absolute term frequencies were reduced; frequently appearing terms are less relevant compared to infrequent terms; terms-CF matrix contains weighted term frequencies.

However, according to [26], and during a number of author's experiments, the solutions were found: TF-IDF approach does not work well because when a CF contains only a 100-150 words, there are seldom terms that occur more than once within a document; but, the most common words occurred within one CF are the so-called key terms, which determine the topic's label of analysed CF in a large scale; it is more important to focus on the allocation of stop words and most significant part-of-speech, to maximise the weight of keywords of the CF by excluding consideration of the terms that have no semantic weight.

LSA-based Analyzing of Latent Semantic Relations

LSA-based Analysis of LSR is the stage, which *aims* to identify the patterns in the relationships between the terms and latent semantic topics. As we already stated, LSA method is based on the principle that terms that are used in the same contexts tend to have similar meanings. For revealing this information about LSR between topics and CF/terms, we need: to assess the degree of semantic correlation relationship between CF/terms via building the reduced model of LSR; to form the semantic clusters of CF *via* determining the cosine distance between the CF in order to identify the LSR between topics and CF; to form the contextual dictionary of semantic clusters of CF *via* determining the cosine distances between the terms in order to identify the LSR between k terms and topics.

Step III. Identifying the Hidden Semantic Connection Within the Documents

Mathematically the Reduced model, as the instrument of preliminary LSR presence identification, is the process of multiplying of SVD transformation results with chosen k-dimension $X_{K_{t\times d}} = U_{K_{t\times d}} \Sigma_{K_{t\times d}} (V_{K_{t\times d}})^T$. The fragment of *PCS* results of Reduced model is presented in Table 5.

Table 5. The fragment of PCS results of the reduced model for identifying the LSR

Terms (Polish/ English)	CF_0	CF_1	CF_2	CF_3	CF_4	CF_5	CF_6
bohater / character	1.115	2.785	2.974	3.535	1.676	2.907	1.636
dobry / good	0.162	0.406	0.401	0.481	0.234	0.338	0.225
film / movie	0.384	0.964	0.888	1.071	0.537	0.626	0.508
główny / main	0.479	1.211	0.687	0.882	0.542	-0.369	0.459
kino / cinema	0.963	2.431	1.512	1.915	1.129	-0.384	0.978
kobieta / woman	0.569	1.440	0.725	0.950	0.617	-0.687	0.508

Via comparison of the red numbers in Table 5 with zero's values in the same places of Table 4 could be, as an example, identified the existence of the following phenomena of LSR:

– the term "*Film/Movie*" seems to have the presence in all CF where the word "*Bohater/Character*" appears;

– the term "*Kobieta/Woman*" seems to have the presence in the CF where the word "*Kino/Cinema*" appears.

At the same time, we can observe the increasing of the values of the correlation coefficient (CC) between terms, compared the results of Tables 4 and 5 (Table 6):

Table 6. Example of PCS results of the comparison of the CC between terms

Source terms	Absolute frequency terms-CF matrix	Reduced model for identifying the hidden connection
Bohater. Film	–0.33391154	0.984754769
Kino. Kobieta	0.64162365	0.984405802

Steps IV-V. Identifying the Degree of Closeness Between the CF/Terms in the Semantic Dimensions of Topics

For measuring the level of LSR, identified on the previous step, the matrix of cosine distance between the vectors of CF and terms should be built. The *PCS* results of this estimation are presented in the Tables 7 and 8

Table 7. PCS results of the matrix of cosine distance between the vectors of CF

	CF_0	CF_1	CF_2	CF_3	CF_4	CF_5	CF_6
CF_0	1	0.9998	0.8052	0.8403	0.9537	–0.3376	0.8764
CF_1	0.9998	1	0.8164	0.8505	0.9592	–0.3196	0.8855
CF_2	0.8052	0.8164	1	0.9981	0.9463	0.2863	0.9912
CF_3	0.8403	0.8505	0.9981	1	0.9645	0.2266	0.9975
CF_4	0.9537	0.9592	0.9463	0.9645	1	–0.0387	0.9807
CF_5	–0.3376	–0.3196	0.2863	0.2266	–0.0387	1	0.1573
CF_6	0.8764	0.8855	0.9912	0.9975	0.9807	0.1573	1

Table 8. The fragment of PCS results of the matrix of cosine distance between the vectors of terms

	akcent/accent	akcja/action	bohater/character	...	łatwo/easily	osiągać/reach
akcent	1	0.9938	0.6136	...	0.873	0.1269
akcja	0.9938	1	0.6978	...	0.8132	0.2367
bohater	0.6136	0.6978	1	...	0.1506	0.8611
...						
łatwo	0.873	0.8132	0.1506	...	1	-0.373
osiągać	0.1269	0.2367	0.8611	...	-0.373	1

Step VI. LSA Clustering of CF/Terms in the Semantic Dimensions of Topics
    Based on the matrices of cosine distances between the vectors of CF and terms, in this step the Semantic clustering process should be realized. An example of the implementation of *k*-means clustering [12, 22] algorithm for CF and terms (in the condition of LDA-based number of SC) is presented in the Tables 9 and 10 and Fig. 3.

Table 9. PCS results of the labels of contextual fragments' clustering

CF	CF_0	CF_1	CF_5	CF_2	CF_3	CF_4	CF_6
Cluster	0	0	1	2	2	2	2

Table 10. PCS results of the contextual dictionary of semantic clusters

Terms (Polish/English)	Cluster	Terms (Polish/English)	Cluster	Terms (Polish/English)	Cluster
fabuła/story	0	reżyser/director	1	bohater/caracter	2
akcent/accent	0	kino/cinema	1	dobry/good	2
scenariusz/script	0	kobieta/woman	1	film/movie	2
akcja/action	0	główny/main	1	intryga/intrigue	2
ksiazka/book	0	obsada/cast	1	sposób/method	2
scena/scene	0	efekt/effect	1	typowy/typical	2
obraz/image	0	schemat/scheme	1	gra/playing	2
historia/history	0	stworzyć/create	1	rola/role	2

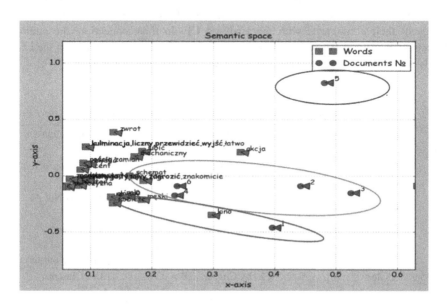

Fig. 3. The example of the graphical presentation of the results of CF semantic clustering
Source: own research results

3.4   Adjustments of the Results of the Two Levels of Analysis

On the VII step of *Author's Algorithm*, it is supposed to combine the results of the implementation of LSA and LDA levels for analysis, namely:

1. Forming the table of the Comparison of the numerical labels of Latent Semantic Clusters of a set of CF, obtained on two levels of research (Table 11). As we can see, the results of clustering for CF_4 and CF_6, obtained in LSA- and LDA-analysis levels, do not match.

Table 11. PCS results of the comparison of the semantic clusters as a set of CF labels

LDA-level			LSA-level	
CF	# Topic (Cluster)	Probability	CF	Cluster
CF_0	1	0.6228	CF_0	0
CF_1	1	0.8022	CF_1	0
CF_2	2	0.7957	CF_2	2
CF_3	2	0.6603	CF_3	2
CF_4	**1**	**0.7039**	**CF_4**	**2**
CF_5	0	0.8411	CF_5	1
CF_6	**1**	**0.4800**	**CF_6**	**2**

2. Formulation and implementation the *Rules of Adjustments* of the results obtained in the LSA- and LDA-analysis levels.

As stated above, LDA method implementation presupposes the assignment of the corresponding topics to CF based on the largest (from existing) *probability* (P) of degree of their compliance with the analysed CF. In this connection, the author's concept of *Rules of Adjustments* (RA) of the results of Semantic Clustering of the LSA- and LDA-analysis levels for each particular CF is proposed (Table 12).

Table 12. Rules of adjustments of CF clustering results

# of rule	LSA-analysis result	Result of comparison	LDA-analysis result	LDA Probability (P)	Assignable cluster
1	LSA Cluster	=	LDA Cluster	$P > 0.3$	LSA Cluster = LDA Cluster
2	LSA Cluster	=	LDA Cluster	$P \leq 0.3$	Cluster is Not recognized
3	LSA Cluster	$\neq$	LDA Cluster	$P \leq 0.3$	LSA Cluster
4	LSA Cluster	$\neq$	LDA Cluster	$0.3 < P \leq 0.7$	LSA Cluster/Re-clustering
5	LSA Cluster	$\neq$	LDA Cluster	$P > 0.7$	LDA Cluster

These rules allow:

– to improve the quality of LDA-method recognizing the CF's topics (rules 3, 4) due to the possibility of correcting the results of clustering, which are characterized by the low level of probability of a CF belonging to a particular topic. Suggested instrument – latent semantic specificity of the LSA method;

– to improve the quality of LSA-method recognition of hidden relations between the CF (rules 2, 5) due to the possibility of correcting the results of clustering, which characterize by situations, when CF coordinates located on the cluster's boundary. Suggested instrument – the probabilistic characteristics of the LDA method.

The *PCS* results of the implementation of RA are presented in Table 13.

Table 13. PCS results of the of final version of the labels of the CF's semantic clusters

CF	CF_5	CF_0	CF_1	CF_4	CF_2	CF_3	CF_6
# topic	0	1	1	1	2	2	2

4 Case Study Results and Discussion

For the process of verification of the *author's Algorithm* was formed the sentimental structure of FRC via classification of the reviews collection on the Subjectively Positive (SPSC) and Subjectively Negative Sentiment Corpuses (SNSC). This procedure is realized on the basis of information on the subjective assessment (SA) of films by the reviewers (measured by 10-point scale).

As a condition of sentimental structure of FRC building, the following *Heuristic 1.3* was adopted: *to consider the SPSC, if the SA is more than 5 points, and SNSC – if it is equal or less than 5 points.*

During the verification, the 30 reviews from each reviews collection were analysed. Totally 208 paragraphs from SPSC and 260 paragraphs from SNSC were studied. The recommended number of clusters (identified in LDA-level of analysis) is equal to 4. The structure (percentage of paragraphs, belonging to the topic) of the Semantic Clusters in each separate level and after adjustment (LSA&LDA) is presented in Table 14. The Contextual labels (CL) of the Topics were assigned automatically on the bases of the terms with the highest frequency in each topic.

Table 14. The structure of the semantic clusters

SPSC				SNSC			
CL of the Topics	LSA, %	LDA, %	LSA & LDA, %	CL of the Topics	LSA, %	LDA, %	LSA&LDA, %
Bohater/Character	19.71	18.75	19.23	Bohater/Character	11.54	13.46	12.31
Reżyser/Director	32.21	36.06	33.65	Aktor/ Actor	30.00	30.38	29.23
Scenariusz/Scenario	17.79	12.50	16.83	Widz/Spectator	28.85	26.92	28.08
Fabuła/Story	30.29	32.69	30.29	Fabuła/Story	29.62	29.23	30.38

The quantitative indicators of the adjustments process of the Latent Semantic Relations Analysis results: percentage of not recognized CF inside the Topic (*Indicator 1*); percentage of CF, which changed the Cluster (*Indicator 2*) and as well as final qualitative characteristic of research (*Recall rate*) are given in Table 15.

Table 15. The quality of the of LSR analysis results

SPSC			SNSC		
Labels of the topics	Indicator 1	Indicator 2	Labels of the topics	Indicator 1	Indicator 2
Bohater/Character	7.50	5.56	Bohater/Character	9.23	6.25
Reżyser/Director	2.82	5.48	Aktor/Actor	1.27	5.13
Scenariusz/Scenario	3.17	12.00	Widz/Spectator	5.52	9.09
Fabuła/Story	6.11	7.81	Fabuła/Story	2.61	2.70
Recall rate		95.19	Recall rate		96.15

5 Conclusions

In this paper authors presented the complex two-level Algorithm of Modelling and Analysis of TMP, aimed at elimination the *Limitations* characterizing the of two mathematical frameworks and taking into account the Document Type *Specificity*. The answers for the main scientific research question were found: the combination of the Discriminant and Probabilistic Methods (*Hypothesis* H2.1) as well as Specificity of the Argumentative Type Document oriented approach (*Hypothesis* H1.2), gave the opportunity to improve the following qualitative characteristics of LSR Analysis:

– recall rate (the ratio of the number of semantically clustered/recognized paragraphs to the total number of paragraphs in the corpora) to 90 95%;

– precision indicator (the average probability of significantly clustered/recognized paragraphs) from 62 to 70–75%.

In the *future research*, these results are planned to be used: to evaluate the Algorithm effectiveness for processing the English language Documents; to develop the algorithm of forming the hierarchical structure of the Latent topics of Corpora with taking into account the Sentiment specificity.

References

1. Baeza-Yates, R., Ribeiro-Neto, B.: Modern Information Retrieval. Addison-Wesley, Wokingham (2011). Second edition (1999)
2. Bahl, L., Baker, J., Jelinek, E., Mercer, R.: Perplexity – a measure of the difficulty of speech recognition tasks. In: Program, 94th Meeting of the Acoustical Society of America, vol. 62, p. S63 (1977)
3. Blei, D., Ng, A., Jordan, M.: Latent Dirichlet allocation. J. Mach. Learn. Res. **3**, 993–1022 (2003)

4. Blei, D.: Introduction to probabilistic topic models. Comm. ACM **55**(4), 77–84 (2012)
5. Ali, D., Juanzi, L., Lizhu, Z., Faqir, M.: Knowledge discovery through directed probabilistic topic models: a survey. In: Proceedings of Frontiers of Computer Science in China, pp. 280–301 (2010)
6. Blei, D.: Topic modeling, http://www.cs.princeton.edu/ ∼ blei/topicmodeling.html
7. Dumais, S.T., Furnas, G.W., Landauer, T.K., Deerwester, S.: Using latent semantic analysis to improve information retrieval. In: Proceedings of CHI 1988: Conference on Human Factors in Computing, pp. 281–285. ACM, New York (1988)
8. Deerwester, S., Dumais, S.T., Harshman, R.: Indexing by Latent Semantic Analysis (1990), http://lsa.colorado.edu/papers/JASIS.lsi.90.pdf
9. Eden, L.: Matrix Methods in Data Mining and Pattern Recognition. SIAM (2007)
10. Furnas, G.W., Deerwester, S., Dumais, S.T., Landauer, T.K., Harshman, R.A., Streeter, L. A., Lochbaum, K.E.: Information retrieval using a singular value decomposition model of latent semantic structure. In: Proceedings of ACM SIGIR Conference, pp. s.465–s.480. ACM, New York (1998)
11. Salton, G., Michael, J.: McGill Introduction to Modern Information Retrieval. McGraw-Hill Computer Science Series, vol. XV, 448 p. McGraw-Hill, New York (1983)
12. Jain, A.K., Murty, M.N., Flynn, P.J.: Data clustering: a review. ACM Comput. Surv. **31**(3) (1999)
13. Gramacki, J., Gramacki, A.: Metody algebraiczne w zadaniach eksploracji danych na przykładzie automatycznego analizowania treści dokumentów. In: XVI Konferencja PLOUG, pp. 227–249 (2010)
14. Kapłanski, P., Rizun, N., Taranenko, Y., Seganti, A.: Text-mining similarity approximation operators for opinion mining in bi tools. In: Proceeding of the 11th Scientific Conference "Internet in the Information Society-2016", pp. 121–141. University of Dąbrowa Górnicza (2016)
15. Canini, K.R., Shi, L., Griffiths, T.: Online inference of topics with latent dirichlet allocation. J. Mach. Learn. Res. Proc. Track **5**, 65–72 (2009)
16. Tomanek, K.: Analiza sentymentu – metoda analizy danych jakościowych. Przykład zastosowania oraz ewaluacja słownika RID i metody klasyfikacji Bayesa w analizie danych jakościowych, Przegląd Socjologii Jakościowej, pp. 118–136 (2014), www. przegladsocjologiijakosciowej.org
17. Aggarwal, C., Zhai, X.: Mining Text Data. Springer, New York (2012)
18. Leticia, H.A.: Comparing Latent Dirichlet Allocation and Latent Semantic Analysis as Classifiers, Doctor of Philosophy (Management Science), 226 p. (2011)
19. Papadimitrious, C.H., Raghavan, P., Tamaki, H., Vempala, S.: Latent semantic indexing: a probabilistic analysis. J. Comput. Syst. Sci. **61**, 217–235 (2000)
20. Rizun, N., Kapłanski, P., Taranenko, Y.: Development and research of the text messages semantic clustering methodology. In: 2016, Third European Network Intelligence Conference, vol. # 33, pp. 180–187. ENIC (2016)
21. Rizun, N., Kapłanski, P., Taranenko, Y.: Method of a Two-Level Text-Meaning Similarity Approximation of the Customers' Opinions. Economic Studies – Scientific Papers. University of Economics in Katowice, vol. 296, pp. 64–85 (2016)
22. Rizun, N., Taranenko, Y.: Development of the algorithm of polish language film reviews preprocessing. In: Proceeding of the 2nd International Conference on Information Technologies in Management, Rocznik Naukowy Wydziału Zarządzania WSM (in print) (2017)
23. Rui, X., Donald, C., Wunsch, I.I.: Survey of clustering algorithms. IEEE Trans. Neural Netw. **16**(3), 645–678 (2005)

24. Salton, G., Wong, A., Yang, C.S.: A vector space model for automatic indexing. Commun. ACM 18(11), s.613–s.620 (1975)
25. Hofman, T.: Probabilistic latent semantic analysis. In: UAI 1999, pp. 289–296 (1999); Hofmann, T.: Probabilistic Latent Semantic Indexing. In: SIGIR, pp. 50–57 (1999)
26. Mika, T.: Term Weighting in Short Documents for Document Categorization, Keyword Extraction and Query Expansion. PhD Thesis, Series of Publications A, Report A-2013-1 (2013)

Discovering Relational Phrases for Qualia Roles Through Open Information Extraction

Giovanni Siragusa[✉], Valentina Leone, Luigi Di Caro, and Claudio Schifanella

Department of Computer Science, University of Turin, Turin, Italy
{siragusa,dicaro,schi}@di.unito.it, valentina.leone@edu.unito.it

Abstract. In *Generative Lexicon* [17], Pustejovsky defined the *Qualia Structure* which organizes the semantic meaning carried by nouns through four roles: *formal, telic, agentive* and *constitutive*. Despite their expressive power, to the best of our knowledge no actual NLP system uses qualia structures possibly due to the large effort needed to construct such knowledge bases. Some researchers have tried to circumvent this obstacle using lexico-syntactic patterns based on Hearst idea [11]. In this paper, we propose an Open Information Extraction method to automatically acquire a set of relational phrases from a large corpus, starting with a small set of nouns and their qualia elements. Our idea is that the relational phrases unveil the relations between the nouns and their qualia elements. We compared our method with Reverb [10], Ollie [18] and ClausIE [9] in terms of patterns quality and the relative qualia elements extraction.

Keywords: Generative Lexicon · Qualia structure · Open information extraction · Template-patterns · Word sense disambiguation · Natural language processing

1 Introduction

In *Generative Lexicon* [17], Pustejovsky defined four levels of representation to capture lexical meaning: *Argument Structure, Event Structure, Qualia Structure* and *Inheritance Structure*. In particular, qualia structure organizes the semantic meaning carried by nouns through four roles:

Formal: the set of terms that distinguish the object within a domain (orientation, shape, colour, position, dimension, taxonomic categorization, etc.);

Telic: function or purpose of the object:

- the purpose that an agent has in using the object or performing an activity (e.g., the act of *cutting* with a *knife*);

The major part of this work has been carried out by the first two authors, equally. The work has been funded by the project Semantic Burst: Embodying Semantic Resources in Vector Space Models, financed by Compagnia di San Paolo.

© Springer International Publishing AG 2017
P. Różewski and C. Lange (Eds.): KESW 2017, CCIS 786, pp. 69–84, 2017.
https://doi.org/10.1007/978-3-319-69548-8_6

– object built-in function or aim of specific activity (e.g., *cut* for *knife*);

Agentive: the reason that brought the object into being, i.e., the origin of the object (agent, artifact, etc.);

Constitutive: the parts that compose the object (material, weight and parts).

Despite their expressive power in terms of describing nouns meaning, to the best of our knowledge no actual NLP system uses qualia structures. Usually, systems tend to rely lexical semantics resources such as WordNet [12], BabelNet [15] or FrameNet [2], to name a few. One of the possible reasons is the huge human effort required to manually construct qualia-based lexical resources. To avoid this issue, researchers tried to define lexico-syntactic patterns aiming at automatically harvesting nouns and verbs for qualia roles. These methods mainly used the Hearst's template-based strategy [11] whereby hand-crafted patterns are employed to match with specific relations of interest. In particular, Yamada and Baldwin [22] tried to extract qualia elements for *telic* and *agentive* roles from the British National Corpus, while Cimiano and Wenderoth [7,8] extracted qualia elements for all four roles. Caselli and Russo [6] adopted the same method of Cimiano and Wenderoth to extract qualia elements for Italian nouns from the glosses of the ontological class *ARTIFACT* of Senso Comune ontology [16]. Table 1 reports some patterns defined by these three works.

Table 1. Some patterns defined in the works of Yamada and Baldwin [22], Cimiano and Wenderoth [7,8], and Caselli and Russo [6]. In the patterns, N refers to the target Noun, V to the target Verb and ADJ to the adjective; V[+nom] is the nominalization of the Verb, V[+ing] is the *-ing* construction and V[+ed] is the past participle of the Verb. For Cimiano and Wenderoth, NP stands for Noun Phrase, NP_C and NP_F stand for the Noun Phrase of the constitutive role and the formal role respectively.

Authors	Patterns
Yamada and Baldwin	N BE worthy of V[+nom]; V[+ing] Noun; N BE Adverb-V[ed]; N (deserves|merits) V[+nom]
Cimiano and Wenderoth	NP comprises NP_C; NP are made up of NP_C; NP_F ,? such as NP
Caselli and Russo	used to (make|put) V; made of N; of ADJ color; produced by N; a kind of N

In contrast to manual-definition of lexico-syntactic patterns, Open Information Extraction systems [3] automatically extract sets of triples of the form (*argument1*; *relational phrase*; *argument2*), where *argument1* and *argument2* are words (or multi-word expressions) and *relational phrase* is a phrase excerpt that describes the semantic relation between the two arguments. For instance, the triple from *"Faust made a deal with the devil"* is (Faust, made a deal with, the devil). In this paper, we propose an OIE system to automatically acquire a set of relational phrases from a large corpus, starting with a small set of nouns and their qualia elements. Those relational phrases can be used either to extend and create qualia structures or to create lexico-syntactic patterns with less effort.

In our work, we pose the constraint that *argument1* and *argument2* must contain the noun of interest and the qualia elements[1], respectively. In detail, we adopted a bootstrapping method: first we define an initial qualia structure, disambiguating the words with BabelNet to capture only those that are related to the noun we are interested in, forming our initial *seeds*. We used those nouns and their qualia elements to retrieve all sentences containing them from 1,000,000 Wikipedia pages. Then, an OIE system is applied on those sentences to extract a set of patterns for each role. Finally, we use the extracted patterns to obtain new qualia elements for the roles. Our contribution is thus twofold:

- to construct a set of OIE relation phrases for the qualia roles;
- to build a multilingual disambiguated qualia structure through the BabelNet IDs associated to the words.

2 Related Works

Open Information Extraction (OIE) was born to solve the problems of *Information Extraction*, which does not scale well in large corpora where a huge set of relations is present. OIE have reached notable results in extracting relational phrases in large corpora such as Wikipedia and the Web [3,21]. To the best of our knowledge, such systems are based on two steps: a tagging step where a Part-Of-Speech tagger or a dependency parser is applied to the sentence, and an extraction step that unravels the relational phrases. However, those systems may suffer from uninformative (relations which omit relevant information - for example, the triple (faust; made; a deal with the devil)) and incoherent (relations with no meaningful interpretation) extractions. Some research works have tried to solve this issue using heuristics. For instance, Reverb [10] uses syntactical constraints, while Moro et al. [13] use a dependency parser and checks if one of the arguments is marked as subject or object of a word in the relational phrase.

Differently from the previous systems, DefIE [4] constructs a syntactic-semantic graph by merging the output of the dependency parser with a Word Sense Disambiguation system. It extracts the relational phrases only between disambiguated words. KrankeN [1], instead, focused on extracting N-ary relations. The idea was that triples of the form (*argument1*; *relational phrase*; *argument2*) suffer of information loss. In the paper, they argue that classical OIE systems extract only a small part of the information present in the sentence "*In the 2002 film Bubba Ho-tep, Elvis lives in a nursing home.*", such as (*Elvis*; *lives in*; *nursing home*), missing the year (2002) and the title of the film (Bubba Ho-tep). A system similar to KrakeN is ClausIE [9], which uses a rich set of rules (based on the dependency parser) to extract a large variety of N-ary relations.

---

[1] Note that we have two possible cases: (argument1 = noun, argument2 = qualia element) and (argument1 = qualia element, argument2 = noun).

Other works used OIE systems to create or to populate ontologies and taxonomies. Nakashole et al. [14] applied OIE to automatically build a taxonomy, while Carlson et al. [5] and Speer and Havasi [20] used OIE to extend an existing ontology.

3   Initial Qualia Structures and Manual Annotation

In [7], Cimiano and Wenderoth defined lexico-syntactic patterns to match noun phrases and verbs in order to capture qualia elements for the four roles using the web as corpus. They tested those patterns on seven nouns (*knife, book, beer, computer, conversation, data mining* and *natural language processing*) extracting candidate qualia elements which are then evaluated by human-annotators through a score between 0 (not related to the noun and role) and 3 (tightly related to the noun and role).

```xml
<qualiastructure name="knife" babelids="bn:00049322n,bn:00049323n">
    <role name="formal">
        <element babelids="bn:00074123n">steel</element>
        <element babelids="bn:00005704n">weapon</element>
        <element babelids="bn:00047730n,bn:00047731n">item</element>
        <element babelids="bn:00049259n">kitchenware</element>
        <element babelids="bn:00058442n">object</element>
        <element babelids="bn:00046961n">instrument</element>
        <element babelids="bn:00079388n">utensil</element>
        <element babelids="bn:00031322n">equipment</element>
        <element babelids="bn:00046962n,bn:00077585n">tool</element>
        <element babelids="">cutting instrument</element>
        <element babelids="">cutting weapon</element>
        <element babelids="">emergency item</element>
    </role>
    <role name="agentive">
        <element babelids="bn:00086013v,bn:00084198v">produce</element>
        <element babelids="bn:00086013v,bn:00084198v,bn:00090565v">make</element>
        <element babelids="bn:00086013v,bn:00084198v,bn:00086009v">create</element>
    </role>
    <role name="constitutive">
        <element babelids="bn:00011086n">blade</element>
        <element babelids="bn:00054550n,bn:00002936n">metal</element>
        <element babelids="bn:00074123n">steel</element>
        <element babelids="bn:00081492n">wood</element>
        <element babelids="bn:00041820n">handle</element>
        <element babelids="bn:01213505n">tang</element>
        <element babelids="bn:00002936n">alloy</element>
        <element babelids="">rotating disc</element>
    </role>
</qualiastructure>
```

Fig. 1. An excerpt of the XML for the noun *knife*. The attribute *babelids* contains the synset IDs. Note that some babelids attributes are empty.

We used their qualia structures as our initial ones, selecting only the qualia elements with a score greater than 0 and that we thought as relevant according to the name and the role. Then, we manually linked the nouns and the qualia elements to one or more synsets of BabelNet [15]. During the annotation, we discovered that some words do not have a synset ID in BabelNet (e.g., rotating disc). The advantage provided by the manual annotation is twofold:

1. It is possible to translate the qualia structure in different languages using the BabelNet synsets structure;
2. It can be used to discard those sentences in which the noun or the qualia element does not have the correct meaning.

We encoded the resulting structure into an XML file. Figure 1 shows an excerpt for the noun *knife*.

4 Qualia-Based Pattern Extraction

Common Open Information Extraction (OIE) systems are developed to extract a large set of relations in an unsupervised fashion. However, these systems are not made to extract particular relations such as "steel in knife" (for the constitutive role) or "computer creates" (for the agentive role), because they are developed to extract triples[2] with the constraint that the relational phrases contain a verb. This means that these techniques may loose some relevant relations. For this reason, we developed an OIE system, named *Qualia open Information Extraction* (QIE), which merges the output of a Dependency Parser with that of a Word Sense Disambiguation method, taking inspiration from [4].

The remainder of this section is structured as follows: in Subsect. 4.1 we present QIE, while in Subsect. 4.2 we describe a cleaning method to filter those relations that do not have the synset ID of the noun (or the qualia element) equals or related to the one of the qualia structure. Finally, in Subsect. 4.3 we describe the probability-based formula used to score the relations and unveil those representative for a specific role.

4.1 QIE: OIE for Qualia Roles

The aim of *Qualia open Information Extraction* (QIE) is to extract phrase excerpts that unveil the relations between a noun and its qualia element.

In details, QIE takes as input a sentence and two lemmatized arguments (the noun and the qualia element) and returns a phrase excerpt. First, it generates all possible lexical variants of the lemmatized words (the word itself, its plural form, its past simple form, its -ing form, and so forth) by using Unimorph English Corpus[3]. Then, it processes the sentences through two steps: a Dependency Parser step and a Merging step. In the first step, a Dependency Parser[4] is applied on the sentences to generate a dependency graph. The graph is passed as input to the second phase, where it is merged with the output of a Word

---

[2] An exception is ClausIE [9] which can extract pairs of the form *(argument, relational phrase)* and N-ary relations composed by a triple and a set of additional information.

[3] http://www.unimorph.org.

[4] We used Mate-Tools parser (http://code.google.com/p/mate-tools).

Sense Disambiguation (WSD) method which assigns a sense to each word of the sentence according to the context. QIE can also modify some edges of the graph to deal with conjunctions (e.g., "and") and coordination (list of words separated by commas). If a node m is connected to a node n with an edge labelled as conjunction (or coordination), QIE splits the connection and links m to a neighbour of n which has the edge labelled as object (or modifier).

Finally, QIE uses the lexical variants of the two arguments to extract the shortest path that connects them in the indirect version of the graph. In case the two arguments are adjacent or they are included in the same node, QIE returns the character "*" to indicate that the arguments are separated only by a space. The output of the system is a set of triples of the form *(argument1, shortest path, argument2)* in which *argument1* and *argument2* are two (possibly disambiguated) words belonging to the lexical variant sets.

4.2   Relation Phrases Cleaning

By Analyzing the triples extracted by QIE we noticed that some of them contain an argument (or both) whose sense is not related to the one in the qualia structure. This is due to the last phase of the method, which uses the lexical variants of the nouns and qualia elements to extract the relational phrases. For instance, the method may extract *"knife maker"* for the noun *"knife"*. To solve this problem, we applied a cleaning phase. For each extracted triple in a role, we checked the sense associated to the arguments to remove those triples in which at least an argument has a different sense with the one within the qualia structure. More in detail, given an argument, we first check if it is a noun or a qualia element in order to retrieve the list of senses (the elements in the *babelids* attribute) associate to it in the qualia structure. Then, we compare the sense of the argument, assigned by the WSD software, with the retrieved ones: if the sense associate to the argument appears in the sense-list or it is an hyponym of a sense in the sense-list, we mark the argument as correct[5].

4.3   Relation Phrases Scoring

Once the extracted triples are cleaned, QIE discovers those relational phrases which are relevant for a specific role. Specifically, for each relation phrase p and role r, we computed $score(r, p)$ using the following formula:

$$score(r, p) = \frac{freq_r(p)}{\sum_{p' \in RelsPhrases} freq_r(p')} \qquad (1)$$

where $freq_r(p)$ is the frequency of a relational phrase p in a role r, and *RelsPhrases* is the set containing the relational phrases. An high value of $score(r, p)$ means that the relational phrase p is relevant for the role r. We did not normalized the score in each role with respect to the frequencies in the

---

[5] If the term referring to the argument has an empty *babelnetids* attribute in the qualia structure, we consider the argument as correct.

other roles since, theoretically, qualia elements may overlap across some of the roles. For instance, the triple (*knife*; *is made of*; *steel*) could be extracted both for the formal and the constitutive role.

5   Evaluation

Our experiment focused on two main points:

1. the extraction of relational phrases to unveil the semantic relations between nouns and qualia elements.
2. the acquisition of qualia elements for the nouns *Beer, Book, Computer, Knife, Conversation, Data Mining* and *Natural Language Processing* using the relational phrases extracted in point (1).

For both the two points we evaluated the results of QIE comparing them with those obtained with Reverb [10], Ollie [18] and ClausIE [9].

This section is composed as follows: in Subsect. 5.1 we discuss the extraction of sentences from Wikipedia using the augmented qualia structures described in Sect. 3; in Subsect. 5.2 we evaluate the quality of the acquired patterns. Finally, in Subsect. 5.3 we describe the evaluation of the automatically extracted qualia elements.

5.1   Wikipedia Extraction

In order to collect a corpus of sentences for building the patterns, we retrieved all sentences containing both a noun and a qualia element contained in our initial qualia structures. We also considered the sentences containing the plural forms for the nouns, and third person, past simple and present continuous for the verbs. Sentences have been separated by qualia role. For the extraction, we used the first 1,000,000 pages of a disambiguated version of Wikipedia [19][6], obtaining 9,714 sentences.

In general, the presence of some punctuation between the noun and the qualia element may indicate that they do not refer to each other. As an example, consider the noun *knife*, its qualia element *steel* and the sentence "*He carried a sword made of steel, a knife and a book*". In this case, the noun and the qualia element are separated by a comma and the qualia element *steel* does not refer to the noun *knife*. Obviously, some exceptions are possible as it happens with syntactic coordinations, as in the sentence "*He carried a knife made of wood, leather and steel*". Even if these dynamics can be identified with the use of a syntactic parser, we decided to foster as much as possible the precision of the extracted relational phrases by filtering out all sentences having the noun and the qualia element separated by some punctuation character. After this procedure, we obtained a new corpus of 4,210 sentences. Figure 2 shows the number of

---

[6] The   resource   is   available   at   the   following   url: http://lcl.uniroma1.it/ babelfied-wikipedia/.

sentences per role before and after the application of the punctuation filter, while Fig. 3 shows the distribution of sentences to the seven nouns in the four roles in the cleaned corpus.

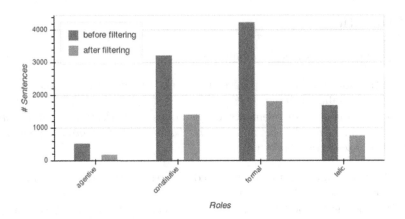

Fig. 2. Number of sentences per role, before and after the application of the filter.

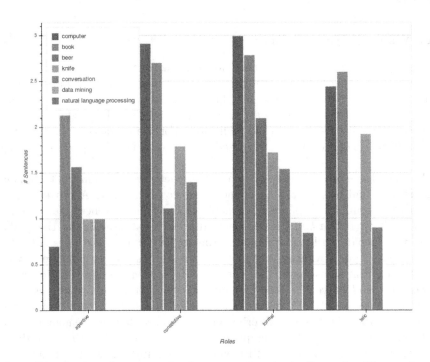

Fig. 3. Distribution of sentences for the seven nouns in the four roles. The number of sentences is in logarithmic scale.

5.2   Evaluation of the Relational Phrases

Given the sentences of a role and their noun-qualia element pairs, we used the OIE systems on those sentences to extract a set of triples. A particular case is ClausIE which can extract pairs of the form (*argument*; *relation*) (e.g., (book; fallen) for the sentence "The book fallen."). We transformed these pairs in triples moving the *relation* to the second argument and adding the character * to the relational phrase (e.g., (book; *; fallen) for the pair (book; fallen)). Unfortunately, ClausIE did not finish the extraction of the relational phrases for the four roles due to some internal problem in the code[7].

After the extraction, we noticed three problems during the evaluation and comparison among the methods:

1. arguments may contain pronouns;
2. arguments may not contain the target noun, the qualia element, or both;
3. arguments may contain words with different senses with respect to those annotated in the initial qualia structures.

We removed all triples falling in one of the first two cases. Then, we filtered the remaining ones checking the BabelNet ID of the arguments (as in Subsect. 4.2 for the corpus cleaning).

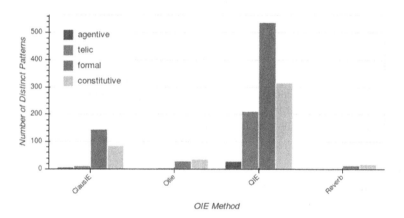

Fig. 4. Distinct relational phrases extracted by Ollie, Reverb, ClausIE and QIE for the four roles.

Figure 4 shows the number of unique relational phrases extracted by the OIE systems for the four roles. Their rank (using the score formula of Subsect. 4.3) is then shown in Table 2 with the top-5 highly-scored patterns, for each method and qualia role. Since the number of the extracted patterns for each role is different from one OIE method to another, we decided to consider the top-20 relational phrases per role in order to equally compare the methods.

---

[7] We could not manage to solve this issue with the help of the available documentation.

Table 2. The table shows the top 5 relational phrases extracted by the OIE systems for the four roles. The symbol "*" means that the noun and its qualia element are separated by a space. The cells containing less than 5 relational phrases report the totality of the extracted patterns.

OIE system	Formal	Telic	Agentive	Constitutive
Reverb	is; have; 's using; prevents; connects printers to			is; are the first chapters of; is a type of; is also available for; is the ability of
Ollie	is; be uses by; be interconnects by; is a part of;)is	be lent in; were still given; was established by; are converted to		is; reported; had been created to manage; are the first chapters of; is a performance feature of
ClausIE	is; was; uses; be connected; are used	is; was used; *; are intended; can be used	took; were used; had; easy; led	is; are; allow; has; had
QIE	*; is; of; for; on	*; to; was; were; used to	to; took years to; took months to; While working on ran out and had; with purpose of building library	*; with; has; is composed of; was divided into

The relational phrases thus selected, were given to three annotators, together with the initial qualia structure, asking them to express their judgment on the relatedness of the pattern to the role.

For this evaluation, the annotators used a numerical scale which comprises three values:

score 0: the relational phrase is not related to the role where it appears;
score 1: the annotator has doubts if the relational phrase is related to the role.
score 2: the relational phrase is strictly related to the role.

For instance, *"is used to"* is not related to the role *constitutive* (score 0), while it is related to role *telic* (score 2).

In order to make the annotation work easier, we provided the annotators with the following guidelines:

1. a relational phrase is good if a full-sense sentence of (*noun*; *pattern*; *qualia*) type can be issued;
2. if a relational phrase is considered suitable for a role but it contains very specific words, then a score between 0 and 1 might be assigned. In fact, the presence of these words suggests the idea of a relational phrase that can be applied on few concepts only. An example is the relational phrase *"are the first*

chapters of" extracted by Reverb for the constitutive role, where the presence of the word *chapters* evokes a literary domain out of which the application of the relational phrase on heterogeneous concepts would be difficult.

Figure 5 shows the inter-annotator agreement for each role and method[8]. The value of Fleiss Kappa ranges from 0.13 (slight agreement) to 0.48 (moderate agreement), showing that the scoring task is difficult for annotators.

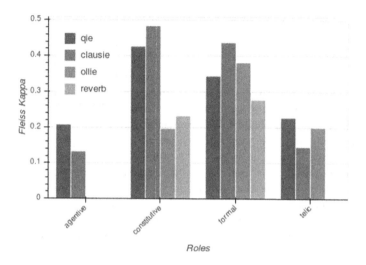

Fig. 5. The figure shows Fleiss Kappa for each role and method.

Each relational phrase received three different scores assigned by the different annotators. Since our evaluation needs a unique score, we used the majority voting to assign a final score to the relational phrases, i.e., we assigned the score with the highest number of votes. For instance, suppose that a relational phrase has the following votes for the three scores: 0 votes for score 0, 2 votes for score 1 and 1 vote for score 2. In this case, the pattern will have a final score of 1. We used the scores to compute the accuracy of each method in the roles, treating patterns with score 2 as the correct ones and patterns with score 0 and 1 as the incorrect ones. The results are reported in Table 3, from which we can notice that QIE achieved the highest accuracy compared to the other systems. Furthermore, all methods achieved an accuracy equals to 0 for the agentive role. This result, combined with the low value of Fleiss Kappa in that role, highlights the fact that good relational phrases for this role are difficult to acquire.

---

[8] In case of a missing score, we assumed it as 0.

Table 3. The table shows the accuracy of the top-20 extracted patterns. Values in bold represent the highest results.

OIE Method	Formal	Telic	Constitutive	Agentive
Reverb	0.31	0.0	0.28	0.0
Ollie	0.2	0.0	0.15	0.0
QIE	**0.4**	**0.65**	**0.6**	0.0
ClausIE	0.3	0.5	0.5	0.0

5.3 Qualia Element Extraction

As mentioned above, our aim was not only to extract a set of relational phrases unveiling the relations between the nouns and their qualia elements for the four roles, but also to use them to acquire old and new qualia elements.

We made a cartesian-product between the seven nouns and the top 20 relational phrases of each role, generating a set of queries of the form *"<noun> <relational phrase>"* for the four OIE systems. For example, given the noun *"knife"* and the pattern *"is used to"*, we generate the query *"knife is used to"*, taking the syntactic chunk that follows in the text as an extracted qualia element, as later detailed. These queries have been then used to retrieve new qualia elements from 100,000 Wikipedia pages (not used in the creation of the corpus of Subsect. 5.1). In detail, we used Clips[9] to bring the noun and the pattern to their lemmatized forms. Then, we also lemmatized the sentence in order to catch all possible inflections of the words. For example, given the above-mentioned noun *"knife"* and the pattern *"is used to"*, we covered other inflections, such as the noun's plural form, and different conjugation of the verbs inside the pattern e.g., *"was used to"*, *"are used to"*, etc.

Subsequently, we extracted the syntactic chunk located at the right of the pattern having the type:

- NP (noun phrase) for the constitutive and formal role;
- VP (verbal phrase) for the agentive role;
- NP or VP for the telic role.

The identification of the types is provided by Clips. Finally, we took and lemmatized the head of NP and/or VP, which is considered as a candidate qualia element.

We scored the candidates of a role using the *Conditional Probability*, described in [8], which ranks the qualia elements on the base of their degree of membership to the role, given the target noun. We took the first 100 qualia elements for each one of the seven nouns of the initial qualia structures, obtaining 700 candidates to annotate for each role; considering the four methods under comparison we had approximately $700 * 4 * 4 = 11,200$ qualia elements to manually validate. Such high number of candidates was unfeasible to annotate, thus

---

[9] http://www.clips.ua.ac.be/pages/pattern-en.

we randomly sampled 100 candidates for each role. The extracted candidates were given to three annotators which expressed if a given qualia element belongs (or not) to the role. Finally, we assigned the final result by using the majority voting. The tagged data were used to compute an accuracy value for each system and role.

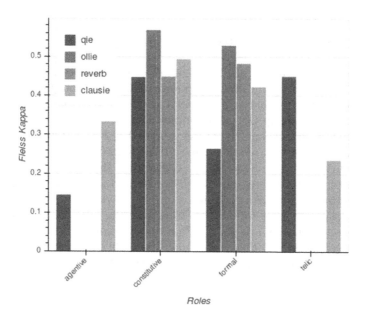

Fig. 6. The figure shows Fleiss Kappa of the qualia elements annotation task.

Table 4. The table shows the accuracy of the methods for the extraction of new qualia elements.

OIE method	Formal	Telic	Constitutive	Agentive
Reverb	0.26	0.0	0.17	0.0
Ollie	0.3	0.0	0.11	0.0
QIE	**0.48**	**0.23**	**0.23**	**0.15**
ClausIE	0.24	0.21	0.20	0.13

Figure 6 reports the inter-annotators agreement, while Table 4 shows the results for this experiment[10]. From the table we can see that QIE achieves the highest results for all the four roles compared to the other three OIE systems.

---

[10] In case of a missing evaluation of a qualia element, we assumed that it does not belong to the role.

Reverb and Ollie reached a 0-accuracy for Telic and Agentive since they did not extract any candidate qualia elements.

Sometimes, annotators found difficult to judge if a candidate qualia element belongs or not to the role, given the noun. This is also reflected in the scores of the qualia elements in [7], as well as the reason behind the low accuracy levels in the table.

6   Conclusion and Future Work

In this paper, we presented a novel method to obtain lexico-syntatic patterns to acquire qualia elements using an Open Information Extraction (OIE) approach. Our contribution was twofold: (i) we created a set of disambiguated qualia structures; (ii) we developed an Open Information Extraction system called *"Qualia open Information Extraction"* (QIE). As validated in Sect. 5, QIE obtained the highest accuracy compared to other three OIE systems on both the experiments.

One of the main problems we had to deal with is the size of the initial data set [7] which is composed by only seven nouns and a small set of qualia elements associated to them. An enrichment of this knowledge would be necessary to retrieve a larger corpus from which to extract the relational phrases. This could be done starting from the information contained in existing resources such as WordNet (or BabelNet) then trying to create qualia structures for the sister terms, hyponyms and hypernyms related to a term. Moreover the qualia structure, as proposed by Cimiano and Wenderoth [7], lacks completely of adjectives to use as qualia elements and, consequently, also the results extracted by QIE do not include them. Nevertheless, we believe that some roles are suitable to be expressed by adjectives. For instance, some adjectival qualia elements for the *formal* role could be: *round, white* or *small*.

Finally, our aim was to find a solution for the above-mentioned problems in order to come up to a method for the automatic creation of Qualia Structures and a consequent large-scale extension of the existing Qualia Base. The Qualia Base will have an impact on Natural Language Processing (e.g., Information Retrieval, Semantic analysis, etc.) and Robotics field (e.g., the robot will know the purpose and the functionality of objects). However, this task should be considered as a very challenging one, and needs further research.

References

1. Akbik, A., Löser, A.: Kraken: N-ary facts in open information extraction. In: Proceedings of the Joint Workshop on Automatic Knowledge Base Construction and Web-scale Knowledge Extraction, pp. 52–56. Association for Computational Linguistics (2012)
2. Baker, C.F., Fillmore, C.J., Lowe, J.B.: The berkeley framenet project. In: Proceedings of the 36th Annual Meeting of the Association for Computational Linguistics and 17th International Conference on Computational Linguistics, vol. 1, pp. 86–90. Association for Computational Linguistics (1998)

3. Banko, M., Cafarella, M.J., Soderland, S., Broadhead, M., Etzioni, O.: Open information extraction from the web. IJCAI **7**, 2670–2676 (2007)
4. Bovi, C.D., Telesca, L., Navigli, R.: Large-scale information extraction from textual definitions through deep syntactic and semantic analysis. Trans. Assoc. Comput. Linguist. **3**, 529–543 (2015)
5. Carlson, A., Betteridge, J., Kisiel, B., Settles, B., Hruschka Jr., E.R., Mitchell, T.M.: Toward an architecture for never-ending language learning. In: AAAI, vol. 5, p. 3 (2010)
6. Caselli, T., Rise, T., Russo, I.: From glosses to qualia: qualia extraction from senso comune. In: 6th International Conference on Generative Approaches to the Lexicon, p. 37 (2013)
7. Cimiano, P., Wenderoth, J.: Automatically learning qualia structures from the web. In: Proceedings of the ACL-SIGLEX Workshop on Deep Lexical Acquisition, pp. 28–37. Association for Computational Linguistics (2005)
8. Cimiano, P., Wenderoth, J.: Automatic acquisition of ranked qualia structures from the web. In: ACL 2007, Proceedings of the 45th Annual Meeting of the Association for Computational Linguistics (2007)
9. Del Corro, L., Gemulla, R.: Clausie: clause-based open information extraction. In: Proceedings of the 22nd International Conference on World Wide Web, pp. 355–366. ACM (2013)
10. Fader, A., Soderland, S., Etzioni, O.: Identifying relations for open information extraction. In: Proceedings of the Conference on Empirical Methods in Natural Language Processing, pp. 1535–1545. Association for Computational Linguistics (2011)
11. Hearst, M.A.: Automatic acquisition of hyponyms from large text corpora. In: Proceedings of the 14th Conference on Computational Linguistics, vol. 2, pp. 539–545. Association for Computational Linguistics (1992)
12. Miller, G.A.: Wordnet: a lexical database for english. Commun. ACM **38**(11), 39–41 (1995)
13. Moro, A., Navigli, R.: Integrating syntactic and semantic analysis into the open information extraction paradigm. In: IJCAI (2013)
14. Nakashole, N., Weikum, G., Suchanek, F.: Patty: a taxonomy of relational patterns with semantic types. In: Proceedings of the 2012 Joint Conference on Empirical Methods in Natural Language Processing and Computational Natural Language Learning, pp. 1135–1145. Association for Computational Linguistics (2012)
15. Navigli, R., Ponzetto, S.P.: Babelnet: the automatic construction, evaluation and application of a wide-coverage multilingual semantic network. Artif. Intell. **193**, 217–250 (2012)
16. Oltramari, A., Vetere, G., Chiari, I., Jezek, E., Zanzotto, F.M., Nissim, M., Gangemi, A.: Senso comune: a collaborative knowledge resource for italian. In: Gurevych, I., Kim, J. (eds.) The Peoples Web Meets NLP. Theory and Applications of Natural Language Processing, pp. 45–67. Springer, Heidelberg (2013). doi:10.1007/978-3-642-35085-6_2
17. Pustejovsky, J.: The generative lexicon. Comput. Linguist. **17**(4), 409–441 (1991)
18. Schmitz, M., Bart, R., Soderland, S., Etzioni, O., et al.: Open language learning for information extraction. In: Proceedings of the 2012 Joint Conference on Empirical Methods in Natural Language Processing and Computational Natural Language Learning, pp. 523–534. Association for Computational Linguistics (2012)

19. Scozzafava, F., Raganato, A., Moro, A., Navigli, R.: Automatic identification and disambiguation of concepts and named entities in the multilingual wikipedia. In: Gavanelli, M., Lamma, E., Riguzzi, F. (eds.) AI*IA 2015. LNCS, vol. 9336, pp. 357–366. Springer, Cham (2015). doi:10.1007/978-3-319-24309-2_27
20. Speer, R., Havasi, C.: Representing general relational knowledge in conceptnet 5. In: LREC, pp. 3679–3686 (2012)
21. Wu, F., Weld, D.S.: Open information extraction using wikipedia. In: Proceedings of the 48th Annual Meeting of the Association for Computational Linguistics, pp. 118–127. Association for Computational Linguistics (2010)
22. Yamada, I., Baldwin, T.: Automatic discovery of telic and agentive roles from corpus data. In: PACLIC (2004)

Probabilistic Topic Modelling for Controlled Snowball Sampling in Citation Network Collection

Hennadii Dobrovolskyi$^{(\boxtimes)}$, Nataliya Keberle, and Olga Todoriko

Department of Computer Science, Zaporizhzhya National University, Zhukovskogo
St. 66, Zaporizhzhya 69600, Ukraine
gen.dobr@gmail.com , nkeberle@gmail.com , o-sun@rambler.ru

Abstract. The paper presents a probabilistic topic model (PTM) application to citation network collection. Snowball sampling method is moderated with the selection of the most relevant papers by means of the PTM. The PTM used in the paper is modified to treat collections of short texts. It is constructed from the titles of seed papers collection united with the papers obtained through unrestricted snowball sampling. The objective of the research is to propose and to experimentally verify the approach of application of PTM of short text documents for improvement of a citation network collection. The preliminary analysis has shown that the method is robust: seed paper collection variations do not affect the most influencing papers subset in the collected citation network.

Keywords: Citation network · Snowball sampling · Text mining · Short text document · Topic modelling

1 Introduction

The completeness of a related work review is a problem well known to each scientist. The main question that should be answered is "If the collected set of the scientific publications contain all significant knowledge of the domain of interest?".

The approaches designed to analyse an existing collection of scientific publications include citation analysis [1,20], the study of the co-authorship [28,29], elaboration of keywords and topics [27], examination of terminology [5]. The evaluation and adjustment of the mentioned methods are performed using different

H. Dobrovolskyi and N. Keberle—The work has been partially done in frame of the EU FP7 Marie Curie IRSES SemData project (http://www.semdata-project.eu/), grant agreement No PIRSESGA-2013-612551.

P. Różewski and C. Lange (Eds.): KESW 2017, CCIS 786, pp. 85–100, 2017.
https://doi.org/10.1007/978-3-319-69548-8_7

and thoughtfully prepared data sets[1][2][3], but the collection of a representative set of publications related to the particular topic is still the task of current interest. First, the existing collections of scientific papers do not cover all the research topics and, second, gathering the publications manually is a time-consuming task which demands the expert involvement. In the actively evolving domains of research, manual search is obstructed with the fast growth of a number of publications. Another issue is the thesaurus diversity of different researchers which limit the completeness of the collected set of publication.

Individual variability of researchers knowledge and terminology affects the search because the authors of each scientific paper use their personal dictionary of terms which can be slightly different from terms used by the researcher looking for publications by keywords. As a result, the set of collected articles does not cover the whole domain of interest and the literature review is biased [32].

The objective of the presented paper is the implementation of restricted snowball sampling to build representative citation network of scientific publications on a domain of interest. To prevent infinite inflation of the sampled set we keep only the papers similar to the seed ones which are manually selected. The main question to be answered is if the sampling can collect most of the seminal publications concerning the domain of interest.

To estimate the publication similarity our algorithm uses probabilistic topic model. One of the challenges of similarity evaluation is that in most of the cases only titles and abstracts of the publications are available and the common methods like latent Dirichlet allocations [4] lose their precision. So the special modification combining word-word co-occurrence [40,44], sparse symmetric nonnegative matrix factorization and principal component approximation [14] is suggested. It is demonstrated that the developed method of probabilistic topic modelling provides the natural estimation of the number of topics and allows calculation of the short text semantic similarity.

This paper is organized as follows. In Sect. 2 we review existing advances in building citation network and estimating similarity of the short texts. Section 3 describes the used restricted sampling method along with a short discussion regarding the employed topic model, term dictionary reduction techniques, description of the suggested sparse symmetric nonnegative matrix factorization approach and the application of ideas of principal component approximation. In Sect. 4 the experiment details are described and Sect. 5 contains results of experiments. Section 6 is devoted to short conclusions and directions of the future studies.

---

[1] AAAI Digital Library Conference Proceedings. https://aaai.org/Library/conferences-library.php.

[2] Journals: Free Texts: Download & Streaming: Internet Archive. https://archive.org/details/journals.

[3] Stanford Large Network Dataset Collection. https://snap.stanford.edu/data/.

2   Related Work

2.1   Building the Citation Network

Scientific search engines like Google Scholar[4], Microsoft Academic[5], Semantic Scholar[6] etc. provide different search facilities but do not answer the question concerning the completeness of keyword search results with respect to a topic of interest.

One of the successful alternatives to keywords search is citation analysis [20] that creates and explores a directed graph which is a citation network. The nodes of the graph represent publications and the edge from node A to node B means that the publication A references publication B. References in a scientific article are selected thoroughly by the authors and advanced search engines can follow citations. It was shown [7] that citation analysis allows the creation of the more comprehensive list of publications than manual keyword search, it enables some formal description and smoothes individual differences of researchers.

The theoretical study of citation patterns was started by Price [34,38] and Garfield [8]. It was demonstrated that detection of hubs simplifies the discovery of other parts of citation network [10,15,36]. The percentage of hubs which are the most cited papers is small because about 90% of publications are never cited [24].

The completeness of the citation network depends on the database coverage and on the quality of the search algorithm. In 2006 Meho and Kiduk [42] show that a single scientific search engine cannot provide enough data to collect the complete set of publications and several databases should be queried. But since then the scientific databases have been significantly extended and the coverage should be much better.

The high quality of citation-based search algorithm is ensured with phenomena of "small world" which is a proved property of scale-free networks [3,28]. Newman [28] demonstrated that in the most of the cases it is enough to do three following iterations: each publication from the current queue is considered then all the papers it references and all papers referencing to the publication are added to the next level queue. The critical point of the algorithm is a selection of the initial queue which is called a seed collection. It should contain the papers that match the domain of interest and are widely cited.

The described iterations are the essence of the snowball sampling algorithm that is widely used to study social relations [17]. But it cannot be applied directly to publication crawling because the scientific texts often refer to the areas that are not directly related to the investigated domain. Therefore the straightforward implementation of snowball sampling causes infinite collection inflation [1] and some restrictions should be introduced while constructing the citation network. For instance, Ahad et.al in their approach uses [1] vector document model and cosine similarity to filter out the most relevant papers while sampling, however,

---

[4] Google Scholar https://scholar.google.com.ua/.

[5] Microsoft Academic https://academic.microsoft.com/.

[6] Semantic Scholar https://www.semanticscholar.org/.

when sampling the scientific abstracts the document vector model doesnt provide enough precision. Lecy et al. [17] apply PageRank calculated by Google Scholar as a measure of paper significance. But PageRank is a property of a global citation network including all topics of knowledge, so it cannot be calculated from its small subset.

In this paper, we show that the restricted snowball sampling which utilises short text similarity can provide a set of publications which is small enough and contains most of the seminal papers concerning the domain of interest.

2.2   Similarity of Short Texts

In most of the scientific databases, the publication title, abstract, author names and the book or journal title are available, sometimes the database-specific keywords and topics are presented while full text is often protected by copyright. Therefore the only information we can rely on is the title and the abstract and we should decide if the paper matches the topic of interest where the topic is defined as a set of seed documents.

Natural approach suitable for such kind of matching is a short text similarity which can be calculated in many ways. String matching methods test whether words in two short texts are similar sequences of characters calculating largest common substring [12], edit distance [25] or lexical overlap [13]. However, string matching fails if the similar texts contain synonyms or have different word order.

Other methods of short text comparison are based on converting texts into syntactic trees [37]. A drawback of the syntactic parser is that it can process only texts having the well-defined structure and works only at the sentence level.

Semantic similarity calculation involves external sources of semantic knowledge such as WordNet [35]. However, such lexical databases are not available for all languages and even if they are, their dictionaries do not contain proper names, domain-specific terms, slang etc.

Recently developed deep learning approaches [26] only require a large amount of unlabeled data and represent words as vectors in a high-dimensional semantic space. Such a representation is referred to as embedding. The challenge is a transition from word level to sentence level [37]. While reducing the sequence of word vectors to a sentence vector of a fixed size some information is necessarily lost and it is crucial to decide which information to keep. Le and Mikolov suggested a variation of the word2vec algorithm for paragraph embedding [16]. Yang et al. [43] suggest an attention mechanism that maps the sequence of the word vectors to a single sentence vector. Also, the straightforward applications of neural networks to short text matching were developed [22]. Another method of word embedding, called GloVe, is proposed in [31]. It is based on word-word co-occurrence matrix and uses global matrix factorization, so it is close to BTM [40] and WNTM [44] statistical topic modelling. But all embedding methods require a lot of data that is not available in a restricted domain of knowledge. Moreover, the neural network embeddings are hard to interpret and adjust.

Distributional semantic approaches assume that words having similar context have the similar meaning. The famous LDA algorithm [4] uses this guess by building word-document co-occurrence matrix to get the probabilistic topic model (PTM). PTM allows mapping a text, sentence or separate word to a low-dimensional vector of topic probabilities. The distance between vectors can be used as the text similarity measure. However the critical drawback is the low precision when handling short texts. To overcome the last obstacle the word-word co-occurrence matrix [40,41,44] or other topic model modifications [30] can be used.

Below we use the word-word co-occurrence frequencies to build the probabilistic topic model and calculate similarity of scientific abstracts. We apply the sparse symmetric nonnegative matrix factorization together with the principal component analysis as a mean to naturally define the number of topics.

3   Restricted Snowball Sampling Method Description

The general workflow of the restricted snowball sampling contains the following steps:

1. Collect a set of seed papers;
2. Start from the seed papers and run several iterations of the unrestricted snowball sampling to gather baseline documents;
3. Create the PTM using baseline documents:
   1 extract title and abstract from each document of the collection;
   2 split all the titles and abstracts into sentences;
   3 create reduced dictionary containing all the significant words occurring in the sentences;
   4 combine all words from the reduced dictionary occurring in the same sentence into pairs and build the joint probability matrix;
   5 detect the collection specific stop-words and exclude them from the reduced dictionary;
   6 perform sparse symmetric nonnegative matrix factorization (Sparse SNMF) to create PTM;
   7 map each of the seed papers to a vector of topic probabilities.
4. Perform the batch restricted snowball sampling:
   1 get a portion of papers from queue;
   2 download the papers referenced by the portion;
   3 download the papers referencing the portion;
   4 extract bag of stemmed words from each of downloaded papers;Word-Word Co-Occurrence and Probabilistic Topic Model
   5 map each of the downloaded papers to a vector of topic probabilities;
   6 calculate distance from each downloaded paper to the seed papers;
   7 add to the next level queue only those of downloaded papers which are close to the seed papers.

Some of the listed steps requiring detailed explanation are discussed below.

3.1   Seed Paper Selection

Selecting the seminal publications is important for creating a comprehensive citation network. In [17] the authors recommend that the seed papers should be the seminal papers of the knowledge domain pointed by experts or the papers selected by the researcher. Valid seed papers should be 5–10 years old and have to be widely cited. The best seeds are the reviews, foundational or framing articles on the topic of interest. Proper analysis of the dependence of the collected citation network recall on the seed publication properties is still an open question.

3.2   Word-Word Co-Occurrence and Probabilistic Topic Model

Let us consider a set of documents D and a dictionary W containing all terms used in D. Each document $d \in D$ is a sequence of n_d terms $(w_1, \ldots, w_{n_d})$ where "term" stands for a word or group of words. As well as common document topic model [39] the method used assumes that

1. Each term w in the document d is related to a topic t from a set of topics T. The document d is formed as a set of pairs (w, t), independently selected in a random way from discrete probability $p(d, w, t)$ defined over set $D \times W \times T$. The document d and the term w are observable and the topic t is the hidden parameter.
2. Order of terms in the document doesn't affect the topic model.
3. Order of documents in the collection doesn't affect the topic model.
4. Conditional probability $p(w|d, t)$ is independent on the document d, i.e. $p(w|d, t) = p(w|t)$.
5. The number of significant topics is far less than the number of words and the number of documents.

Like Biterm Topic Model [40] and Word Network Topic Model [44] the suggested method utilizes the joint probability $p(w_i, w_k)$ that both word w_i and word w_k occur in the same document or document fragment

$$p(w_i, w_k) = \sum_{t=1}^{T} p(w_i|t)\, p(t)\, p(w_k|t) \tag{1}$$

where t identifies a topic. The probability $p(w_i, w_k)$ is estimated as relative number of term pairs (w_i, w_k). To count the pairs each document d_k is mapped to the set of short sequences of terms $S(d_k) = (s_{k1}, s_{k2}, \ldots)$, where $s_{kq} = (w_{kq1}, \ldots, w_{kqr})$. Next, each sentence s_{kq} is mapped to pairs (w_i, w_k), $w_i \in s_{kq}$, $w_k \in s_{kq}$, $w_i \neq w_k$.

To build a topic model we need to evaluate the probabilities $p(w_i|t)$ and $p(t)$. Then detection of the document topics $p(t|d)$ is performed using the expression

$$p(t|d) = \sum_{i=1}^{|W|} p(t|w_i)\, p(w_i|d) \tag{2}$$

where $p(t|w_i)$ is found from the Bayes equation

$$p(t|w_i) = \frac{p(w_i|t)p(t)}{p(w_i)} \tag{3}$$

$p(w_i)$ is the probability of word w_i in the collection,

$$p(w_i) = \sum_{j=1}^{|W|} p(w_i, w_j) \tag{4}$$

and $p(w_i|d)$ is the relative frequency of word w_i in document d.

Because the documents are represented through their topic probabilities the difference between them can be measured with Kullback-Leibler divergence, Fisher's χ^2 or other statistical distances [23].

3.3 Dictionary Reduction

In the presented work, the size of joint probability matrix is reduced with stemming[7], keeping only nouns and adjectives, omitting stop-words and rare words.

Words which are not nouns and not adjectives can be excluded with part-of-speech tagger[8] because of small contribution to document topic assignment [33].

Stop-words are the terms that do not affect topic detection. Various lists of common stop-words are available online[9] but the collection-specific list has to be constructed.

To extract a set of collection-specific stop-words the probability $p(w_i, w_j)$ is employed. The background idea is that the stop-word w_i can co-occur with any other word so it has a large value of the Shannon information entropy

$$I(w_i) = \sum_{j=1}^{|W|} p(w_i, w_j) \, log \, [p(w_i, w_j)] \tag{5}$$

So the N_s terms w_i having the largest values of $I(w_i)$ are considered as stop-words and N_s can serve as the additional parameter of the algorithm.

Rare words are detected with the comparison of the single word probability $p(w_i)$ and a threshold value P_r. To define P_r we require that the kept terms cover $\alpha\%$ of occurrences.

---

[7] See NLTK Stemmers http://www.nltk.org/howto/stem.html.

[8] See NLTK part-of-speech tagger http://www.nltk.org/book/ch05.html.

[9] For instance list of English stop words is available at Snowball stemmer site http://snowball.tartarus.org/algorithms/english/stop.txt.

$$\alpha = \sum_{p(w_i) \geq P_r} p(w_i) \tag{6}$$

We need to exclude the rare words from further consideration because of the statistical nature of our topic model. The small number of the rare words don't allow the reliable estimate of joint probabilities and keeping them we decrease the accuracy of the PTM.

After all the surplus words are excluded the joint probability matrix P becomes much smaller and should be decomposed into product of three matrixes according to (1).

3.4 Sparse Symmetric Nonnegative Matrix Factorization and Principal Component Analysis

The decomposition (1) can be simplified by defining matrix H such that

$$H_{it} = p\left(w_i | t\right) \sqrt{p\left(t\right)} \tag{7}$$

Dimensionality of the factor H_{ij} is $|W| \times |T|$, where $|W|$ is number of words in the reduced dictionary and $|T|$ is a suggested number of topics.

After the substitution (7) performed the Eq. (1) becomes well known Symmetric Nonnegative Matrix Factorization (SNMF) problem [9].

$$P \approx HH^T, H_{it} \geq 0 \tag{8}$$

Typically SNMF is formulated as optimization problem

$$\left\|P - HH^T\right\|_F^2 \to min, H_{it} \geq 0 \tag{9}$$

where $\|Z\|_F^2$ denotes squared Frobenius norm of a matrix Z.

Non-negative sparse coding [11] is a decomposition in which the factor H is sparse - it depends only on a few significant parameters improving human interpretability of the results. Usually [11] the sparsity is achieved with adding extra term to objective function (9):

$$\left\|P - HH^T\right\|_F^2 + \lambda \sum_{i,t} |H_{it}| \to min, H_{it} \geq 0 \tag{10}$$

where the parameter λ affects both sparsity level and factorization accuracy. The particularity of the Sparse Symmetric Nonnegative Matrix Factorization (SSNMF) defined with (10) is the simultaneous requirements of symmetricity, non-negativity and sparsity.

Similarly to other optimization problems (10) can be solved with projected gradient descent approach [19] which consists of the following update rule

$$H_{it}^{(n+1)} = max\left(0, H_{it}^{(n)} + \delta\nabla_{it}\right) \tag{11}$$

where ∇_{it} is a gradient of objective function (10) that takes into account symmetry of the matrix P:

$$\nabla_{it} = 4\left[\sum_j\left(\sum_p H_{ip}H_{jp} - P_{ij}\right)H_{jt}\right] + \lambda \tag{12}$$

and δ is a variable step size which is gradually decreased during iterations.

The main point of the presented method is that the number of topics T can be determined from the solution of (10). Below we demonstrate that sparsity condition leads to very small values of some topic probabilities. So their values can be set to zero and corresponding topics can be neglected in the resulting topic model. The used approach differs from the well known method of Principal Component Analysis [14] with the method of matrix decomposition.

After the factor H_{it} is calculated, the topic probabilities can be found as squared norm of matrix columns

$$p(t) = \left(\sum_{i=1}^{|W|} H_{it}\right)^2 \tag{13}$$

and the normalized columns are the topic word probabilities

$$p(w_i|t) = H_{it}/\sqrt{p(t)} \tag{14}$$

4   Experiment Details

In our experiments, seed papers are collected with searching for keywords "automatic pronunciation assessment" in the Google Scholar database. The set of the seed papers is incomplete and imbalanced but it matches the main objective to test if we can collect the seminal papers starting from such a bad seed.

The current implementation of the restricted snowball sampling uses Academic Knowledge API[10] to search for scientific publication. The API allows to follow the citations in any direction and we can select both the papers referenced by current publication and the papers referencing the current publication. Also the API provided title, abstract, author names and list of topics.

To get the difference between two vectors of topic probabilities v_1 and v_2 we use the Kullback-Leibler divergence

$$D(v_1|v_2) = \sum_i v_1(i)\,\log\frac{v_1(i)}{v_2(i)}. \tag{15}$$

---

[10] https://azure.microsoft.com/en-us/services/cognitive-services/academic-knowledge/.

and distance to the set of seed papers is calculated as KL-divergence with the closest entry of the set. Then the distance is compared to a threshold to decide if the publication will be ignored or added to the snowball.

The parameters used in the experiment are: the upper bound of the topic number is set to 200; the percentage of stop words to exclude equals to 0.02; the percentage of rare words to exclude is 0.05; the sparsity parameter λ is 0.005; the threshold distance is 0.2; the number of seed papers is chosen to be 100 and random subset of seed papers has size 50; the number of levels of snowball sampling is 3; the number of runs is set to 12.

For our experiments, we selected a large set of seed papers from "pronunciation quality assessment" domain and run the sampling method starting from random subsets of that set. Then the following average measures are calculated: the number of citation inside the collected network and the probability of detection with sampling. The number of citations is used instead of the total number of citations because the citation in the relevant publication is more valuable than the same citation in other domain of knowledge. The probability of a paper detection is a percentage of sampling runs discovered the paper. This measure shows how closely is the paper linked to other relevant papers.

5   Results and Discussion

5.1   Stop-Words Detection

Figure 1 shows dictionary terms ranked by values of information entropy calculated using the results of the run of unrestricted snowball sampling which is necessary to create the probabilistic topic model. The Shannon information entropy as function of the rank does not have some special points. It demonstrates the smoothly increasing dependency and we cannot point out some natural threshold. Thus the number of omitted stop-words remains an arbitrary external parameter of the developed algorithm allowing some adjustments.

The top 5 of collection-specific stop-words having the largest values of the entropy are "article", "need", "effect", "main", and "aspect" which are a common terms specific to the scientific writing style. Those terms can be dropped out without loss of precision of the topic model.

5.2   SSNMF and PCA for Topic Modelling

In our experiments each of SSNMF runs starts from random initial state and in general case produces a different number of topics. However Fig. 2 demonstrates that in each case the number of topics can be determined in natural way because probabilities of some topics are too small to be significant and there is a critical topic rank where the topic probability sharply decreases to very small values.

However, the smaller number of topics does not mean the better model quality. Table 1 shows that when number of topics gets bigger the average topic coherence [2] increases i.e. the accuracy of the topic model increases.

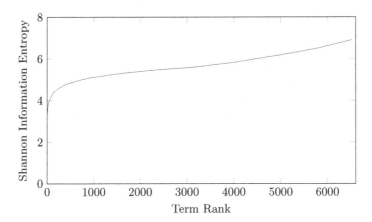

Fig. 1. Shannon information entropy as function of the term rank where the terms are ranked by value of the entropy.

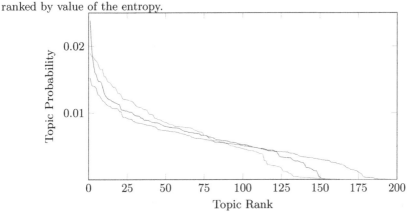

Fig. 2. Topic probabilities for different SSNMF runs for sparsity parameter $\lambda = 0.005$ and random initial states as function of topic rank. Topics are ranked by decrease of values of topic probability.

Table 1. Dependence of average topic coherence on number of topics

Snowball run	Number of topics	Average coherence
1	135	0.471
2	151	0.481
3	180	0.487
4	182	0.519

5.3   Restricted Snowball Sampling

In the restricted snowball sampling, the mathemathically grounded recall evaluation is still an open question because the analogue of central limit theorem is not known for citation sub-networks as well as their statistical properties [17]. The most promising observed feature of the restricted snowball sampling related to the recall is the saturation. The Fig. 3 shows the probability that the publication will be accepted and added to the snowball. More detailed insight shows that the publications added at last stages of the sampling represent the related topics but do not directly concern the domain of interest. So we can conclude that the right part of the plot is generated with topic model inaccuracy rather than actual data. Hence we can stop sampling much earlier than we did.

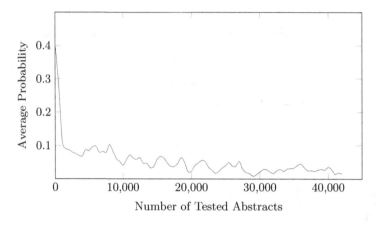

Fig. 3. Average probability of the situation that the publication will be added to the restricted snowball as a function depending on the number of tested abstracts.

It should be noted that new papers in the field will be added to the snowball as referencing already published relevant papers but they will not appear in the list of the most important articles because it takes time for others to read, understand and refer to such papers.

Next feature of interest is the level of precision. To measure the precision we extract top 50, top 100, top 150 etc. most cited papers, then mark the most relevant ones with hands and calculate their percentage as the function of detection probability and size of the most cited papers set. The Table 2 shows that most of the relevant publications were sampled at each snowball run and multiple sampling runs slightly increase the method precision.

The more effective way to increase the precision is the topic model justification which can provide the more precise estimation of semantic distance between the scientific abstracts.

5.4   Expert Selected Papers Evaluation

Another way to evaluate the restriction method is the test of publications selected by experts to check if the publications will be accepted or rejected.

Table 2. The percentage of valid publications as function of the probability of paper detection where N_{top} is the number of the most cited papers extracted and P_{detect} is probability of paper detection

N_{top}	Interval of P_{detect}					
	0–100%	0–20%	20–40%	40–60%	60–80%	80–100%
50	0.50	0.04	0.02	0.02	0.06	0.36
100	0.39	0.02	0.01	0.04	0.05	0.27
150	0.36	0.02	0.01	0.05	0.05	0.25
200	0.29	0.01	0.01	0.04	0.04	0.19
250	0.27	0.02	0.01	0.04	0.03	0.17

For the preliminary tests, we have chosen the publications of Ann Lee [18], Jonás Fouz-González [6], and Victoria López et al. [21]. The first and second papers contain reviews of the domain of interest and the last one describes one of machine learning topic which does not mention the automatic pronunciation assessment. Then for each of the publications we calculate the average distance between the list of references and the set of seed papers and obtain the following estimates: average distance from references selected by Jonás Fouz-González [6] to seed papers is 0.1199 ± 0.0136, the references of Ann Lee [18] are slightly further (distance is 0.1465 ± 0.0267), and review of Victoria López et al. [21] concerning different topic has the largest distance 0.1744 ± 0.0186. Regarding the last paper as baseline corresponding to different area of knowledge we can see that its' references are clearly separated from relevant ones.

6   Conclusions and Future Studies

The main objective of the paper was to develop a method of collecting scientific publications on a domain of interest. The method relies only on the information contained in most of databases, namely paper titles, abstracts, author names and, sometimes, keywords. It provides data for future analysis, including mainstream outline, detection of cutting edge ideas and emerging subfields.

In the presented study we propose the application of probabilistic topic model to restricted snowball sampling which is an intelligent crawling the paper citation network and collecting the papers similar to the seed collection. The starting point of the crawler is a collection of the seed papers found with keywords search in any of scientific search engine. The advantage of the proposed method is its independence on the external information like Google Scholar page rank or full text of publication which may be not available.

The used probabilistic topic model is adapted to handle collections of a short texts with a known approach utilizing the word-word co-occurrence probability. The proposed modification of the topic model uses sparse symmetric nonnegative matrix factorization and provides a natural way to determine the number of topics which is similar to principal component analysis.

The experiments show that the snowball sampling demonstrates saturation and is tolerant to the collection of the seed papers where both features are restricted with topic model inaccuracy. So the main direction of future works is increasing the precision of the topic model with parameter justification or running multiple iterations. A complete analysis of the collected citation network using known methods like PageRank or Search Path Analysis is also planned.

References

1. Ahad, A., Fayaz, M., Shah, A.S.: Navigation through citation network based on content similarity using cosine similarity algorithm. Int. J. Database Theory Appl. **9**(5), 9–20 (2016)
2. Aletras, N., Stevenson, M.: Evaluating topic coherence using distributional semantics. IWCS **13**, 13–22 (2013)
3. Barabási, A.L.: Scale-free networks: a decade and beyond. Science **325**(5939), 412–413 (2009)
4. Blei, D.M., Ng, A.Y., Jordan, M.I.: Latent dirichlet allocation. J. Mach. Learn. Res. **3**, 993–1022 (2003)
5. Ermolayev, V., Batsakis, S., Keberle, N., Tatarintseva, O., Antoniou, G.: Ontologies of time: Review and trends. Int. J. Comput. Sci. Appl. **11**(3), 57–115 (2014)
6. Fouz-González, J.: Trends and directions in computer-assisted pronunciation training. In: Mompean, J.A., Fouz-González, J. (eds.) Investigating English Pronunciation, pp. 314–342. Palgrave Macmillan UK, London (2015). doi:10.1057/9781137509437_14
7. Garfield, E.: From computational linguistics to algorithmic historiography. In: Symposium in Honor of Casimir Borkowski at the University of Pittsburgh School of Information Sciences (2001)
8. Garfield, E., Merton, R.K.: Citation Indexing: Its Theory and Application in Science, Technology, and Humanities, vol. 8. Wiley, New York (1979)
9. Gillis, N.: Introduction to nonnegative matrix factorization. arXiv preprint arXiv:1703.00663 (2017)
10. Harris, J.K., Beatty, K.E., Lecy, J.D., Cyr, J.M., Shapiro, R.M.: Mapping the multidisciplinary field of public health services and systems research. Am. J. Prev. Med. **41**(1), 105–111 (2011)
11. Hoyer, P.O.: Non-negative sparse coding. In: Proceedings of the 2002 12th IEEE Workshop on Neural Networks for Signal Processing, pp. 557–565. IEEE (2002)
12. Islam, A., Inkpen, D.: Semantic text similarity using corpus-based word similarity and string similarity. ACM Trans. Knowl. Discov. Data (TKDD) **2**(2), 10 (2008)
13. Jijkoun, V., de Rijke, M.: Recognizing textual entailment: is word similarity enough? In: Quiñonero-Candela, J., Dagan, I., Magnini, B., d'Alché-Buc, F. (eds.) MLCW 2005. LNCS, vol. 3944, pp. 449–460. Springer, Heidelberg (2006). doi:10.1007/11736790_25
14. Jolliffe, I.T.: Principal component analysis and factor analysis. Principal component analysis. Springer Series in Statistics, pp. 115–128. Springer, New York (1986). doi:10.1007/978-1-4757-1904-8_7
15. Kajikawa, Y., Ohno, J., Takeda, Y., Matsushima, K., Komiyama, H.: Creating an academic landscape of sustainability science: an analysis of the citation network. Sustain. Sci. **2**(2), 221 (2007)

16. Le, Q., Mikolov, T.: Distributed representations of sentences and documents. In: Proceedings of the 31st International Conference on Machine Learning (ICML 2014), pp. 1188–1196 (2014)
17. Lecy, J.D., Beatty, K.E.: Representative literature reviews using constrained snowball sampling and citation network analysis (2012)
18. Lee, A., et al.: Language-independent methods for computer-assisted pronunciation training. Ph.D. thesis, Massachusetts Institute of Technology (2016)
19. Lin, C.J.: Projected gradient methods for nonnegative matrix factorization. Neural Comput. **19**(10), 2756–2779 (2007)
20. Liu, J.S., Lu, L.Y., Lu, W.M., Lin, B.J.: Data envelopment analysis 1978–2010: a citation-based literature survey. Omega **41**(1), 3–15 (2013)
21. López, V., Fernández, A., García, S., Palade, V., Herrera, F.: An insight into classification with imbalanced data: empirical results and current trends on using data intrinsic characteristics. Inf. Sci. **250**, 113–141 (2013)
22. Lu, Z., Li, H.: A deep architecture for matching short texts. In: Advances in Neural Information Processing Systems, pp. 1367–1375 (2013)
23. MacKay, D.J.: Information Theory. Inference and Learning Algorithms. Cambridge University Press, Cambridge (2003)
24. Meho, L.I.: The rise and rise of citation analysis. Phys. World **20**(1), 32 (2007)
25. Mihalcea, R., Corley, C., Strapparava, C., et al.: Corpus-based and knowledge-based measures of text semantic similarity. AAAI **6**, 775–780 (2006)
26. Mikolov, T., Chen, K., Corrado, G., Dean, J.: Efficient estimation of word representations in vector space. arXiv preprint arXiv:1301.3781 (2013)
27. Moya-Anegón, F., Vargas-Quesada, B., Herrero-Solana, V., Chinchilla-Rodríguez, Z., Corera-Álvarez, E., Munoz-Fernández, F.: A new technique for building maps of large scientific domains based on the cocitation of classes and categories. Scientometrics **61**(1), 129–145 (2004)
28. Newman, M.E.: The structure of scientific collaboration networks. Proc. Natl. Acad. Sci. **98**(2), 404–409 (2001)
29. Newman, M.E.: Coauthorship networks and patterns of scientific collaboration. Proc. Natl. Acad. Sci. **101**(suppl 1), 5200–5205 (2004)
30. Pang, J., Li, X., Xie, H., Rao, Y.: SBTM: topic modeling over short texts. In: Gao, H., Kim, J., Sakurai, Y. (eds.) DASFAA 2016. LNCS, vol. 9645, pp. 43–56. Springer, Cham (2016). doi:10.1007/978-3-319-32055-7_4
31. Pennington, J., Socher, R., Manning, C.D.: Glove: global vectors for word representation. EMNLP **14**, 1532–1543 (2014)
32. Petticrew, M., Gilbody, S.: Planning and conducting systematic reviews. In: Health Psychology in Practice, pp. 150–179 (2004)
33. Popova, S., Khodyrev, I., Egorov, A., Logvin, S., Gulyaev, S., Karpova, M., Mouromtsev, D.: Sci-search: academic search and analysis system based on keyphrases. In: Klinov, P., Mouromtsev, D. (eds.) KESW 2013. CCIS, vol. 394, pp. 281–288. Springer, Heidelberg (2013). doi:10.1007/978-3-642-41360-5_24
34. Price, D.: Citation measures of hard science, soft science, technology, and non-science. In: Nelson, C.E., Pollack, D.K. (eds.) Communication Among Scientists and Engineers. Heath Lexington Books Massachusetts (1970)
35. Ramage, D., Rafferty, A.N., Manning, C.D.: Random walks for text semantic similarity. In: Proceedings of the 2009 Workshop on Graph-based Methods for Natural Language Processing, pp. 23–31. Association for Computational Linguistics (2009)
36. Small, H.: Visualizing science by citation mapping. J. Associat. Inf. Sci. Technol. **50**(9), 799 (1999)

37. Socher, R., Huang, E.H., Pennin, J., Manning, C.D., Ng, A.Y.: Dynamic pooling and unfolding recursive autoencoders for paraphrase detection. In: Advances in Neural Information Processing Systems, pp. 801–809 (2011)
38. de Solla Price, D.J.: Networks of scientific papers. Science **149**(3683), 510–515 (1965)
39. Vorontsov, K., Potapenko, A.: Tutorial on probabilistic topic modeling: additive regularization for stochastic matrix factorization. In: Ignatov, D.I., Khachay, M.Y., Panchenko, A., Konstantinova, N., Yavorskiy, R.E. (eds.) AIST 2014. CCIS, vol. 436, pp. 29–46. Springer, Cham (2014). doi:10.1007/978-3-319-12580-0_3
40. Yan, X., Guo, J., Lan, Y., Cheng, X.: A biterm topic model for short texts. In: Proceedings of the 22nd International Conference on World Wide Web, pp. 1445–1456. ACM (2013)
41. Yan, X., Guo, J., Liu, S., Cheng, X., Wang, Y.: Learning topics in short texts by non-negative matrix factorization on term correlation matrix. In: Proceedings of the 2013 SIAM International Conference on Data Mining, pp. 749–757. SIAM (2013)
42. Yang, K., Meho, L.I.: Citation analysis: a comparison of Google Scholar, Scopus, and web of science. Proc. Am. Soc. Inf. Sci. Technol. **43**(1), 1–15 (2006). http://dx.doi.org/10.1002/meet.14504301185
43. Yang, Z., Yang, D., Dyer, C., He, X., Smola, A.J., Hovy, E.H.: Hierarchical attention networks for document classification. In: HLT-NAACL, pp. 1480–1489 (2016)
44. Zuo, Y., Zhao, J., Xu, K.: Word network topic model: a simple but general solution for short and imbalanced texts. Knowl. Inf. Syst. **48**(2), 379–398 (2016)

Russian Tagging and Dependency Parsing Models for Stanford CoreNLP Natural Language Toolkit

Liubov Kovriguina[1]([✉]), Ivan Shilin[1], Alexander Shipilo[1,2],
and Alina Putintseva[1]

[1] ITMO University, Saint-petersburg, Russia
{lyukovriguina,shilinivan}@corp.ifmo.ru,
alexandershipilo@gmail.com, aaputintseva@niuitmo.ru
[2] SPbSU, Saint-petersburg, Russia

Abstract. The paper concerns implementing maximum entropy tagging model and neural net dependency parser model for Russian language in Stanford CoreNLP toolkit, an extensible pipeline that provides core natural language analysis. Russian belongs to morphologically rich languages and demands full morphological analysis including annotating input texts with POS tags, features and lemmas (unlike the case of case-, person-, etc. insensitive languages when stemming and POS-tagging give enough information about grammatical behavior of a word form). Rich morphology is accompanied by free word order in Russian which adds indeterminacy to head finding rules in parsing procedures. In the paper we describe training data, linguistic features used to learn the classifiers, training and evaluation of tagging and parsing models.

Keywords: Dependency parsing · Neural net dependency parsing · Dependency parsing for Russian language · Russian models for Stanford CoreNLP · Maxent tagger · Label embeddings

1 Introduction

The paper describes the process of development of the Russian language package for POS-tagging, lemmatization and dependency parsing in Stanford CoreNLP pipeline, which has not previously included any linguistic resources for Russian[1]. Implemented models for morphology and syntax are available on github[2]. The paper is structured as follows: Sect. 2 gives a brief description of Stanford CoreNLP toolkit, in Sect. 3 we provide general information about Universal Dependencies representation, a contemporary standard for describing morphology and syntax of world languages and show some use-cases for Russian language. Sections 4–6 are the central sections where details on algorithms, models

---

[1] https://github.com/MANASLU8/CoreNLP.
[2] https://github.com/MANASLU8/CoreNLPRusModels.

© Springer International Publishing AG 2017
P. Różewski and C. Lange (Eds.): KESW 2017, CCIS 786, pp. 101–111, 2017.
https://doi.org/10.1007/978-3-319-69548-8_8

implementation and evaluation are provided. We preferred to include a general overview of original algorithms used for training tagger and parser for Russian language to maintain coherence of the paper and ensure results understanding. Conclusion contains some ideas on future work and possible improvements. Undertaken efforts on model implementation may seem to have clear practical contribution, however, as it can be seen from Sects. 4–6 it was important to consider theoretical aspects of language typology to build sound models and get better performance.

2   Stanford CoreNLP Toolkit

Stanford CoreNLP is a Java annotation pipeline framework, which provides common natural language processing steps including tokenization, POS-tagging, parsing, NER, etc. [1]. It is organized as a pipeline of annotators. An Annotator has a uniform interface that adds some analysis information to the input text. It takes in an Annotation object to which it can add extra information. An Annotation is stored as a typesafe heterogeneous map [1]. This architecture provides a light weight and easy to understand and use solution (in comparison to UIMA, DKPro Core and several other packages), which also benefits from a vast documentation.

3   Universal Dependencies Representation

Universal Dependencies[3] is a cross-linguistic standard for annotating treebanks and representing surface syntax relations in the sentence. Its first version as Stanford typed dependencies appeared in 2008 (acc. to [2]). Stanford Dependencies adopts the lexicalist hypothesis in syntax, whereby grammatical relations should be between whole words or lexemes ([2]), which holds in Universal Dependencies. UD representation has been extended to other genres, including social media [3] and spoken language [4]. Universal Dependencies initiative promoted building treebanks for most known world languages and its datasets are widely used in parsing competitions [9] and dependency parsing applications (i.e., UDPipe[4]).

3.1   Universal POS Tags and Morphological Features

Universal POS Tags. Universal Dependencies representation distinguishes 17 core part-of-speech categories, which are divided into open class words, closed class words and miscellaneous "other" category, which includes punctuation marks, symbols and words that for some reason cannot be assigned a real part-of-speech category. Open class words include adjectives, adverbs, interjections, nouns, proper nouns and verbs. The closed class words have a larger set of

---

[3] http://universaldependencies.org/.
[4] https://ufal.mff.cuni.cz/udpipe.

tags, including prepositions, determiners, numerals, particles, pronouns, auxiliary verbs (which are distinguished from other verbs) and coordinate and subordinate conjunctions which also have separate tags[5].

Universal Morphological Features. UD morphological features unit traditional grammar categories defined for world languages (i.e. gender, number, case, definiteness for nouns; tense, aspect, verb form, polarity for verbs, etc.). Adverbial and adjectival participles, which are often considered separate part-of-speech classes in Russian, are tagged as verbs in universal dependencies and distinguished by their morphological features (i.e., VerbForm feature).

3.2 Universal Dependency Relations

Universal Dependencies relations aim to describe syntactic relations between content words. First of all, the clear distinction between core arguments and other dependents is made. The group of core arguments is split into nominals (subject, which can be active or passive, direct object and indirect object) and clauses (clausal subject, clausal complement and open clausal complement). Oblique modifiers belong to the group of non-core dependents. Taxonomy of UD relations can be seen in Fig. 1.

	Nominals	Clauses	Modifier words	Function Words
Core arguments	nsubj obj iobj	csubj ccomp xcomp		
Non-core dependents	obl vocative expl dislocated	advcl	advmod* discourse	aux cop mark
Nominal dependents	nmod appos nummod	acl	amod	det clf case
Coordination	**MWE**	**Loose**	**Special**	**Other**
conj cc	fixed flat compound	list parataxis	orphan goeswith reparandum	punct root dep

Fig. 1. Classification of universal dependencies relations (from universaldependencies. org)

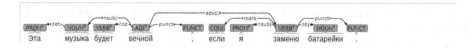

Fig. 2. Example of a sentence with adverbial subordinate clause. (English translation: "This music will last forever, if I replace the batteries".)

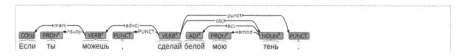

Fig. 3. Example of a sentence with adjectival clause and adverbial subordinate clause.(English translation: "If you can, turn my shadow white".)

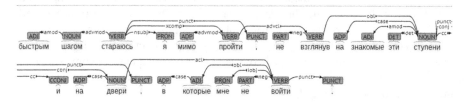

Fig. 4. Example of a sentence with open clausal complement and coordination.(English translation: "I try to walk by quickly, not having looked at these close stairs and at the door, which I shall not enter".)

Examples of Russian sentences[6] annotating according to UD representation can be seen in Figs. 2, 3 and 4.

4 Implementing Stanford CoreNLP Annotators for Russian Language

In this section we describe the process of development of Stanford CoreNLP annotators for morphological analysis, lemmatization and parsing of Russian language. Basic guidelines can be found at StanfordNLP site[7] [8], and CoreNLP javadoc documentation[9] and a number of publications.

4.1 Datasets

SynTagRus Treebank. SynTagRus is the largest and most elaborated dependency treebank for Russian language and provides a good basis for experimental studies using data-driven methods [11]. SynTagRus is an integral but fully

---

[6] Examples were taken from Ilya Kormiltsev poetry, therefore, English translations in the figures footnotes are approximate and do not preserve the author's syntax.

[7] https://nlp.stanford.edu/software/tagger.html

[8] https://nlp.stanford.edu/software/nndep.shtml

[9] https://nlp.stanford.edu/nlp/javadoc/javanlp/.

autonomous part of the Russian National Corpus developed in a nationwide research project[10]. Texts of SynTagRus treebank belong to a variety of genres (contemporary fiction, popular science, newspaper and journal articles dated between 1960 and 2008, texts of online news, etc.) [11].

SynTagRus treebank, converted to Universal Dependencies representation was included in experimental evaluations such as the CoNLL-2017 shared task [9]. The treebank version in CoNLL-U format, presented for parsers' evaluation track, contains 48,214 sentences (871,125 tokens) in the training set and 6,584 sentences (118,646 tokens) in the development set. These data are freely available as Universal Dependencies v.2[11].

4.2   CoNLL-U Format

Treebanks, annotated in CoNLL-U format (a revised version of CoNLL-X format), include sentences consisting of one or more word lines, and word lines contain the following fields[12]:

1. ID: Word index, integer starting at 1 for each new sentence; may be a range for tokens with multiple words.
2. FORM: Word form or punctuation symbol.
3. LEMMA: Lemma or stem of word form.
4. UPOSTAG: Universal part-of-speech tag drawn from our revised version of the Google universal POS tags.
5. XPOSTAG: Language-specific part-of-speech tag; underscore if not available.
6. FEATS: List of morphological features from the universal feature inventory or from a defined language-specific extension; underscore if not available.
7. HEAD: Head of the current token, which is either a value of ID or zero (0).
8. DEPREL: Universal Stanford dependency relation to the HEAD (root if HEAD = 0) or a defined language-specific subtype of one.
9. DEPS: List of secondary dependencies (head-deprel pairs).
10. MISC: Any other annotation.

Treebanks, published in CoNLL-U format, contain annotations, which can be used for training both tagging and parsing algorithms.

5   Part-of-Speech Tagger Implementation

POS-tagger was trained and evaluated on datasets of SynTagRus treebank (training and development sets correspondingly, see Sect. 5.1). The first model (**model POS** in Table 2) was trained to predict POS-tags only, the second model (**model POS+MF** in Table 2) was trained to predict POS-tags and morphological features.

---

[10] http://www.ruscorpora.ru/.
[11] https://lindat.mff.cuni.cz/repository/xmlui/handle/11234/1-1983.
[12] http://universaldependencies.org/docs/format.html.

Algorithm. POS-tagger uses maximum entropy approach to predict part-of-speech tags [5]. The idea of maximum entropy modeling is to choose the probability distribution p that has the highest entropy out of those distributions that satisfy a certain set of constraints. The constraints restrict the model to behave in accordance with a set of statistics collected from the training data. The statistics are expressed as the expected values of appropriate functions defined on the contexts h and tags t. Knowledge sources for the tagger can be found in properties files ud-ru-tagger.props, ud-ru-mf-tagger.props.

5.1   Lemmatization

Mapping Table and Homonymy Resolution. To perform lemmatization, a LemmatizationAnnotator class was developed. This class adds lexeme of the input word form to the class CoreAnnotations.LemmaAnnotation.class. Lemmas are looked up in the mapping table which contains word form, its lexeme and part-of-speech tag. Below is excerpt from the mapping table showing how the right lexeme can be chosen if the tagger faced ambiguous word forms.

```
берегу берег NOUN
берегу беречь VERB
```

The mapping table was compiled using SynTagRus training set and open-sourse lexical dictionaries[13]. Lemmatization accuracy is 93.58.

6   Neural Net Dependency Parser

SynTagRus training and developments sets were used for training and evaluating the parsing model. Results can be seen in Table 2.

Algorithm. CoreNLP neural net dependency parser employs arc-standard transition-based parsing algorithm [6]. Transition-based dependency parsing aims to predict a transition sequence from an initial configuration to some terminal configuration, which derives a target dependency tree [7]. This algorithm uses a neural net classifier to predict the correct transition based on features extracted from the configuration.

According to [7,8] transition system for dependency parsing is a quadruple $S = (C, T, c_s, C_t)$, where

1. C is a set of configurations, each of which contains a buffer b of (remaining) nodes and a set A of dependency arcs,
2. T is a set of transitions, each of which is a (partial) function $t : C \to C$,
3. c_s is an initialization function, mapping a sentence $x = (w_0, w_1, ..., w_n)$ to a configuration with $b = [1, ..., n]$,
4. $C_t \subseteq C$ is a set of terminal configurations.

---

[13] http://odict.ru/.

A configuration is required to contain at least a buffer b, initially containing the nodes $[1, ..., n]$ corresponding to the real words of a sentence $x = (w_0, w_1, ..., w_n)$, and a set A of dependency arcs, defined on the nodes in $V = 0, 1, ..., n$, given some set of dependency labels L [7].

The Stanford neural net dependency parser uses a stack-based variant of transition-based parsing. The stack is used to store partially processed tokens, that is, tokens that have been removed from the input buffer but which are still considered as potential candidates for dependency links, either as heads or as dependents. A parser configuration is therefore defined as a triple $c = (s, b, A)$ with a stack s, a buffer b and a set of dependency arcs A. Initial parser configuration is c_0 with the stack $s = [ROOT]$, buffer containing the whole sentence to be parsed and empty set A of dependency arcs (see an example in Table 1).

Denoting s_i $(i = 1, 2, ...n)$ as the ith top element on the stack, and $b_i(i = 1, 2, ...n)$ as the ith element on the buffer, the arc-standard system defines three types of transitions:

Table 1. An example of transition-based dependency parsing.

Transition	Stack	Buffer	A
	[ROOT]	[К сожалению, арабский оригинал книги не сохранился.]	
[SHIFT]	[ROOT К]	[сожалению, арабский оригинал книги не сохранился.]	
[SHIFT]	[ROOT К сожалению]	[, арабский оригинал книги не сохранился.]	
[LEFT-ARC (case)]	[ROOT сожалению]	[, арабский оригинал книги не сохранился.]	$A \cup case$(сожалению, К)
[SHIFT]	[ROOT сожалению ,]	[арабский оригинал книги не сохранился.]	
[RIGHT-ARC (punct)]	[ROOT сожалению]	[арабский оригинал книги не сохранился.]	$A \cup punct$(сожалению, ,)
[SHIFT]	[ROOT сожалению арабский]	[оригинал книги не сохранился.]	
[LEFT-ARC (amod)]	[ROOT сожалению оригинал]	[книги не сохранился.]	$A \cup amod$(оригинал, арабский)
[SHIFT]	[ROOT сожалению арабский книги]	[не сохранился.]	
[RIGHT-ARC (nmod)]	[ROOT сожалению оригинал]	[не сохранился.]	$A \cup nmod$(оригинал, книги)
...	...	...	...
[RIGHT-ARC (root)]	[ROOT]	[]	$A \cup root$(ROOT, сохранился)

1 The example sentence in Russian and its English translation:

```
К сожалению   , арабский   оригинал   книги       не сохранился    .
Unfortunately , the Arabic  version    of the book  is not preserved .
```

- *LEFT − ARC(l)*: adds an arc $s_1 \rightarrow s_2$ with label l and removes s_2 from the stack. Precondition: $|s| \geq 2$.
- *RIGHT − ARC(l)*: adds an arc $s_2 \rightarrow s_1$ with label l and removes s_1 from the stack. Precondition: $|s| \geq 2$.
- *SHIFT* moves b_1 from the buffer to the stack. Precondition: $|b| \geq 1$ [6].

Original feature templates described in [6] employ wordforms, its leftmost and rightmost children and POS-tags. Firstly, we reproduced training with this features and then added morphological information to feature templates. Templates taking morphological features into account, have shown better parsing accuracy. Neural net classifier used embeddings of words, POS-tags and relation labels for training. Word embeddings has become a well-known method to track semantic similarity, however developers of the parser have shown [6] that part-of-speech embeddings and arc label embeddings excibit many semantic similarities like words (see "Implementation" paragraph for details).

Implementation. Following the guidelines on training neural network dependency parser, we wrote RussianMorphoFeatureSpecification, RussianTreebankLanguagePack and several HeadFinders. The last RussianHeadFinder was written for finding heads in trees following UD representation and distinguishing heads for main and subordinate clauses, noun and prepositional phrases in Russian has become a problem. So, this version of the RussianHeadFinder is just a prototype where the structure of only the most frequent phrases is described. Figure 5 shows t-SNE visualization of label embeddings. It can be seen that most of the labels are grouped according to their syntactic roles: i.e., group A contains labels denoting direct object, indirect object, oblique modifiers and clausal complement, group B includes labels used for headless and exocentric constructions.

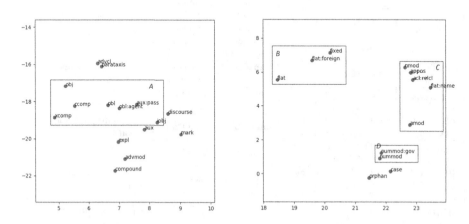

Fig. 5. t-SNE visualization of label embeddings (enlarged).

Evaluation. Results of parser training with different options (sample size, embeddings dimensionality, attached headFinder and morphological features)

Table 2. Parser performance with different training options

Sample size	Training options							
	Head finder	Tagger model	Tagging accuracy	Embedd. model	Embedd. dimens.	Hidden layer size	UAS	LAS
12,000 sentences	+	POS	97.92	SMT	100	200	69.55	61.27
48,814 sentences	+	POS	97.92	SMT	100	200	77.81	72.63
48,814 sentences	-	POS+MF	86.30	SMT	100	200	80.66	75.99
48,814 sentences	-	POS+MF	86.30	SMT	100	200	81.02	76.40
48,814 sentences	-	POS+MF	86.30	Aranea	100	300	81.63	77.12
48,814 sentences	-	POS+MF	86.30	Aranea	100	400	81.73	77.30

can be seen in Table 2. Two different corpora were used for building word embeddings: Russian segment of news parallel corpora (*SMT* in Table 2)[14] and Aranea Maius Russian Web coprus (*Aranea* [10] in Table 2). Parsing quality was evaluated using unlabeled attachment score (UAS) and labeled attachment score (LAS). The best registered performance was on the run when morphological features were used for building templates with word embeddings calculated on Aranea corpus (81.73 UAS, 77.30 LAS).

This is not the best result obtained on SynTagRus. Stanford NLP group presented a new algorithm based on LSTM a couple of months ago on CoNLL shared task [9]. It reaches 94.0 UAS and 92.6 LAS on SynTagRus. Results of other parsers (evidence from [9]) on SynTagRus can be seen in Table 3. However, these results might not be directly comparable with our model because of differences in tokenizing procedures in the evaluation script.

Table 3. Parsers evaluation on the SynTagRus corpus

No.	Name	UAS	LAS
1	Stanford (Stanford)	94.00	92.60
2	C2L2 (Ithaca)	92.15	90.03
3	UParse (Edinburgh)	91.90	89.18
4	IMS (Stuttgart)	91.81	89.80
5	HIT-SCIR	91.71	89.77
..	..	..	..
26	OpenU NLP Lab (Ra'anana)	84.20	75.63
27	**Stanford CoreNLP (Rus)**	**81.73**	**77.30**
28	TRL (Tokyo)	69.49	61.66

---

[14] http://statmt.org/.

7   Conclusion and Future Work

The main contribution described in the paper are open-source Russian tagging and parsing models developed for Stanford CoreNLP toolkit that allow fast and scalable processing of large text collections and can be easily integrated with other NLP tools in the CoreNLP pipeline. The best registered tagging quality was 97.92 for POS-tags and 86.30 for predicting whole morphological information for a word form. We assume that training and testing (both for tagger and parser) was performed in a methodologically correct way, using development set of SynTagRus corpus in Universal Dependencies format. However, it should be admitted, that performed evaluation is of limited extent and some additional testing has to be undertaken, i.e. evaluating the tagger on social media texts and spoken language and performing a neat comparison with the baseline and existing taggers using, i.e., Dialogue Evaluation datasets[15]. To have a competitive tagger, it's also necessary to implement annotators for predicting lexemes of non-vocabulary word forms. Highest scores obtained for the parsing model are 81.73 for unlabeled attachment score and 77.30 for labeled attachment. This is a good result in general, but still a lot of work has to be done in respect of training set representativeness, checking the steadiness of training set annotations, learning low-frequency trees and distant syntax relations.

Acknowledgment. This work was financially supported by the Russian Fund of Basic Research (RFBR), Grant No. 16-36-60055.

References

1. Manning, C.D., et al.: The standford CoreNLP natural language processing toolkit. In: ACL (System Demonstrations), pp. 55–60 (2014)
2. de Marneffe, M.-C., et al.: Universal Dependencies: A cross-linguistic typology. In: Language Resources and Evaluation Conference (LREC), European Language Resources Association (ELRA), Iceland, Reykjavik, pp. 4585–4592 (2014). ISBN:978-2-9517408-8-4
3. de Marneffe, M.-C., et al.: Extending stanford dependencies. In: Proceedings of the 13th International Conference on Dependency Linguistics, pp. 187–196 (2013). ISBN:978-2-9517408-9-1
4. Dobrovojc, K., Nivre, J.: The universal dependencies treebank of spoken slovenian. In: Proceedings of LREC Conference, European Language Resources Association (ELRA), Portorož, Slovenia, pp. 1566–1573 (2016)
5. Toutanova, K., Manning, C.D.: Enriching the knowledge sources used in a maximum entropy part-of-speech tagger. In: Joint SIGDAT Conference on Empirical Methods in Natural Language Processing and Very Large Corpora (EMNLP/VLC-2000), vol. 13, pp. 63–70 (2000)
6. Chen, D., Manning, C.D.: A fast and accurate dependency parser using neural networks. In: Empirical Methods in Natural Language Processing (EMNLP), pp. 740–750 (2014)

---

[15] https://github.com/dialogue-evaluation/morphoRuEval-2017.

7. Nivre, J.: Algorithms for deterministic incremental dependency parsing. Comput. Linguist. **34**(4), 513–553 (2008). doi:10.1162/coli.07-056-R1-07-027

8. Nivre, J., et al.: Labeled pseudo-projective dependency parsing with support vector machines. In: Proceedings of the 10th Conference on Computational Natural Language Learning, CoNLL 2006, pp. 221–225 (2006)

9. Zeman, D., Popel, M., Straka, M., Hajic, J., Nivre, J., et al.: CoNLL 2017 shared task: multilingual parsing from raw text to universal dependencies. In: Proceedings of the CoNLL 2017 Shared Task: Multilingual Parsing from Raw Text to Universal Dependencies, Vancouver, Canada, August 3–4, 2017, pp. 1–19 (2017). doi:10. 18653/v1/K17-3001

10. Benko, V., Zakharov, V.P.: Very large russian corpora: new opportunities and new challenges. In: Proceedings of the International Conference "Dialogue 2016" (2016)

11. Nivre, J., Boguslavsky, I.M., Iomdin, L.L.: Parsing the SynTagRus treebank of russian. In: Proceedings of the 22nd International Conference on Computational Linguistics, vol. 1, pp. 641–648 (2008). ISBN: 978-1-905593-44-6

Investigating the Relationship Between Tweeting Style and Popularity: The Case of US Presidential Election 2016

Farideh Tavazoee$^{(\boxtimes)}$, Claudio Conversano, and Francesco Mola

University of Cagliari, Cagliari, Italy
farideh.tavazoee@unica.it

Abstract. Predicting popularity from social media has been explored about a decade. As far as the number of social media users is soaring, understanding the relationship between popularity and social media is really beneficial because it can be mapped to the real popularity of an entity. The popularity in social media, for instance in Twitter, is interpreted by drawing a relationship between a social media account and its followers. Therefore, in this paper, to understand the popularity of candidates of the US election 2016 in social media, we verify this association in Twitter by analyzing the candidates' tweets. More specifically, our aim is to assess if candidates put efforts to improve their style of tweeting over time to be more favorable to their followers. We show that Mr. Trump could wisely exploit Twitter to attract more people by tweeting in a well-organized and desirable manner and that tweeting style has increased his popularity in social media.

Keywords: Popularity · US election 2016 · Sentiment analysis · Twitter · Retweet · Classification · Machine learning

1 Introduction

Most of people tend to do more and more of their daily activities online such as shopping, trading and communicating, thanks to the increasing Internet access opportunities. This huge amount of data makes the Internet an invaluable source of information for analyzers to extract knowledge. A prominent example of knowledge extraction from the Internet is social media analysis, which can lead to understanding what people think about a subject or how they make a decision. This kind of analysis is usually known as opinion mining or sentiment analysis [21].

Among the existing social media platforms, Twitter is widely used to share opinions on a variety of topics, especially by political parties. This popularity has motivated scholars to extract knowledge from Twitter although it has its own limitations which make it challenging. Some examples of its limitations are that people have just 140 characters to describe their opinions as well as tweets can be sarcastic and ironical [25]. Considering all of the characteristics of Twitter,

© Springer International Publishing AG 2017
P. Różewski and C. Lange (Eds.): KESW 2017, CCIS 786, pp. 112–123, 2017.
https://doi.org/10.1007/978-3-319-69548-8_9

different opinion mining approaches have been proposed by researchers. There are many studies that have exploited Twitter to predict the future trends, such as predicting the box-office revenues for movies [1] or predicting citations of articles [6]. Other studies have shown how it is feasible to predict the popularity of tweets from the content of the tweets [14,22,23,26]. Some other studies are orientated towards predicting the results of election by detecting the polarity or emotions of tweets [2,8,11,13,18,20,24,25].

The effectiveness of the analysis of Twitter data motivated us to consider this valuable source of information to investigate if it was beneficial for the candidates of the US election 2016 to advertise themselves properly. It is well known that the use of this social media grown exponentially during the campaigns. The management of Twitter reported that more than 1 billion tweets related to the campaigns were sent from August 2016, when the primary debates began, till the end of the campaign[1]. This grown is not surprising, since the use of social media, including Twitter, was also crucial for the election of the former president Barak Obama. As documented in [12], the previous presidential campaign was the first in which in which the Internet, the electorate, and political campaign strategies for the White House successfully converged to propel a candidate to the highest elected office in the nation. Thus, there is a strong belief that social media played a key role also in the Mr. Trump's victory. The president-elect argued that "his victory offers proof that his social media presence had greater leverage on the election than hundreds of millions of dollars spent by the Democrats to influence its outcome, social media played a key role in his victory and he will continue to use it in the White House"[2]. For instance, Evan Williams, the co-founder of Twitter, apologized for any effect of Twitter on Trump's victory and he said "If it's true that he wouldn't be president if it weren't for Twitter, then yeah, I'm sorry"[3].

Taking all the above-mentioned considerations into account, in this paper we evaluate how much candidates of the US presidential election 2016 invested on the content of their tweets to make them more favorable to their followers. To this end, we predict the popularity of tweets from their content by considering a set of well-known syntactic features from the tweets, as well as by proposing some semantic features like sentiment of tweets in order to improve the accuracy of the prediction. Moreover, we show that predicting the popularity of candidates' tweets reveals interesting knowledge on the style of tweeting of candidates during the pre-election period. It allows them to attract more people, and it is likely that a proper tweeting style increases their winning chance. Overall, the main contributions of our study are:

---

[1] See, for example: https://blog.twitter.com/official/en_us/a/2016/how-election2016-was-tweeted-so-far.html.

[2] See, for example: http://www.cbsnews.com/news/president-elect-trump-says-social-media-played-a-key-role-in-his-victory/.

[3] http://www.foxnews.com/tech/2017/05/21/twitter-co-founder-apologizes-for-helping-elect-trump.html.

(a) To find a connection between the number of favorites/retweets a tweet of a candidate receives and the popularity of the candidate itself;

(b) To derive an index of popularity based on social media information. This is done by training a classifier in order to predict that a specific tweet has to be considered as popular and by observing how sensitivity of the classifier outcome evolves over time.

The structure of our paper is as follows. Our approach is described in Sect. 2. Section 3 presents the results of the empirical analysis and Sect. 4 ends the paper with some concluding remarks.

2   System Overview

Our system consists of four main steps. First, a substantial number of tweets are extracted from candidates profiles in Twitter (2.1). Second, the gathered tweets are cleaned and prepared for the statistical analysis (2.2). Next, to construct the features vector, the sentiment of tweets along with some well-known characteristics of tweets are extracted (2.3). Finally, a classifier is trained on the candidates' tweets based on the extracted features to evaluate the amount of efforts each candidate put in social media to increase his popularity (2.4).

2.1   Data Gathering

There are two ways to gather data from Twitter, either through Twitter API or by crawling the Twitter website. Extracting data via Twitter API has a limitation that does not let us gathering tweets prior to two weeks before. As we need to provide a comprehensive evaluation, we should access tweets for longer time periods so we resort to a crawler implemented in Python [10].

Crawlers are known as automated programs or scripts to reach the depth of a web page with the purpose of gathering a specific type of information or data. By the aid of crawling, we monitored tweets since January 2016 until a day before the election, namely 7th of November. In total, we gathered 9,025 tweets from the profile of both the candidates. Of these, 5,106 tweets come from Clinton's timeline and 3,919 tweets from Trump's timeline. The distribution of gathered tweets in each month is shown in Fig. 1.

2.2   Data Preparation

One of the most important steps after data gathering is cleaning the data as there might be many meaningless characters. Especially in Twitter, tweets are different from the normal text because there are lots of useless elements that might affect some experimental parts like sentiment analysis [25]. To this end, we removed punctuation, numbers, stop words, URLs, and white spaces. Moreover, we convert all the uppercase letters to lower case letters and run stemming on words.

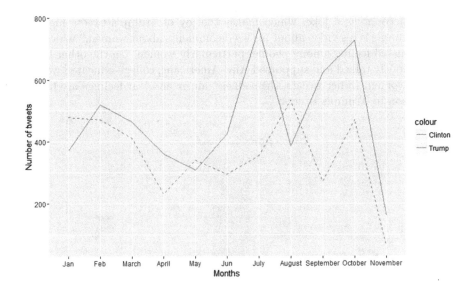

Fig. 1. Tweets extracted from candidates' timeline.

2.3 Feature Extraction

The heart of any learning-based system is feature engineering. We resort to a set of well-known features that have been used successfully in previous research, such as frequency of frequent words (191 features from Clinton's tweets and 137 features from Trump's tweets) and the count of some important punctuation like @, ! and # in a tweet.

Moreover, we use sentiment analysis [16] to obtain a set of 9 features and amend them to the feature vector. This task has been accomplished by using the Sentiment package in R programming language[4]. The set of sentiment features consists of 6 emotional features and 3 features describing the polarity (i.e., positive and negative) of tweets. In total, the feature vector contains 203 and 149 features from Clinton and Trump's tweets respectively. All these features are used to train a classifier, as described in Sect. 2.4.

The main motivation behind using sentiment features as predictors in a classification model is that they should reveal a degree of popularity of the tweets since, in general, sentiment analysis is able to describe the potential reaction of people to the candidates' tweets. The six emotions deriving from tweets are anger, disgust, joy, sadness, fear, and surprise.

A comparison of the distribution of emotional features between both candidates is shown in Fig. 2, which shows a reflection of the personality of candidates from their tweets. For instance, during the election period, Trump surprised people frequently by talking about building a wall around US, about preventing the entrance of immigrants or about nuclear deals. Some people took

---

[4] https://cran.r-project.org/src/contrib/Archive/sentiment/.

them into account as a joke, which makes the joy of Trump's tweets higher[5]. Another example is his video about his lewd comments about women,[6] which has spread disgusted feeling among people, particularly women. On the other hand, Clinton seriously talked and supported Afro-Americans, college-educated women and single women. In her tweets, the sadness, anger and fear feelings are higher in comparison to Trump's tweets.

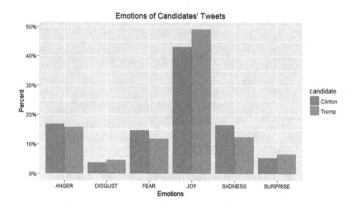

Fig. 2. Emotions of candidates' tweets.

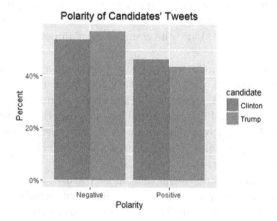

Fig. 3. Polarity of candidates? tweets.

To compare the polarity of tweets, the percentage of positive and negative tweets is shown in Fig. 3. Figure 3 shows that both the candidates tweeted more

---

[5] See, for example: http://www.snopes.com/donald-trump-sentence/.

[6] http://www.smh.com.au/world/us-election/donald-trump-recorded-having-extremely-lewd-conversation-about-women-20161007-grxrwp.html.

negative words rather than positive ones. Nevertheless, if we compare the two candidates, Trump used more negative words as well as less positive words. However, it doesn't necessarily mean that using negative words has a negative effect on people, and vice versa. As a conclusion, features derived from sentiment of tweets seems to be appropriate in describing the reaction of people following a tweet.

2.4 Prediction

As we mentioned earlier, we aim to understand if a tweet can be considered as popular, or not. This problem is innately a binary classification task. We define a tweet as popular if both the number of favorites (like) and retweets obtained from Twitter users are higher than a specific threshold. In this respect, the response variable describes the popularity of each single tweet. It equals one when the number of favorites and the number of retweets a single tweet has received are both above two specific pre-specified thresholds, otherwise it is set to zero. The threshold is, therefore, a specific quantile of the distribution of the variable of interest (number of favorites or number of retweets). As for the choice of a suitable value for this quantile, we run several experiments by varying this value in the 0–1 range. Results of these experiment (not reported) indicate that small values of the quantile or large values of it (say less than 0.25 or more than 0.75) lead to classifiers that perform rather poorly since the distribution of the response variable in each experiment is unbalanced. Values of the threshold between 0.25 and 0.75 lead to classifiers that are very similar in terms of accuracy. For these reasons, we conjectured that although the thresholds for favorites and retweets could be different, in our analysis it was worth considering the median of all favorites and that of all retweets as the most common representative measures.

To classify tweets, we trained many of the classifiers commonly used in the statistical learning framework [9], such as Logistic Regression, SVM and Random Forests (RF) classifiers. Results indicate that the RF classifier outperforms competitors on our data in terms of accuracy, probably because of the strength of the individual trees in the forest and the correlation between them, which cause the overall classification engine to be more robust with respect to noise. This finding is consistent with previous literature on classifiers' performance, where it has proven that RF perform very well in addressing complex classification type decision problems in many fields. Specifically, the performance of Random Forests has been evaluated on different data sets, and it has shown that RF outperforms more than a hundred of classifiers (see, for instance, [4,5,7,17]).

3 Evaluation

In the following, we describe the experimental setup and the evaluation metric (3.1). We then address the problem about understanding which candidate intelligently invested more on social media (3.2).

3.1 Experimental Setup

We train a RF classifier separately for each candidate in order to observe how their social media popularity evolves in time. Following standard supervised classification experiments, we partition our data into two different sets: a training set and a test set. The latter is composed of all the tweets of each specific month, whose class label (i.e., "popular" or "not popular") should be estimated by training a RF classifier on data containing the tweets of the previous months. In other words, we consider the tweets in a specific month as the test set and all the older tweets as the training set. For instance, to predict the popularity of tweets in April 2016, we include all the tweets of the period January–March 2016 in the training set and all the tweets of April 2016 in the test set. To define the popularity index described in Sect. 3.2 data has been partitioned in this way for 10 times in order to consider 10 different classification experiments related to the period from February 2016 to November 2016.

We used a laptop with 2.7 GHz processor and 4 GB of memory to run the experiments, and the RandomForest package [15] implemented in the R environment to train the RF classifier and to predict the class label for test set instances. The performance of the RF classifier is evaluated by calculating the sensitivity from the confusion matrix. The reason of considering this metric is because we are interested in estimating $P(Y = popular|\mathbf{x})$, which corresponds to the conditional probability for a tweet to be popular based on specific values of the set of predictors $\mathbf{x}$.

3.2 Results

To recap our logical assumption, if candidates follow a structure to tweet in a way that their tweets are more favorable to their followers, they might be able to increase their chance of winning although this correlation might not be causality. This assumption should be observable by using the sensitivity metric, as mentioned in Sect. 3.1.

Furthermore, we consider the variation of the sensitivity in time in order to define a "popularity index" based on a proper rescaling of the sensitivity metric. We set a base value of 100 in February 2016 and computed the relative variation of monthly changes in sensitivity from the base period to the last month (i.e., November 2016). The "popularity index" for both candidates is plotted in Fig. 6.

At the same time, we are interested in understanding which of the two proxies of popularity (i.e., number of favorites and number of retweets) is more representative in defining the index. As the results in Figs. 6 and 5 show, either retweets or the combination of retweets and favorites could be considered as more realistic in terms of popularity if compared to the favorites alone. If we follow the "popularity index" represented in Figs. 4, 5 and 6, the trend of popularity for Donald Trump is increasing in time. We believe that this increasing trend for Trump is not obtained randomly and either himself or his staff managing his Twitter account wisely selected the tweets that were considered as more favorable to people. In contrast, the "popularity index" for Hilary Clinton is decreasing. This

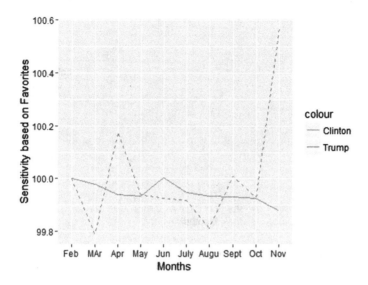

Fig. 4. Based on the number of Favorites

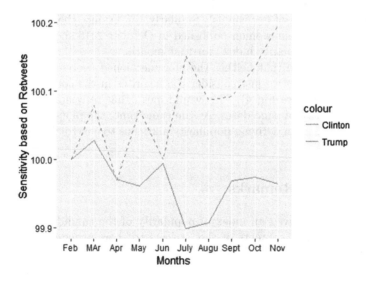

Fig. 5. Based on the number of Retweets

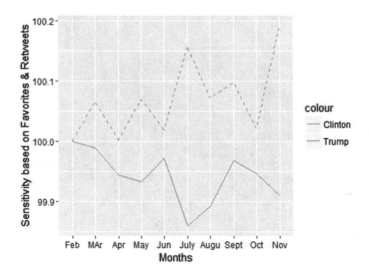

Fig. 6. Predictions based on the number of favorites and retweets.

could indicate that she probably spoiled the advantage of exploiting social media for advertisement. Interestingly, there are two large variations in the index, that are observed in July for Clinton and in October for Trump. These two changes can be mapped to two important events occurred during the pre-election period which affect the people opinion at least for a short period. The news that should affect the "popularity index" of Hillary Clinton in July 2016 is about her criminality in a new batch of her emails[7]. Similarly, for Trump, the news about the lewd conversation about women published in October 2016 should be a reason of dropping the "popularity index" for that month.

It is worth to note that neither the aforementioned evolving trend nor the drops are observable if we just consider the favorites in the construction of the "popularity index" (see Fig. 4). This would mean that retweets actually reflect more the popularity of candidates. At the same time, the plots show that considering both factors in deriving popularity allows us to control for the volatility of the index.

4   Concluding Remarks

In this paper, we derived an index of popularity of the candidates in 2016 US election based on the contents of their tweets to show a possible relationship between tweeting style and popularity. More specifically we showed that, if candidates take advantages of social media and learn how to tweet in a proper way they can increase their popularity. Our results enforce the Trump's opinion that

---

[7] https://vault.fbi.gov/hillary-r.-clinton.

the key reason of his victory was social media. In the future, we will use a similar approach to understand the power of social media in advertising different businesses.

Our research topic could be lead to the current phenomena of Organized Social Media Manipulation and white populism, that are very debated nowadays. As for the first one, some scholars (as, for example [3]) proved empirically that some "cyber troops" like government, military or political party teams committed to manipulating public opinion over social media. They report on specific organizations created, often with public money, to help define and manage what is in the best interest of the public and conclude that cyber troops are a pervasive and global phenomenon. Many different countries employ significant numbers of people and resources to manage and manipulate public opinion online, sometimes targeting domestic audiences and sometimes targeting foreign publics. As for populism, it is documented in [19] that populism is an appropriate descriptor of the 2016 election and that Donald Trump stands out in particular as the populist par excellence. Historical data reveal a large "representation gap" that typically accompanies populist candidates and content analysis of campaign speeches shows that Trump, more so than any other candidate, employs a rhetoric that is distinctive in its simplicity, anti-elitism, and collectivism. At the same time, original survey data in [19] show that Trump's supporters are distinctive in their unique combination of anti-expertise, anti-elitism, and pronationalist sentiments. Together, these findings highlight the distinctiveness of populism as a mechanism of political mobilization and the unusual character of the 2016 race.

Going back to our "popularity index", there are a few considerations that might affect our results. First, there might be fake accounts among the followers that can influence the index construction: discriminating them from the set of normal tweets needs further investigation. Moreover, it was not possible to consider the geographical location of the followers. This is another important issue that will be addressed in the future, since it should be noted that just the US citizens can vote for the US election. In addition, the problem of predicting popularity could be addressed as a multiple classification model instead of a binary classification. However it needs some quantitative evaluation to understand its effectiveness and benefits.

References

1. Asur, S., Huberman, B.A.: Predicting the future with social media. In: Proceedings of the 2010 IEEE/WIC/ACM International Conference on Web Intelligence and Intelligent Agent Technology, WI-IAT 2010, vol. 01, pp. 492–499. IEEE Computer Society, Washington, DC (2010). doi:10.1109/WI-IAT.2010.63
2. Barbosa, L., Feng, J.: Robust sentiment detection on twitter from biased and noisy data. In: Proceedings of the 23rd International Conference on Computational Linguistics: Posters, COLING 2010, pp. 36–44. Association for Computational Linguistics, Stroudsburg, PA, USA (2010). http://dl.acm.org/citation.cfm?id=1944566.1944571

3. Bradshaw, S., Howard, P.N.: Troops, trolls and troublemakers: a global inventory of organized social media manipulation (2017). http://comprop.oii.ox.ac.uk/wp-content/uploads/sites/89/2017/07/Troops-Trolls-and-Troublemakers.pdf. Working paper
4. Breiman, L.: Random forests. Mach. Learn. **45**(1), 5–32 (2001). doi:10.1023/A:1010933404324
5. Emanet, N., Öz, H.R., Bayram, N., Delen, D.: A comparative analysis of machine learning methods for classification type decision problems in healthcare. Decis. Analytics **1**(1), 6 (2014). doi:10.1186/2193-8636-1-6
6. Eysenbach, G.: Can tweets predict citations? metrics of social impact based on twitter and correlation with traditional metrics of scientific impact. J. Med. Internet. Res. **13**, e123 (2011)
7. Fernández-Delgado, M., Cernadas, E., Barro, S., Amorim, D.: Do we need hundreds of classifiers to solve real world classification problems? J. Mach. Learn. Res. **15**(1), 3133–3181 (2014). http://dl.acm.org/citation.cfm?id=2627435.2697065
8. Gayo-Avello, D.: Don't turn social media into another 'literary digest' poll. Commun. ACM **54**(10), 121–128 (2011). doi:10.1145/2001269.2001297
9. Hastie, T., Tibshirani, R., Friedman, J.: The Elements of Statistical Learning: Data Mining, Inference, and Prediction. Springer, Heidelberg (2003). http://www.worldcat.org/isbn/0387952845
10. Henrique, J.: Getoldtweets by python. https://github.com/Jefferson-Henrique/GetOldTweets-python
11. Jahanbakhsh, K., Moon, Y.: The predictive power of social media: On the predictability of U.S. presidential elections using twitter. CoRR abs/1407.0622 (2014). http://arxiv.org/abs/1407.0622
12. Jarvis, S.E.: Communicator-in-chief: How barack obama used new media technology to win the white house edited by John Allen Hendricks and Robert Denton Jr. Presidential Stud. Q. **40**(4), 800–802 (2010). doi:10.1111/j.1741-5705.2010.03815.x
13. Jiang, L., Yu, M., Zhou, M., Liu, X., Zhao, T.: Target-dependent twitter sentiment classification. In: Proceedings of the 49th Annual Meeting of the Association for Computational Linguistics: Human Language Technologies, HLT 2011, vol. 1, pp. 151–160. Association for Computational Linguistics, Stroudsburg, PA, USA (2011). http://dl.acm.org/citation.cfm?id=2002472.2002492
14. Kupavskii, A., Ostroumova, L., Umnov, A., Usachev, S., Serdyukov, P., Gusev, G., Kustarev, A.: Prediction of retweet cascade size over time. In: Proceedings of the 21st ACM International Conference on Information and Knowledge Management, CIKM 2012, NY, USA, pp. 2335–2338 (2012). doi:10.1145/2396761.2398634
15. Liaw, A., Wiener, M.: Classification and regression by randomforest. R News **2**(3), 18–22 (2002). http://CRAN.R-project.org/doc/Rnews/
16. Liu, B.: Sentiment analysis and opinion mining. Synth. Lect. Hum. Lang. Technol. **5**(1), 1–167 (2012)
17. Näppi, J.J., Regge, D., Yoshida, H.: Comparative performance of random forest and support vector machine classifiers for detection of colorectal lesions in CT colonography. In: Yoshida, H., Sakas, G., Linguraru, M.G. (eds.) ABD-MICCAI 2011. LNCS, vol. 7029, pp. 27–34. Springer, Heidelberg (2012). doi:10.1007/978-3-642-28557-8_4
18. O'Connor, B., Balasubramanyan, R., Routledge, B.R., Smith, N.A.: From tweets to polls: linking text sentiment to public opinion time series. In: Proceedings of the Fourth International Conference on Weblogs and Social Media, ICWSM 2010, Washington, DC, USA, 23–26 May 2010 (2010)

19. Oliver, J.E., Rahn, W.M.: Rise of the trumpenvolk. Ann. Am. Acad. Polit. Soc. Sci. **667**(1), 189–206 (2016). doi:10.1177/0002716216662639
20. Pak, A., Paroubek, P.: Twitter as a corpus for sentiment analysis and opinion mining. In: Chair, N.C.C., Choukri, K., Maegaard, B., Mariani, J., Odijk, J., Piperidis, S., Rosner, M., Tapias, D. (eds.) Proceedings of the Seventh International Conference on Language Resources and Evaluation (LREC 2010). European Language Resources Association (ELRA), Valletta, Malta, May 2010
21. Pang, B., Lee, L.: Opinion mining and sentiment analysis. Found. Trends Inf. Retr. **2**(1–2), 1–135 (2008). doi:10.1561/1500000011
22. Petrovic, S., Osborne, M., Lavrenko, V.: RT to win! predicting message propagation in twitter. In: Proceedings of the Fifth International Conference on Weblogs and Social Media, Barcelona, Catalonia, Spain, 17–21 July 2011 (2011)
23. Suh, B., Hong, L., Pirolli, P., Chi, E.H.: Want to be retweeted? large scale analytics on factors impacting retweet in twitter network. In: Proceedings of the 2010 IEEE Second International Conference on Social Computing, SOCIALCOM 2010, pp. 177–184 (2010). doi:10.1109/SocialCom.2010.33
24. Tumasjan, A., Sprenger, T., Sandner, P., Welpe, I.: Predicting elections with twitter: What 140 characters reveal about political sentiment. In: Proceedings of the Fourth International AAAI Conference on Weblogs and Social Media, pp. 178–185 (2010)
25. Wang, H., Can, D., Kazemzadeh, A., Bar, F., Narayanan, S.: A system for real-time twitter sentiment analysis of 2012 U.S. presidential election cycle. In: Proceedings of the ACL 2012 System Demonstrations, ACL 2012, pp. 115–120. Association for Computational Linguistics, Stroudsburg, PA, USA (2012). http://dl.acm.org/citation.cfm?id=2390470.2390490
26. Zaman, T., Fox, E.B., Bradlow, E.T.: A bayesian approach for predicting the popularity of tweets. CoRR abs/1304.6777 (2013). http://arxiv.org/abs/1304.6777

Knowledge Representation and Reasoning

Temporal Reasoning with Non-convex Intervals

Miguel Bento Alves[1,2(✉)], Carlos Viegas Damásio[2], and Nuno Correia[2]

[1] ESTG, Instituto Politécnico de Viana do Castelo,
4900-348 Viana do Castelo, Portugal
mba@estg.ipvc.pt

[2] NOVA-LINCS, Universidade Nova de Lisboa, 2829-516 Caparica, Portugal
{cd,nmc}@fct.unl.pt

Abstract. The time dimension is fundamental to Semantic Web applications for supporting commercial transactions or allowing the retrieval of resources contextualised in time. In this work, we propose an ontology that is an extension to the OWL-Time ontology defining non-convex intervals. We also present temporal operators for non-convex intervals and how to reason with them in our approach. We present a temporal reasoner that is able to reason both with temporal entities encoded with OWL-Time ontology and with non-convex temporal entities.

1 Introduction

Time is a major dimension around which human life cycles evolve [13]. Commercial transactions occur in a given time. Its importance can be seen in business intelligent systems, where the time dimension is present in almost all systems. Events and historical facts are something that occur in time. Time is also important in document retrieval as well as in image and video retrieval, where not only answers about the *When* dimension but also about the *Where* dimension can be given to users when querying spatio-temporal events.

Despite the power of Semantic Web technologies, they have very limited support for temporal information modeling [21]. OWL [9], for example, provides no temporal support beyond allowing data values to be typed as basic XML Schema dates, times or durations [14]. There are no standard high-level mechanisms to consistently represent and reason with temporal information in OWL [21]. OWL-Time [4] proposes an ontology that provides rich descriptions of temporal instants, intervals, durations, and calendar terms. However, reasoning over OWL-Time cannot be done directly with the existing mechanisms of the Semantic Web.

Time intervals can be classified in intervals which have no gaps (i.e. convex intervals) and periods of time that have gaps inside them (i.e. non-convex intervals [18]). Non-Convex time intervals are unions of convex time intervals which often occur in the real world. Ladkin defined a taxonomy of relations between non-convex intervals [18] and a set of operators over such intervals. The former kind only allows to represent singular events, continuous in time. The latter allows to represent events with gaps (e.g., a swimming competition that holds

© Springer International Publishing AG 2017
P. Różewski and C. Lange (Eds.): KESW 2017, CCIS 786, pp. 127–142, 2017.
https://doi.org/10.1007/978-3-319-69548-8_10

on Saturday, from 14 h to 18 h and on Sunday, from 9 h to 12 h) and periodic (or recurring) events (e.g., Christmas day). The existing OWL-Time ontology is restricted to represent convex intervals. Thus, in this work we propose an extension to OWL-Time ontology to represent non-convex intervals.

We also describe a OWL reasoner for temporal data, capable to deal both with temporal entities encoded in OWL-Time and with non-convex intervals, encoded in our proposed ontology. The temporal reasoner was integrated in a Semantic Web framework, namely, Jena [1,20]. OWL-Time, even if it is not a W3C standard, is the mostly used and studied ontology of temporal concepts with rich descriptions for capturing the temporal content and is currently in development. Jena framework is a free and open source Java framework for building Semantic Web applications. It provides a programmatic environment for RDF, RDFS, OWL, a query engine for SPARQL and it includes a rule-based inference engine. Jena is widely accepted for Semantic Web applications because it offers an "all-in-one" Java solution. Consequently, the integration of our temporal reasoner in Jena framework allows a seamless integration with semantic web projects. Our ultimate goal is that the knowledge base can be temporal queried in a high level fashion.

This document is structured as follows. In **Sect. 2** we introduce the OWL-Time ontology. In **Sect. 3** we present the NCO (non-convex) ontology, an extension to OWL-Time ontology to deal with non-convex intervals. The presentation of our approach is supported by several real world examples. In **Sect. 4** we detail the temporal reasoning in NCO ontology and we present the temporal reasoner developed in this work and its system architecture in **Sect. 5**. In **Sect. 6** we present related work and we finish with conclusions in **Sect. 7**.

2   Time Ontology in OWL

OWL-Time [4] is an ontology that has been developed for describing the temporal content of Web pages and the temporal properties of Web services. It intends to capture its essential features and make them, and their associated resources, easily available to a large group of Web developers and users. OWL-Time defines two main types of temporal entities, instants and intervals. Intervals are things with extent and instants are point-like, having no interior points. It is generally safe to think of an instant as an interval with zero length, where the beginning and end are the same instant. These two types of temporal entities are, in OWL-Time, the classes *Instant* and *Interval*, sub-classes of the class *TemporalEntity*. The class *ProperInterval* is a sub-class of *Interval*, which corresponds with the common understanding of intervals, in that the beginning and end are distinct, and is therefore disjoint from *Instant*. The properties *hasBeginning* and *hasEnd* are relations between temporal entities and instants, and the beginnings and ends of temporal entities, if they exist, are unique. Instead of the end of a temporal entity, a duration can also be defined through the property *hasDurationDescription*. In infinite intervals, a positively infinite interval has no end, and a negatively infinite interval has no beginning. An instant can be represented either by a XMLSchema datetime or by an OWL-Time datetime description. A datetime description has the following properties: *unitType*, *year*, *month*, *week*, *day*, *dayOfWeek*, *dayOfYear*, *hour*, *minute*, *second*, and *timeZone*. The property unitType specifies the temporal unit type of the datetime description, and its range is a *TemporalUnit*, i.e., one of the following values: *Second*, *Minute*, *Hour*, *Day*, *Week*, *Month* and *Year*. The properties *inXSDDateTime* and *inDateTime* allows to encode an *Instant* by a XMLSchema datetime or a datetime description. In **Example** 1 we give an example of an event defined by using these two properties.

```
exa:PortugalDay2017                      time:hasEnd [
   time:hasBeginning [                      time:inDateTime [
      time:inXSDDateTime                        time:year "2017"^^xsd:gYear;
         "2017-06-10T00:00:00"^^xsd:dateTime    time:month "--06"^^xsd:gMonth;
   ] ;                                          time:day "---10"^^xsd:gDay;
                                                time:unitType :unitDay ] ] .
```

Example 1

Allen [5,6] (**Fig.** 1) developed a calculus of binary relations on intervals for representing qualitative temporal information and address the problem of reasoning about such information (e.g., meets, overlaps). The standard Allen's interval calculus assumes all intervals are proper, i.e., intervals whose beginning and end are different. OWL-Time provides the corresponding interval relations, between two temporal entities: intervalEquals, intervalBefore, intervalMeets, intervalOverlaps, intervalStarts, intervalDuring, intervalFinishes, and their reverse interval relations: intervalAfter, intervalMetBy, intervalOverlappedBy, intervalStartedBy, intervalContains, intervalFinishedBy. In **Example** 2 is given an example using these relations, where it is defined when *Summer of 2016* happens.

```
exa:Summer2016
   :intervalStartedBy [
      :hasDateTimeDescription [
         :unitType :unitDay ;
         :day "---20"^^xsd:gDay ;
         :month "--06"^^xsd:gMonth ;
         :year "2016"^^xsd:gYear ] ] ;
   :intervalFinishedBy [
      :hasDateTimeDescription [
         :unitType :unitDay ;
         :day "---22"^^xsd:gDay ;
         :month "--09"^^xsd:gMonth ;
         :year "2016"^^xsd:gYear ] ] .
```

Example 2

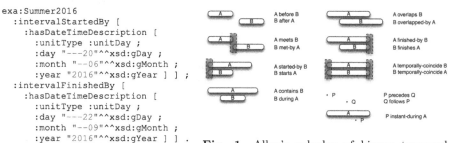

Fig. 1. Allen's calculus of binary temporal relations

3   Ontology for Non-convex Intervals

OWL-Time only allows to represent intervals without gaps, i.e., convex intervals. However, there are several kinds of temporal events that we want to represent that require periods of time with gaps. For example, the Christmas day for every year, the market that happens every second Sunday of each month, the academic year and its semesters, etc. To overcome this limitation, we propose an extension to OWL-Time ontology to represent non-convex intervals, the NCO ontology. Our ontology allows the definition of recurring intervals (e.g. *the summer night*

*market that is held in June, July and August at Wednesdays and Fridays and
starts at 21:00*) and non-recurring intervals (e.g. *The tennis challenge that it
happens in 25th of April, from 15pm to 18pm and in 26th of April, from 9am
to 12am*). To the best of our knowledge this is the only ontology for non-convex
intervals that allows the definition of non-recurring intervals explicitly. In **Sect.** 6
we present some temporal ontologies that support non-convex intervals. However,
they only support recurring intervals.

Formally, a convex interval is represented by $[b_i, e_i]$ where b_i and e_i are
instants, representing, respectively, the beginning and the end of the interval i
and $e_i > b_i$. A non-convex interval is an union of disjoint convex intervals (finite
or arbitrary). To simplify presentation, a non-convex time interval is represented
with the usual set notation.

Example: The year of 2016: $\{[2016\text{-}01\text{-}01, 2016\text{-}12\text{-}31]\}$
Months of February: $\{\ldots, [2015\text{-}02\text{-}01, 2015\text{-}02\text{-}28],$
$[2016\text{-}02\text{-}01, 2016\text{-}02\text{-}29], [2017\text{-}02\text{-}01, 2017\text{-}02\text{-}28], \ldots\}$

In **Fig.** 2 is represented a non-convex interval, a swimming competition that
takes place in 18th and 19th of April of 2016, between 9:00 and 18:00.

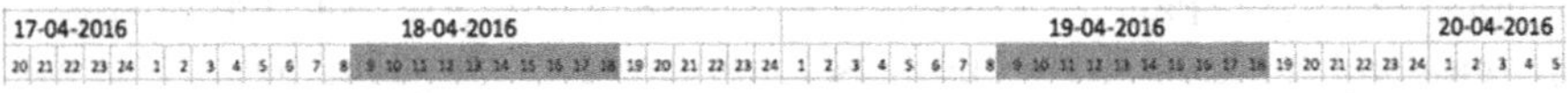

Fig. 2. A non-convex event

On the top of our NCO ontology, that uses the prefix *nco* in the examples,
we have the class *nco:TemporalAggregate* that defines a set of arbitrary temporal
entities. For instance, a *nco:TemporalAggregate* can be all events that happens
in *July of 2017, in the city of Viana do Castelo*. A *nco:TemporalAggregate* can
also be a set of *nco:TemporalAggregate*, for instance, all the *events of the city of
Viana do Castelo in 2017* are the set of events of each month.

The class *nco:TemporalSequence* is a sub-class of *nco:TemporalAggregate* that
defines a set of ordered temporal entities. For instance, a *nco:TemporalSequence*
can be the set of each antiques fair that happens in *July of 2017, in the city
of Viana do Castelo, every Fridays*. The property *nco:hasBeginning* allows to
define the beginning of a temporal sequence whereas the property *nco:hasEnd*
allows to define the end of a temporal sequence. The begging or the end of a
temporal sequence can be defined with a *time:Instant* of the OWL-Time or with
a *nco:GeneralDateTimeLimit* or a *nco:GeneralDateTimeLimitStr* that will be
explained later. The property *nco:hasSpan* is the length of the period of time in
which the temporal sequence happens whereas the property *nco:hasDuration* is
the total time of a given period.

The class *nco:RecurringInterval* represents the ontologies that are non-
convex intervals that defines recurring intervals. The temporal aggregates of
Hobbs and Pan or the RDF Calendar (both detailed in **Sect.** 6) are subclasses of *nco:RecurringInterval*. The core of our approach, that allows the
definition of non-convex intervals, both recurring intervals and non-recurring

intervals, is represented by the class *nco:NCInterval* that is also a sub-class of *nco:RecurringInterval*.

In **Fig.** 3 is presented a sub-model of NCO ontology where is evidencing the top-classes of NCO ontology. Afterwards, our approach to non-convex intervals will be detailed.

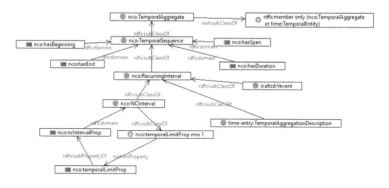

Fig. 3. Sub-model 1 of the NCO ontology

In our NCO ontology the properties *nco:lowerLimit*, *nco:upperLimit* and *nco:hasLimit*, whose domain is a *nco:NCInterval* and whose range is a *nco:TemporalLimit*, allow to make definitions about when a non-convex interval starts, when a non-convex interval ends and a range when a non-convex interval happens. These properties are sub-properties of the property *nco:temporalLimitProp*. The properties *nco:lowerLimit* and *nco:upperLimit* are similar to the properties *time:hasBeginning* and *time:hasEnd* but applicable to non-convex intervals. The property *nco:hasLimit* is used to define that a given temporal entity must be held in a given period or instant in time. The class *nco:TemporalLimit* allows the definition of the temporal limits associated to a *nco:NCInterval*, which is a convex interval or an instant. For instance, 6th *of July of 2016, at 10:00* is a temporal limit, but the year of *2016* is also a temporal limit, as well as a *Wednesday*. The property *nco:hasDurationNCI* allows to define a duration for a *nco:NCInterval* and its range is a *nco:DurationTemporalLimit*. In **Example** 3 is encoded the swimming competition represented in **Fig.** 2. In **Fig.** 4 is represented a sub-model of NCO ontology, evidencing the class *nco:NCInterval* and its relations.

```
exa:SwimmingCompetition                    exa:upperLimit1
   nco:lowerLimit exa:lowerLimit1 ;           time:year "2016"^^xsd:gYear ;
   nco:upperLimit exa:upperLimit1 ;           time:month "--04"^^xsd:gMonth ;
   nco:lowerLimit exa:lowerLimit2 ;           time:day "---19"^^xsd:gDay .
   nco:upperLimit exa:upperLimit2 .        exa:lowerLimit2 time:hour 9 .
exa:lowerLimit1                            exa:upperLimit2 time:hour 18 .
   time:year "2016"^^xsd:gYear ;
   time:month "--04"^^xsd:gMonth ;
   time:day "---18"^^xsd:gDay .
```

Example 3

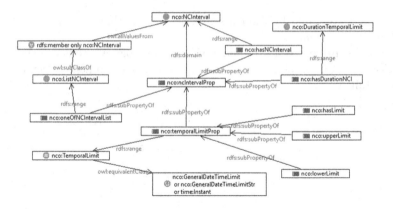

Fig. 4. Sub-model 2 of the NCO ontology

A *nco:TemporalLimit* can be defined in several ways, i.e., it can be a *nco:GeneralDateTimeLimit*, a *nco:GeneralDateTimeLimitStr* (that will be detailed in **Subsect. 3.1**) or a *time:Instant*. Also, it can be defined with a *time:GeneralDateTimeDescription* that is a sub-class of *nco:GeneralDateTimeLimit*. A *nco:GeneralDateTimeLimit* is similar to the *time:GeneralDateTimeDescription* but with two more properties and without a unit type. This class allow us to define elements of a calendar-clock system, i.e., a year, a year and month, a day of the week, etc., to define a lower limit, an upper limit or a range (*hasLimit*) to the non-convex interval. For the sake of compatibility with OWL-Time, we defined a general class for the date time description and a sub-class to define calendar-clock system elements in the gregorian calendar. We present the sub-model of the class *nco:GeneralDateTimeLimit* and the definition of the classes *nco:GeneralDateTimeLimit*, *nco:GeneralDateTimeLimitStr* and *nco:DurationTemporalLimit* in **Fig. 5**. In our ontology, the class *nco:ListNCInterval* is a *rdf:Seq* that defines a sequence of temporal definitions, where only one must be held, allowing a logical disjunction between temporal definitions. The property *nco:oneOfNCIntervalList* associates a *nco:ListNCInterval* to a *nco:NCInterval*. For instance, a match of the champions league takes place on *Tuesdays* or on *Wednesdays*. This example is encoded in **Example 4**.

```
evt:ChampionsLeagueGamesDays nco:oneOfNCIntervalList [
    rdf:_1 [ nco:hasLimit [ time:dayOfWeek time:Tuesday ]] ;
    rdf:_2 [ nco:hasLimit [ time:dayOfWeek time:Wednesday ]] ] .
```
Example 4

As is evidenced by the previous examples, in our approach it is possible to combine all the temporal properties for specifying non-convex temporal intervals. Furthermore, we allow a cardinality bigger than one in these temporal limits properties. That means that we can define more than one lower limit, upper limit or range (*hasLimit*). For example, we can define, for the same temporal entity, a start by date and a start by time or by the day of the week. This feature helps to define more succinct and readable temporal entities. This is clear in the

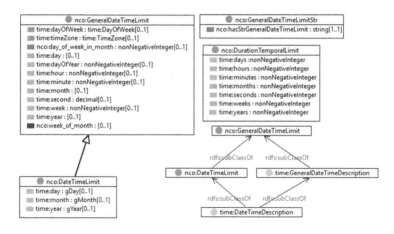

Fig. 5. Sub-model 3 of the NCO ontology

examples of the real world that we give in this work. For example, the summer night market that is held in June, July and August at Wednesdays and Fridays and starts at 21:00, encoded in **Example** 6, needs more and less 25 OWL-Time temporal entities only for one year.

Next, we give several examples of the real world events encoded with our approach. In the next example, we define the Christmas day.

```
exa:ChristmasDay nco:hasLimit [
    time:month "--12"^^xsd:gMonth;
    time:day "---25"^^xsd:gDay ] .
```
Example 5

The next example defines the summer night market that is held in June, July and August at Wednesdays and Fridays. The market starts at 21:00.

```
exa:SummerNightMarket                   nco:lowerLimit [
  nco:lowerLimit [                        time:hour 21 ];
    time:month "--06"^^xsd:gMonth;      nco:oneOfNCIntervalList (
    time:day "---01"^^xsd:gDay ] ;          [nco:hasLimit [
  nco:upperLimit [                            time:dayOfWeek time:Wednesday ]]
    time:month "--08"^^xsd:gMonth;          [nco:hasLimit [
    time:day "---31"^^xsd:gDay ];           time:dayOfWeek time:Friday ]] ) .
```
Example 6

In **Example** 7, we present another way of encoding the example shown in **Example** 3.

```
exa:SwimmingCompetition                 [
  nco:oneOfNCIntervalList ( [           nco:lowerLimit [ time:inXSDDateTime
    nco:lowerLimit [ time:inXSDDateTime   "2016-04-19T09:00:00"^^xsd:dateTime
      "2016-04-18T09:00:00"^^xsd:dateTime    ] ;
      ] ;                               nco:upperLimit [ time:inXSDDateTime
    nco:upperLimit [ time:inXSDDateTime   "2016-04-19T18:00:00"^^xsd:dateTime
      "2016-04-18T18:00:00"^^xsd:dateTime    ] ] ) .
      ] ],
```
Example 7

Looking again to the definition of the class *nco:GeneralDateTimeLimit*, we can evidentiate that the class definition shares almost all properties of date-time description in OWL-Time ontology, but there are two more properties, *weekOfMonth* and *dayOfWeekInMonth*. The property *weekOfMonth* allows to define the week number within a month while the property *dayOfWeekInMonth* allows to define the ordinal number of the day of the week within the month. In **Example** 8 we define the second Sunday of each month. The temporal operator *nco:hasNCInterval* allows the re-using of the non-convex temporal intervals, avoiding repeated encoding. In **Example** 9, we define the Christmas day of 2016, re-using the definition shown in **Example** 5.

```
exa:SecondSundayOfMonth
  nco:hasLimit [
    :dayOfWeekInMonth 2 ] ;
  nco:hasLimit [
    time:dayOfWeek 7 ] .
```
Example 8

```
exa:ChristmasDay2016
  nco:hasNCInterval exa:ChristmasDay ;
  nco:hasLimit [
    time:year "2016"^^xsd:gYear ] .
```
Example 9

The NCO ontology is located at https://github.com/mbentoalves/nco/blob/master/nco.ttl.

3.1 NC Intervals Represented by Literals

In our ontology we also allow the encoding of non-convex temporal intervals using literals. The literal is a string with a construction similar to *xsd:duration*, with the following format: **Y9999M99W99D99WD9WM9DY999DWM9H99MI99S99TZAAAAA**, where Y indicates the year, M indicates the month, W indicates the week of the year, D indicates the day, DW indicates the day of the week, between 1 and 7, WM the week number within a month, DY indicates the day of the year, DWM indicates the day of the week in month, H indicates the hour, MI indicates de minutes, S indicates de seconds and TZ indicates the time zone (by defining a positive or negative time, like $+06:00$ or $-06:00$). For instance, **Example** 6 could be defined as shown in **Example** 10. The advantage of this approach is that it simplifies gently the encoding.

```
exa:SummerNightMarket
  nco:lowerLimit [nco:hasStrGeneralDateTimeLimit "M06D01"] ;
  nco:upperLimit [nco:hasStrGeneralDateTimeLimit "M08D31"] ;
  nco:lowerLimit [nco:hasStrGeneralDateTimeLimit "H21"] ;
  nco:oneOfNCIntervalList [
    rdf:_1 [nco:hasLimit [nco:hasStrGeneralDateTimeLimit "WD4"]] ;
    rdf:_2 [nco:hasLimit [nco:hasStrGeneralDateTimeLimit "WD6"]]
  ] .
```
Example 10

4   Temporal Operators in NCO Ontology

In our work, we extend temporal operators to Allen's relations between time intervals (**Fig.** 1) to relations between non-convex intervals. For instance, the temporal operator *nco:intervalBefore* defines that a *nco:NCInterval* happens before another *nco:NCInterval*. This temporal operator is defined by the **Rule** 1.

i nco:intervalBefore j <-
> *cii is the latest convex interval of i, lcii is the latest instant of cii,*
> *cij is the earliest convex interval of j, ecij is the earliest instant of cij,*
> *lcii < ecij.*

Rule 1 - temporal operator *nco:intervalBefore*

However, the definition of the non-convex operators have some differences to the corresponding in OWL-Time intervals operators. For instance, the operator *nco:intervalAfter* cannot be defined by OWL axioms as is defined to the time interval operator *time:intervalAfter* in OWL-Time, where is the inverse of *time:intervalBefore*. By the **Rule 1**, we induce that the rule will fail for i infinite at right (without latest instant) or for j infinite at left (without earliest instant). Therefore, *nco:intervalBefore* being false does not imply that *nco:intervalAfter* will be true. The property *nco:intervalAfter* should be defined by the **Rule 2**.

i nco:intervalAfter j <- j nco:intervalBefore i.

Rule 2 - temporal operator *nco:intervalAfter*

In [18] is proposed an extensive set of relations for specifying relationships between non-convex intervals that are called non-convex relations. Non-convex relations are combined using relation functors, namely, **mostly**, **always**, **partially**, **sometimes**, and **disjunction**. For instance, considering two non-convex intervals, i and j, and considering R a Allen's relation (convex relation), i **mostly** R j, if, for every convex interval of j there is a convex interval of i that is R to it (can exist other components of i, but not of j), i **always** R j, where R is a convex relation, iff matched pairs of components of i and j are R to each other.

All temporal operators described in [18] can be captured by our approach since we have adapted them to NCO ontology. For instance, we created the temporal operators *nco:mostlyIntervalDuring* and *nco:alwaysIntervalDuring*. In **Rule 3** is defined the temporal operator *nco:mostlyIntervalDuring* whereas in **Rule 4** is defined the temporal operator *nco:alwaysIntervalDuring*.

i nco:mostlyIntervalDuring j <-
> *Si is the list of the convex intervals in i, Sj is the list of the convex intervals in j,*
> *for each CIj in Sj,*
>> *There is a convex interval CIi in Si such as CIi time:intervalDuring CIj.*

Rule 3 - temporal operator *nco:mostlyIntervalDuring*

i nco:alwaysIntervalDuring j <-
> i nco:mostlyIntervalDuring j, j nco:mostlyIntervalContains i.

Rule 4 - temporal operator *nco:alwaysIntervalDuring*

Next, we list the rules to represent Allen's relations between non-convex intervals that weren't listed up to this point.

i nco:intervalContains j <-
    i nco:mostlyIntervalContains j.

Rule 5 - operator $nco{:}intervalContains$

i nco:intervalEquals j <-
    i nco:alwaysIntervalEquals j.

Rule 7 - operator $nco{:}intervalEquals$

i nco:intervalMetBy j <-
    j nco:intervalMeets i.

Rule 9 - operator $nco{:}intervalMetBy$

i nco:intervalOverlappedBy j <-
    j nco:intervalOverlaps i.

Rule 11 - operator $nco{:}intervalOverlappedBy$

i nco:intervalStarts j <-
    j nco:intervalStartedBy i.

Rule 13 - operator $nco{:}intervalStarts$

i nco:intervalFinishes j <-
    j nco:intervalFinishedBy i.

Rule 14 - operator $nco{:}intervalFinishes$

i nco:intervalDuring j <-
    j nco:intervalContains i.

Rule 6 - operator $nco{:}intervalDuring$

i nco:intervalFinishedBy j <-
    j nco:mostlyIntervalFinishes i.

Rule 8 - operator $nco{:}intervalFinishedBy$

i nco:intervalOverlaps j <-
    i nco:sometimesIntervalMeets j.

Rule 10 - temporal operator $nco{:}intervalOverlaps$

i nco:intervalStartedBy j <-
    j nco:mostlyIntervalStarts i.

Rule 12 - operator $nco{:}intervalStartedBy$

i nco:intervalMeets j <-
    cii is the latest convex interval of i,
    $lcii$ is the latest instant of cii,
    cij is the earliest convex interval of j,
    $ecij$ is the earliest instant of cij,
    $lcii == ecij$.

Rule 15 - operator $nco{:}intervalMeets$

The reusing of the semantic rules allows an easy creation of new temporal relations not anticipated yet. For instance, let's consider the temporal relation *nco:dos* that represents the disjunction of relations During, Overlaps and Starts. This relation is presented in **Rule** 16, converted to the respective composition.

i nco:dos z <- i nco:intervalOverlaps j, j nco:intervalDuring z.

Rule 16 - temporal operator *nco:dos*

In the context which we have applied the work described in this paper, we noticed
that we need some temporal operators for non-convex intervals not provided in [18]. We created the *nco:intervalCloseBefore* and the *nco:intervalCloseAfter* temporal operators. The operator *nco:intervalCloseBefore* relates a non-convex temporal entity that holds before but close to other non-convex temporal entity. For instance, consider the non-convex temporal entity of the **Example** 5. An event that happens in 23th of December of 2015 holds after of 25th of December of 2014 and before of 25th of December of 2015 but closer to 25th of December

of 2015, thus we consider that the event that happens in 23th *of December of 2015* is before but close to 25th *of December.*

Taking the operator *nco:intervalCloseBefore*, defined in **Rule** 17, a temporal entity *S1* is close before a temporal entity *S2* iff for each element *M1* of *S1* there are two consecutive moments in *S2*, *M21* and *M22*, where *M1* is closer to *M22* than *M21* considering that *M21* happens before *M22*.

```
i nco:intervalCloseBefore j <-
    S1 is the ordered list of the convex intervals in i,
    S2 is the ordered list of the convex intervals in j,
    for each M1 in S1,
        M21 and M22 are two consecutive convex intervals of S2
        where M21 < M22,
        D1 is the temporal distance between M1 and M21,
        D2 is the temporal distance between M1 and M22,
        D2 < D1.
```

Rule 17 - temporal operator *nco:intervalCloseBefore*

5   Temporal Reasoner

An objective of our work is the development of a temporal reasoner, performing reasoning with temporal entities since encoded using the OWL-Time ontology or NCO ontology. We want to perform temporal queries related with temporal entities in a high-level fashion. For example, let's consider the **Example 3**. We want to be able to perform SPARQL queries like: *All events that take place during the swimming competition*:

```
Select ?Event {
    ?Event nco:intervalDuring exa:SwimmingCompetition . }
```

As described in **Sect.** 4, temporal relations can be supported by semantic rules. A rule language is needed for several reasons, at least because of the limitations of OWL [24]. SWRL [16] is a proposal for representing rules/axioms for the Semantic Web, implemented by several semantic web frameworks. Others semantic web frameworks have their own rule format, e.g., Jena framework with Jena rules [17]. Therefore, we need a Semantic Web framework that supports semantic rules. Since SWRL and Jena rules are an extension of the OWL ontology language, they are restricted to unary and binary DL-predicates. In temporal instants, non-convex intervals or convex intervals, the URI of the temporal entity encompasses a set of definitions about it. These definitions can have different forms since scalar datatypes to compound values and a set of different operators such as *lowerLimit, upperLimit, hasLimit*. Therefore, we cannot resort only to the mechanisms of inference of the semantic rules to define directly interval-based temporal operators because they are restricted to unary and binary DL-predicates. We need to gather all information about a temporal interval and this task may involve several queries to the knowledge base. Consequently, we need a Semantic Web framework that allows the development of custom built-in functions and with the feature of allowing queries to the knowledge base. Because of the previous limitations of existing systems, we developed our temporal reasoner in the Jena framework. In summary, in our system, the object properties

of temporal relations are supported by a (Jena) rule. This rule calls a custom
built-in function that will be responsible for evaluating the relation between two
temporal entities. It is possible to configure the system to run in *backward mode*
or in *forward mode*. Next, we give an example of a Jena rule that supports a
temporal relation property, **Rule 18.**

```
(?x time:intervalBefore ?y) <- (?x rdf:type time:TemporalEntity),     (?y rdf:type
    time:TemporalEntity), TR_intervalBefore(?x, ?y) .
```
Rule 18 - Example of a Jena rule

In this example, *TR_intervalBefore* is a custom built-in function which is
responsible for defining if a given temporal entity occurs before another temporal
entity. We can develop the custom built-in function *TR_intervalBefore* in two
ways. One is developing (java) code to query the knowledge base and perform the
evaluation. In the custom built-in functions developed in Jena framework, the
knowledge base can only be queried by triples statements. SPARQL queries are
not allowed. The other way is resorting to the work developed in [7], a system
that allows the combination of a declarative language with a Semantic Web
framework, namely, the Jena framework and XSB Prolog using InterProlog [10],
a library allowing the development of combined Java+Prolog applications. This
library allows RDF and SPARQL queries in Prolog predicates. Thus, in this
approach, we have the advantage to have the reasoning of temporal entities in
logic rules. Moreover, we can resort to SPARQL to query the knowledge base.
We look to the development of our interval-based temporal operators as a set of
rules. Prolog approach fits better when we implement a set of rules. Furthermore,
is easy to add new rules or change the existing ones. In the **Subsect.** 5.1 we give
more details of the development of the temporal reasoner in this approach.

5.1 Temporal Relations Evaluation System Using RDF Query and Inference in Prolog

In **Example** 11, we list an excerpt of code in Prolog that queries the knowledge
base to extract all statements of an instant defined with a datetime description
(the example assumes that *Instant* is instantiated). In **Fig.** 6 the architecture of
the system is shown.

```
rdfDB(dtd(Property, Value),
    [
    triple(Instant,
        'time:inDateTime',
        '?dtd'),
    triple('?dtd' , Property,
        Value)
    ],
    Ldtd)
```
Example 11 - Excerpt of code in
Prolog that queries the knowl-
edge base

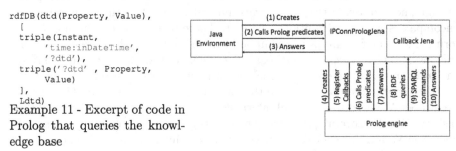

Fig. 6. System architecture of integration of Jena
with Prolog

As we have seen before, the object property *time:intervalBefore* is supported by a rule. This rule calls a built-in function, called *TR_intervalBefore*. In the development of built-in functions, what we can pass to a built-in function as parameter is a literal, an URI or an id of a blank node of our knowledge base. In temporal intervals, we pass the URI or the blank node id of a temporal interval and this encompasses a set of definitions about it. However, in a first step we only have the id of the temporal interval. To get all the information about the temporal interval, the knowledge base must be queried. The custom built-in function will call a rule in Prolog that will evaluate if a given temporal interval occurs before another given temporal interval. This (prolog) rule will query the knowledge base to retrieve the necessary information to perform the reasoning. In **Listing** 1 we list two of the rules that support the before temporal relation. The task of the *rdfDB/1* predicate is to query the knowledge base.

```
tr_intervalBefore(T1, T2):-              tr_beginOfInt(T, InternalValue,
  tr_endOfInt(T1, LST1, PrecisionT1),          Precision):-
  tr_beginOfInt(T2, RST2, PrecisionT2),    rdfDB(triple(T, 'time:hasBeginning',
  compareTR(LST1, RST2, PrecisionT1,           Instant)),
      PrecisionT2, C),                     getInstant(Instant, InternalInstant,
  C == -1 .                                    Precision),
                                                   ...
```

Listing 1

6   Related Work

In [8] is provided a survey on the models and query languages for temporally annotated RDF. Temporal RDF annotation can be characterised by temporal entities translated into RDF as literals instead of literals or resources, as is done in OWL-Time. In [19] is presented an ontology design pattern to represent recurring intervals, a specific case of non-convex intervals in which the period between subintervals and the duration of such subintervals are constant. In [19] only recurring intervals are taken into account, leaving the non-recurring intervals out of the scope. Comparing with our work, we permit all types of non-convex intervals, periodic and non-periodic. Furthermore, is not clear how to represent recurring intervals that do not have a specific beginning/end, like *Every Wednesday*, an example gave in the current paper. The Timeline Ontology [3] is an extension of the OWL-Time ontology, where time intervals have been modelled. The properties of Timeline Ontology related with time are all datatype properties. Thus, rich descriptions as we used in our examples to define moments and intervals are not allowed. Once again, our approach allows logical disjunction in the temporal definitions and we can have explicit non-periodic events. In [15, 22, 23], Hobbs and Pan described an approach for representing temporal aggregates in OWL, as an extension to the OWL-Time. Temporal aggregates are defined as collections/aggregates of temporal entities. The definition is close to non-convex intervals, thus, our NCO ontology is able to represent temporal aggregates. The approach of Hobbs and Pan allows the definition of a set of intervals that our approach does not allow (e.g., four consecutive Sundays). However, in the work of Hobbs and Pan the definition of a set of intervals is complex, resulting in a long descriptions, without a good readability. This drawback is enhanced when we want to encode non-recurring intervals, because the reasoning in the work of Hobbs and Pan is supported by recurrent rules. There are several useful queries that are not easy to do with the encoding of events specified by the work of Hobbs and Pan, namely when we want to query with the time properties. This happen in the Hobbs and Pan approach because of the recurrent rules where is not explicit the time properties like the day of the week, the day of the month and so on. Furthermore, the examples do not illustrate some aspects that are not clear, for instance, how we can deal with the day of the month *vs* day of the week. Our approach allows the definition of events that happen in non-recurring intervals (e.g., *An event that holds in 18th and in 19th of April of 2016, between 9:00 and 18:00*) that is not allowed, (or, at least, in explicitly way) in the work of Hobbs and Pan. Furthermore, our approach allows logical disjunction in the temporal definitions. RDF Calendar [11] applies the Resource Description Framework (RDF) to iCalendar [12] data, a computer file format which allows Internet users to send meeting requests and tasks to other Internet users, in order to integrate calendar data with other Semantic Web data such as social networking data, syndicated content, and multimedia meta-data. RDF Calendar support several declarations using non-atomic data. For instance, *"The BYDAY rule part specifies a COMMA character (US-ASCII decimal 44) separated list of days of the week"* [2]. Non-atomic data leads to difficulties in inference process in Semantic technologies as SPARQL or semantic rules because they assume atomic data. Another example is that in RDF Calendar we cannot create an interval that involves time outside the date. Again due to a problem of non-atomic data. RDF Calendar allows the definition of infinite intervals such as the Christmas day, the independence day, but needs, even if virtual, a start date. RDF Calendar doesn't allow the definition o temporal bounds in rules, we must define each element. This carry to long descriptions that may not be user friendly to read. RDF Calendar supports events that occur every given time, for instance, every three days, that is not foreseen in our work.

7   Conclusion

Time is presented in many facts that evolve humans and it is a very important dimension to support commercial transactions or to retrieve context data, like webpages, multimedia contents, and other kind of resources. Hence, representing time entities in Semantic Web is crucial to support both the representation of the time and reasoning over it. In this work, we propose the NCO ontology, an extension to OWL-Time ontology to represent non-convex intervals, e.g., temporal entities with gaps. In NCO ontology, in addition to recurring intervals, non-recurring intervals can also be represented. We present our approach giving several examples of the real world, proving that our ontology meets the most of uses cases in temporal entities. We provide several temporal operators to deal with non-convex intervals encoded in our approach. The temporal operators are easily re-used to create new temporal operators. In this work, we also developed a temporal reasoner that allows queries in a high-level fashion, encapsulating all the details about temporal reasoning. Our temporal reasoner assumes temporal entities encoded in OWL-Time ontology and in NCO ontology. We developed our system on top of the Jena framework, integrating with Prolog, and using Jena rules. The development of a temporal reasoner in Jena environment allows the integration in a wide range of Semantic Web projects.

References

1. Jena. http://jena.apache.org/
2. RDF calendar documentation. https://www.w3.org/wiki/RdfCalendarDocumentation
3. Timeline ontology. http://motools.sourceforge.net/timeline/timeline.html
4. Time Ontology in OWL, September 2006. http://www.w3.org/TR/owl-time
5. Allen, J.F.: Towards a general theory of action and time. Artif. Intell. **23**(2), 123–154 (1984). doi:10.1016/0004-3702(84)90008-0
6. Allen, J.F., Ferguson, G.: Actions and events in interval temporal logic. Technical report, Rochester, NY, USA (1994)
7. Alves, M.B., Damásio, C.V., Correia, N.: RDF query and inference in prolog. In: Ngonga Ngomo, A.-C., Křemen, P. (eds.) KESW 2016. CCIS, vol. 649, pp. 191–201. Springer, Cham (2016). doi:10.1007/978-3-319-45880-9_15
8. Anastasia Analyti, I.P.: A survey on models and query languages for temporally annotated RDF. Int. J. Adv. Comput. Sci. Appl. (IJACSA) **3**(9), 28–35 (2012)
9. Bechhofer, S., van Harmelen, F., Hendler, J., Horrocks, I., McGuinness, D., Patel-Schneijder, P., Stein, L.A.: Owl web ontology language reference, 10 February 2004. http://www.w3.org/TR/owl-ref/
10. Calejo, M.: InterProlog: towards a declarative embedding of logic programming in Java. In: Alferes, J.J., Leite, J. (eds.) JELIA 2004. LNCS, vol. 3229, pp. 714–717. Springer, Heidelberg (2004). doi:10.1007/978-3-540-30227-8_64
11. Connolly, D., Miller, L.: RDF calendar — an application of the resource description framework to icalendar data. In: World Wide Web Consortium, September 2005
12. Dawson, F., Stenerson, D.: Internet calendaring and scheduling core object specification (icalendar). Technical report 2445, November 1998. http://www.ietf.org/rfc/rfc2445.txt
13. Dodgshon, R.A.: Society in Time and Space: A Geographical Perspective on Change. Cambridge University Press, Cambridge (1998). http://www.worldcat.org/isbn/0521596408
14. Gao, S., Sperberg-McQueen, C.M., Thompson, H.S., Mendelsohn, N., Beech, D., Maloney, M.: W3c XML schema definition language (XSD) 1.1 part 1: Structures. WWW Consortium, Working Draft WD-xmlschema11-1-20080620, June 2008
15. Hobbs, J.R., Pan, F.: An ontology of time for the semantic web. ACM Trans. Asian Lang. Inf. Process. **3**, 66–85 (2004)
16. Horrocks, I., Patel-Schneider, P.F., Boley, H., Tabet, S., Grosof, B., Dean, M.: SWRL: a semantic web rule language combining OWL and RuleML. W3c Member Submission, World Wide Web Consortium (2004)
17. Jena Documentation: Reasoners and rule engines: Jena inference support. http://jena.apache.org/documentation/inference/

18. Ladkin, P.: Time representation: a taxonomy of interval relations. In: Proceedings of the Sixth National Conference on Artificial Intelligence (1986)
19. Poveda-Villalón, M., Mari Carmen Suárez-Figueroa, A.G.P.: A pattern for periodic intervals, October 2014. http://www.semantic-web-journal.net/system/files/swj897.pdf
20. Mcbride, B.: Jena: a semantic web toolkit. IEEE Internet Comput. 6(6), 55–59 (2002)
21. O'Connor, M.J., Das, A.K.: A method for representing and querying temporal information in OWL. In: Fred, A., Filipe, J., Gamboa, H. (eds.) BIOSTEC 2010. CCIS, vol. 127, pp. 97–110. Springer, Heidelberg (2011). doi:10.1007/978-3-642-18472-7_8
22. Pan, F.: A temporal aggregates ontology in owl for the semantic web. In: Proceedings of the AAAI Fall Symposium on Agents and the Semantic Web, Arlington, Virginia, pp. 30–37. AAAI Press (2005)
23. Pan, F.: Representing complex temporal phenomena for the semantic web and natural language. Ph.D. thesis, Los Angeles, CA, USA (2007)
24. Parsia, B., Sirin, E., Grau, B.C., Ruckhaus, E., Hewlett., D.: Cautiously approaching SWRL. Preprint submitted to Elsevier Science (2005). http://www.mindswap.org/papers/CautiousSWRL.pdf

More on the Data Complexity of Answering Ontology-Mediated Queries with a Covering Axiom

Olga Gerasimova[1]([✉]), Stanislav Kikot[2], Vladimir Podolskii[1,3], and Michael Zakharyaschev[2]

[1] National Research University Higher School of Economics, Moscow, Russia
olga.g3993@gmail.com
[2] Birkbeck, University of London, London, UK
[3] Steklov Mathematical Institute, Moscow, Russia

Abstract. We report on our recent results in the ongoing attempts to classify conjunctive queries (CQs) q according to the data complexity of answering ontology-mediated queries of the form $(\{A \sqsubseteq F \sqcup T\}, q)$. In particular, we present new families of path CQs for which this problem is NL-, P- or coNP-complete.

1 Introduction

Ontology-based query answering [5,6,15,19] is a way of organising access to data where, instead of the schemas of data sources, the user is provided with an ontology that serves two purposes: (i) it gives a familiar and convenient vocabulary for formulating end-user queries (e.g., standard geological terms for geologists who want to query a company's databases) and (ii) enriches the data with background knowledge. The key notion in this case is *ontology-mediated query* (OMQ), a pair of the form $Q = (T, q(x))$, where T is an ontology and $q(x)$ a query. The schema of the data is related to the terms in T by means of mappings, $\mathcal{M}$ (say, in R2RML). Now, given a data instance $\mathcal{A}$, we say that a tuple a of constants from $\mathcal{A}$ is a *certain answer* to Q over $\mathcal{A}$ if $q(a)$ holds true in every model of T and $\mathcal{M}(\mathcal{A})$. Whether finding certain answers to OMQs is feasible in practice depends on the languages of T and q. Thus, if T is an *OWL 2 QL*[1] ontology and q a conjunctive query (CQ), then answering Q can be done in AC^0 for data complexity; in other words, there is a first-order query $\Phi(x)$, called an *FO-rewriting* of Q, answers to which over $\mathcal{A}$ are precisely the certain answers to Q over $\mathcal{A}$ [3,7]. Classifying OMQs according to data complexity has become one of the hottest topics in the area of ontology-based data access [8,12,14,17,20].

A systematic investigation of this problem was launched in [5], which, in particular, connected it to constraint satisfaction problems. As shown in [13], answering CQs with basic schema.org ontologies and CQs of qvar-size ≤ 2 is in P for combined complexity, where q is of *qvar-size* n if the restriction of q to

---

[1] https://www.w3.org/TR/owl2-profiles/.

© Springer International Publishing AG 2017
P. Różewski and C. Lange (Eds.): KESW 2017, CCIS 786, pp. 143–158, 2017.
https://doi.org/10.1007/978-3-319-69548-8_11

its quantified variables is a disjoint union of CQs with at most n variables each. Moreover, FO- and datalog-rewritability of OMQs of the form $(\mathcal{T}, \boldsymbol{u})$, where $\mathcal{T}$ is a schema.org ontology and $\boldsymbol{u}$ is a UCQ, are decidable in NExpTime. It has also been recently established in [9] that checking FO-rewritability of OMQs with ontologies formulated in any description logic between $\mathcal{ALCI}$ and $\mathcal{SHI}$ is 2NExpTime-complete. Datalog rewritability of OMQs with ontologies given in disjunctive datalog has been investigated in [14]. An $AC^0/NL/P$ trichotomy of OMQs with $\mathcal{EL}$ ontologies and atomic queries has been established in [18].

In this paper, we report on our ongoing attempts to obtain a complete classification of OMQs of the form $\boldsymbol{Q} = (\mathcal{D}is_A, \boldsymbol{q})$, where $\mathcal{D}is_A = \{A \sqsubseteq F \sqcup T\}$ and $\boldsymbol{q}$ is a CQ. Ontologies with *covering axioms* such as $A \sqsubseteq F \sqcup T$ (saying that, in every model of $\mathcal{D}is_A$, the class A is covered by the union of the classes F and T) are very common in practice: for example, $Animal \sqsubseteq Male \sqcup Female$. The simple examples collected in the table below show how minor tweaks to $\boldsymbol{q}$ can drastically affect the complexity of $\boldsymbol{Q} = (\mathcal{D}is_A, \boldsymbol{q})$ [10]. In the table and elsewhere in the paper, we represent CQs by diagrams. For example, the first CQ below represents $\exists x, y, (F(x) \wedge R(x, y))$ and the second one $\exists x, y, (F(x) \wedge R(x, y) \wedge R(y, x) \wedge T(y))$.[2] (Binary predicates different from R will be shown in diagrams explicitly.)

Complexity	CQ $\boldsymbol{q}$	Explanation
AC^0	$F \circ\!\!\longrightarrow\!\!\circ$	if $\boldsymbol{q}$ has only F but no T, then the F can be ignored
L	$F \circ\!\!\overrightarrow{\underleftarrow{}}\!\!\circ T$	checks undirected reachability: $F \circ\!\!\leftarrow\!\!\circ\!\!\leftarrow\!\!\circ\!\!\leftarrow\!\!\circ T$ the answer to $\boldsymbol{Q}$ is 'yes'
NL	$F \circ\!\!\longrightarrow\!\!\circ T$	checks directed reachability: $F \circ\!\!\rightarrow\!\!\circ\!\!\rightarrow\!\!\circ\!\!\rightarrow\!\!\circ T$ the answer to $\boldsymbol{Q}$ is 'yes'
P	$T \circ\!\!\longrightarrow\!\!\circ\!\!\overset{F}{\longrightarrow}\!\!\circ T$	evaluates monotone circuits
coNP	$\overset{F}{\circ}\!\!\rightarrow\!\!\overset{F}{\circ}\!\!\rightarrow\!\!\overset{T}{\circ}\!\!\rightarrow\!\!\overset{T}{\circ}$	checks CNF satisfiability

The plan of the paper is as follows. Having introduced in Sect. 2 the basic notions we need in what follows, in Sect. 3 we use the $AC^0/NL/P$ trichotomy from [18] to establish a similar trichotomy for the OMQs $(\mathcal{D}is_A, \boldsymbol{q})$ whose CQ $\boldsymbol{q}$ is tree-shaped and the only solitary F-atom in it is at the root. In Sect. 4, we show that the AC^0-criterion for path CQs from [10] collapses for CQs with

---

[2] The OMQ $\boldsymbol{Q} = (\mathcal{D}is_{\mathsf{T}}, \boldsymbol{q})$ with this CQ $\boldsymbol{q}$ can be interpreted as follows, assuming that F stands for 'female', T for 'male', $\top$ for all the individuals of the domain in question, and R for the 'follows' relation: given a graph of Twitter users, in which the gender may be specified for some nodes and missing for the other ones, check whether there certainly exist two people (nodes) in the graph of different gender who follow each other.

loops. In Sect. 5, we present a few classes of path CQs q with a single solitary F, for which answering $(\mathcal{D}is_A, q)$ is NL-complete and P-complete. Finally, in Sect. 6, we give a class of path CQs for which this problem is coNP-complete.

2 Preliminaries

By a *conjunctive query* (*CQ*) we mean in this paper any FO-formula $q(x) = \exists y, \varphi(x, y)$, where φ is a conjunction of unary or binary atoms $P(z)$ with $z \subseteq x \cup y$. Given a data instance—or an *ABox*, in the description logic parlance—$\mathcal{A}$, we denote by $\mathsf{ind}(\mathcal{A})$ the set of individual names that occur in $\mathcal{A}$. A tuple $a \subseteq \mathsf{ind}(\mathcal{A})$ is a *certain answer* to the OMQ $\boldsymbol{Q} = (\mathcal{D}is_A, q(x))$ over $\mathcal{A}$ if $\mathcal{I} \models q(a)$, for every model $\mathcal{I}$ of $\mathcal{D}is_A \cup \mathcal{A}$; in this case we write $\mathcal{D}is_A, \mathcal{A} \models q(a)$. If the set x of *answer variables* is empty, a *certain answer* to $\boldsymbol{Q}$ over $\mathcal{D}$ is 'yes' if $\mathcal{I} \models q$, for every model $\mathcal{I}$ of $\mathcal{D}is_A \cup \mathcal{A}$, and 'no' otherwise. OMQs and CQs without answer variables x are called *Boolean*. We often regard CQs as *sets* of their atoms. For the purposes of this paper, it is enough to assume that all CQs q are *Boolean* and *connected* (in the sense that any two distinct variables in q are connected by a not necessarily directed path of binary atoms from q).

By *answering* a given OMQ $\boldsymbol{Q} = (\mathcal{D}is_A, q(x))$, we understand the problem of checking, given an ABox $\mathcal{A}$ and a tuple $a \subseteq \mathsf{ind}(\mathcal{A})$, whether $\mathcal{D}is_A, \mathcal{A} \models q(a)$. It is easy to see that this problem is always in coNP. It is in the complexity class AC^0 if there is an FO-formula $q'(x)$, called an *FO-rewriting* of $\boldsymbol{Q}$, such that $\mathcal{D}is_A, \mathcal{A} \models q(a)$ iff $q'(a)$ holds in the interpretation given by $\mathcal{A}$, for any ABox $\mathcal{A}$ and any $a \subseteq \mathsf{ind}(\mathcal{A})$.

A *datalog program*, Π, is a finite set of *rules* $\forall z, (\gamma_0 \leftarrow \gamma_1 \wedge \cdots \wedge \gamma_m)$, where each γ_i is an atom $Q(y)$ with $y \subseteq z$ or an equality $(z = z')$ with $z, z' \in z$. (As usual, we omit the prefix $\forall z$.) The atom γ_0 is the *head* of the rule, and $\gamma_1, \ldots, \gamma_m$ its *body*. All the variables in the head must occur in the body, and $=$ can only occur in the body. The predicates in the head of rules are *IDB predicates*, the rest *EDB predicates*.

A *datalog query* is a pair $(\Pi, G(x))$, where Π is a datalog program and $G(x)$ an atom. A tuple $a \subseteq \mathsf{ind}(\mathcal{A})$ is an *answer* to $(\Pi, G(x))$ *over* an ABox $\mathcal{A}$ if $G(a)$ holds in the FO-structure with domain $\mathsf{ind}(\mathcal{A})$ obtained by closing $\mathcal{A}$ under the rules in Π, in which case we write $\Pi, \mathcal{A} \models G(a)$. A datalog query $(\Pi, G(x))$ is a *datalog rewriting* of an OMQ $\boldsymbol{Q} = (\mathcal{D}is, q(x))$ in case $\mathcal{D}is, \mathcal{A} \models q(a)$ iff $\Pi, \mathcal{A} \models G(a)$, for any ABox $\mathcal{A}$ and any $a \subseteq \mathsf{ind}(\mathcal{A})$. The *evaluation problem* for $(\Pi, G(x))$—i.e., checking, given an ABox $\mathcal{A}$ and a tuple $a \subseteq \mathsf{ind}(\mathcal{A})$, whether $\Pi, \mathcal{A} \models G(a)$—is known to be in P. Evaluation of a datalog query with a *linear* program, where the body of any rule has at most one IDB predicate, can be done in NL; see [11] and references therein. The NL upper bound also holds for datalog queries with linear-stratified programs that are defined as follows. A *stratified* program [1] is a sequence $\Pi = (\Pi_0, \ldots, \Pi_n)$ of datalog programs, called the *strata* of Π, such that each predicate in Π can occur in the head of a rule only in one stratum Π_i and can occur in the body of a rule only in strata Π_j with $j \geq i$. If, additionally, the body of each rule in Π contains at most one occurrence of a head predicate from the same stratum, we call Π *linear-stratified*. It is shown in [2] that every linear-stratified program (called there piecewise linear) can be converted in an equivalent linear datalog program.

3 $\mathrm{AC}^0/\mathrm{NL}/\mathrm{P}$ Trichotomy for F-Tree OMQs

By a *solitary occurrence* of F in a CQ q we mean any occurrence of $F(x)$ in q, for some variable x, such that $T(x) \notin q$; likewise, a *solitary occurrence* of T in q is any occurrence $T(x) \in q$ such that $F(x) \notin q$. An F-*tree CQ* is a CQ q with a single solitary $F(x)$ such that the binary atoms in q form a directed tree with root x.

Our first observation is that answering any OMQ $Q = (\mathcal{D}is_A, q)$ with an F-tree CQ q is either in AC^0 or NL-complete or P-complete. We obtain this trichotomy using a recent result of Lutz and Sabellek [18] establishing such a trichotomy for OMQs of the form $(\mathcal{T}, G(x))$, where $\mathcal{T}$ is an ontology formulated in the description logic $\mathcal{EL}$ [4] and G is a concept name (unary predicate).

Theorem 1. *Answering any OMQ* $Q = (\mathcal{D}is_A, q)$ *with an F-tree CQ q is either in* AC^0 *or NL-complete or P-complete.*

Proof. Let Π_Q be the datalog program with the following rules:

$$
\begin{aligned}
G &\leftarrow F(x) \wedge \tilde{q}'(x, y_1, \ldots, y_n) \wedge P(y_1) \wedge \cdots \wedge P(y_n), \\
P(x) &\leftarrow T(x), \\
P(x) &\leftarrow A(x) \wedge \tilde{q}'(x, y_1, \ldots, y_n) \wedge P(y_1) \wedge \cdots \wedge P(y_n),
\end{aligned}
$$

where q' is obtained from q by removing all of its solitary occurrences of T- and F-atoms and $\tilde{q}'$ is the result of omitting all the $\exists$ from q'. As shown in [10, Theorem 7], for any ABox $\mathcal{A}$, we have $\mathcal{D}is_A, \mathcal{A} \models q$ iff $\Pi_Q, \mathcal{A} \models G$.

Denote by $\mathcal{T}_Q$ the $\mathcal{EL}$ TBox with two concept inclusions:

$$
T \sqsubseteq P, \qquad A \sqcap C_q \sqsubseteq P,
$$

where C_q is an $\mathcal{EL}$-concept representing $q \setminus \{F(x)\}$. For example, for

$$
q = F(x) \wedge R_1(x, y_1) \wedge F(y_1) \wedge T(y_1) \wedge R_2(x, y_2) \wedge R_3(y_2, y_3) \wedge T(y_3),
$$

we have

$$
C_q = \exists R_1.(F \sqcap T) \sqcap \exists R_2.\exists R_3.T.
$$

It is readily seen that, for any ABox $\mathcal{A}$ and any $a \in \mathsf{ind}(\mathcal{A})$, we have $\Pi_Q, \mathcal{A} \models P(a)$ iff $\mathcal{T}_Q, \mathcal{A} \models P(a)$.

Finally, we observe that (*i*) answering Q is in AC^0 iff answering $(\mathcal{T}_Q, P(x))$ is in AC^0; (*ii*) answering Q is NL-complete iff answering $(\mathcal{T}_Q, P(x))$ is NL-complete; (*iii*) answering Q is P-complete iff answering $(\mathcal{T}_Q, P(x))$ is P-complete.

Note that [18] gives an EXPTIME algorithm for checking which of the three complexity classes a given $\mathcal{EL}$-OMQ of the form $(\mathcal{T}, G(x))$ falls into. However, applying this algorithm in our case is tricky because the input ontology $\mathcal{T}_Q$ must first be converted to a normal form. In particular, it does not give clear syntactic criteria on the shape of the CQ q that would guarantee that the OMQ $(\mathcal{D}is_A, q)$ belongs to the desired complexity class (see examples below). Note also that the reduction in the proof above does not work for CQs that are not F-trees.

4 AC^0

As shown in [10], answering any CQ $Q = (\mathcal{D}is_A, q)$ is in AC^0 if the CQ q does not have solitary occurrences of F (or T). This sufficient condition becomes also a necessary one if q is a *path* CQ, that is, the variables $x_0, \ldots, x_n$ in q are ordered so that

- the binary atoms in q form a chain $R_1(x_0, x_1), \ldots, R_n(x_{n-1}, x_n)$;
- the unary atoms in q are of the form $T(x_i)$ and $F(x_j)$, for some i and j with $0 \le i, j \le n$.

In fact, we have the following AC^0/NL-dichotomy for OMQs $Q = (\mathcal{D}is_A, q)$ with a path CQ q [10]:

- either q does not contain a solitary F or a solitary T, and answering Q is in AC^0,
- or q contains both solitary F and T, and answering Q is NL-hard.

Here, we give an example showing that this dichotomy collapses for path CQs with loops of the form $R(x, x)$.

Proposition 1. *Answering the OMQ $(\mathcal{D}is_A, q)$, where q is the CQ with a solitary F and a solitary T shown in the picture below, is in AC^0 for data complexity.*

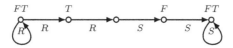

Proof. It suffices to show that $\mathcal{D}is_A, \mathcal{A} \models q$ iff $\mathcal{A} \models q$. The implication ($\Leftarrow$) is trivial.

($\Rightarrow$) Suppose $\mathcal{A} \not\models q$. Let $x_1, \ldots, x_5$ be the consecutive variables in q. We construct a model $\mathcal{I}$ of $\mathcal{D}is_A$ with $\mathcal{I} \not\models q$. Consider the following subsets of $\mathsf{ind}(\mathcal{A})$:

$$B_R = \{a \in \mathsf{ind}(\mathcal{A}) \mid R(a, a) \in \mathcal{A}, \ F(a) \in \mathcal{A}, \ T(a) \in \mathcal{A}\},$$
$$B_S = \{a \in \mathsf{ind}(\mathcal{A}) \mid S(a, a) \in \mathcal{A}, \ F(a) \in \mathcal{A}, \ T(a) \in \mathcal{A}\},$$
$$X = \{a \in \mathsf{ind}(\mathcal{A}) \mid R(b, a) \text{ for some } b \in B_R\},$$
$$Y = \{a \in \mathsf{ind}(\mathcal{A}) \mid S(a, b) \text{ for some } b \in B_S\}.$$

Note that since $\mathcal{A} \not\models q$, the sets X and Y do not intersect. Indeed, if $b \in X \cap Y$, then B_R contains some element a such that $R(a, b) \in \mathcal{A}$, B_S contains some c with $S(b, c) \in \mathcal{A}$, and the map h given by $h(x_1) = h(x_2) = a$, $h(x_3) = b$ and $h(x_4) = h(x_5) = c$ is a homomorphism from q to $\mathcal{A}$. Define a model $\mathcal{I}$ of $\mathcal{D}is_\mathcal{A}$ by extending $\mathcal{A}$ with

- $F(a)$, for all $a \in X$;
- $T(a)$, for all $a \in \text{ind}(\mathcal{A}) \setminus X$.

We claim that $\mathcal{I} \not\models q$. Indeed, suppose there is a homomorphism $h\colon q \to \mathcal{I}$. Clearly, $h(x_1) \in B_R$ and $h(x_5) \in B_S$. It follows that $h(x_2) \in X$ and $h(x_4) \in Y$. Since $T(x_2) \in q$, we have $h(x_2) \in T^{\mathcal{I}}$, and so $T(h(x_2)) \in \mathcal{A}$. Similarly, $F(x_4) \in \mathcal{A}$. It follows that h is a homomorphism from q to $\mathcal{A}$, contrary to our assumption.

Note that the CQ q above is *minimal* (not equivalent to any of its proper sub-CQs).

5   NL vs. P

A path CQ q is called an *F-path CQ* if q has a single solitary occurrence of F at its root; in other words, q is both a path CQ and an F-tree CQ. We represent such a q as shown in the picture below, which indicates *all* the solitary occurrences of F and T:

We know from [10] that

- answering OMQs $(\mathcal{D}is_A, q)$ with F-path CQs can be done in P;
- if $x, y_1, \ldots, y_m$ are all the variables in q, then answering $(\mathcal{D}is_A, q)$ is NL-complete.

There is also a table in [10] with quite a few odd examples of CQs of both kinds. Our next result sheds some light on the left column of this table.

We require the following sub-CQs of the F-path CQ q shown above:

- q_i is the suffix of q that starts at y_i, but without $T(y_i)$, for $1 \le i \le m$;
- q_i^* is the prefix of q that ends at y_i, but without $F(x)$ and $T(y_i)$, for $1 \le i \le m$;
- q_{m+1}^* is q without $F(x)$.

We write $f_i\colon q_i \twoheadrightarrow q$ if f_i is a homomorphism from q_i into q with $f_i(y_i) = x$.

Theorem 2. *If there exist $f_i\colon q_i \twoheadrightarrow q$, for $1 \le i \le m$, then $(\mathcal{D}is_A, q)$ is NL-complete.*

Proof. Let Π be a linear datalog program with the following rules:

$$G \leftarrow F(x) \wedge \tilde{q}^*{}_{m+1}, \tag{r1}$$
$$G \leftarrow F(x) \wedge \tilde{q}^*{}_i \wedge P(y_i), \quad \text{for } 1 \leq i \leq m, \tag{r2}$$
$$P(x) \leftarrow A(x) \wedge \tilde{q}^*{}_{m+1}, \tag{r3}$$
$$P(x) \leftarrow A(x) \wedge \tilde{q}^*{}_i \wedge P(y_i), \quad \text{for } 1 \leq i \leq m. \tag{r4}$$

It suffices to show that, for any ABox $\mathcal{A}$, we have $\mathcal{D}is_A, \mathcal{A} \models q$ iff $\Pi, \mathcal{A} \models G$.
($\Rightarrow$) Suppose $\mathcal{D}is_A, \mathcal{A} \models q$. Let $V_P = \{a \in \text{ind}(\mathcal{A}) \mid \Pi, \mathcal{A} \models P(a)\}$. Define an interpretation $\mathcal{I}$ with domain $\text{ind}(\mathcal{A})$ by taking

$$T^{\mathcal{I}} = \{a \mid T(a) \in \mathcal{A}\} \cup \{a \in V_P \mid A(a) \in \mathcal{A}, \ F(a) \notin \mathcal{A}\},$$
$$F^{\mathcal{I}} = \{a \mid F(a) \in \mathcal{A}\} \cup \{a \notin V_P \mid A(a) \in \mathcal{A}, \ T(a) \notin \mathcal{A}\}.$$

Clearly, $\mathcal{I} \models \mathcal{D}is_A$, and so there is a homomorphism $h\colon q \to \mathcal{I}$. We show now that $\Pi, \mathcal{A} \models G$. Note that we have both $a \in F^{\mathcal{I}}$ and $a \in T^{\mathcal{I}}$ only if $F(a), T(a) \in \mathcal{A}$.
Case 1: $T(h(y_i)) \in \mathcal{A}$, for $1 \leq i \leq m$. Then $\Pi, \mathcal{A} \models G$ by $(r1)$ since $h(x) \in F^{\mathcal{I}}$ can only be because $F(h(x)) \in \mathcal{A}$ (if this is not the case, then $A(h(x)) \in \mathcal{A}$ and we have $h(x) \in V_P$ by $(r3)$, which is a contradiction).
Case 2: $T(h(y_i)) \notin \mathcal{A}$, for some i $(1 \leq i \leq m)$. Let i be minimal with this property. By the definition of $\mathcal{I}$, we then have $h(y_i) \in V_P$ and $A(h(y_i)) \in \mathcal{A}$. Then $\Pi, \mathcal{A} \models G$ by $(r2)$ since $h(x) \in F^{\mathcal{I}}$ can only be because $F(h(x)) \in \mathcal{A}$ (if this is not the case, then $A(h(x)) \in \mathcal{A}$ and we have $h(x) \in V_P$ by $(r4)$, which is a contradiction).
($\Leftarrow$) Suppose there is a derivation of G from Π and $\mathcal{A}$. Then there exist a sequence of homomorphisms

$$h_1\colon q^*_{m+1} \to \mathcal{A}, \quad h_2\colon q^*_i \to \mathcal{A}, \quad \ldots, \quad h_k\colon q^*_j \to \mathcal{A},$$

for some $i, \ldots, j \leq m$, with $h_1(x) = h_2(y_i), \ldots, h_{k-1}(x) = h_2(y_j)$, $F(h_k(x)) \in \mathcal{A}$ and $A(h_n(x)) \in \mathcal{A}$, for $1 \leq n \leq k-1$. Now, consider any model $\mathcal{I}$ of $\mathcal{D}is_A$ extending $\mathcal{A}$ and show that $\mathcal{I} \models q$. If $h_1(x) \in F^{\mathcal{I}}$, then h_1 is a homomorphism from q to $\mathcal{I}$. So, let $h_1(x) \in T^{\mathcal{I}}$. Then the homomorphisms h_2 and f_i give us a homomorphism $h'_2\colon q^*_{m+1} \to \mathcal{I}$ such that $h'_2(x) = h_3(y_l)$. Again, if $h'_2(x) \in F^{\mathcal{I}}$, then h'_2 is a homomorphism from q to $\mathcal{I}$. Otherwise, we combine h'_2 with f_l, and so on. As $h_k(x) \in F^{\mathcal{I}}$, sooner or later we must obtain a homomorphism from q to $\mathcal{I}$.

Example 1. By Theorem 2, the following CQs q give NL-complete OMQs $(\mathcal{D}is_A, q)$:

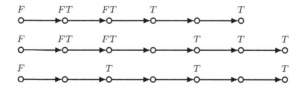

Denote by $\boldsymbol{q}_{TnT}$, for $n \geq 0$, the CQ shown in the picture below, where all the binary predicates are R and the n variables without labels do not occur in F- or T-atoms:

Clearly, Theorem 2 only applies to $\boldsymbol{q}_{T0T}$. Our next results show that, surprisingly, $(\mathcal{D}is_T, \boldsymbol{q}_{T1T})$ is NL-complete, $(\mathcal{D}is_A, \boldsymbol{q}_{T1T})$ is P-complete, and $(\mathcal{D}is_T, \boldsymbol{q}_{TnT})$ is P-complete, for every $n \geq 2$ (where, as usual, $\top$ denotes the class of all domain individuals).

Proposition 2. *Answering the OMQ $(\mathcal{D}is_T, \boldsymbol{q}_{T1T})$ is NL-complete.*

Proof. The NL-hardness follows from [10, Theorem 4]. To establish the matching upper bound, consider the datalog program Π' with the following rules:

$$G \leftarrow F(x) \wedge R(x,y) \wedge P(y) \wedge R(y,z) \wedge R(z,u) \wedge P(u),$$
$$P(x) \leftarrow T(x),$$
$$P(x) \leftarrow R(x,y) \wedge P(y) \wedge R(y,z) \wedge R(z,u) \wedge P(u).$$

As shown in [10, Theorem 7], $\mathcal{D}is_T, \mathcal{A} \models \boldsymbol{q}_{T1T}$ iff $\Pi', \mathcal{A} \models G$. Now, consider a program Π with the single rule

$$T(x) \leftarrow R(x,y) \wedge T(y) \wedge R(y,z) \wedge R(z,u) \wedge T(u). \qquad (r)$$

It is not hard to see that if checking whether $\Pi, \mathcal{A} \models T(a)$, for any given $a \in \mathsf{ind}(\mathcal{A})$, can be done in NL, then checking whether $\Pi', \mathcal{A} \models G$ can also be done in NL. Thus, it suffices to show that checking whether $\Pi, \mathcal{A} \models T(a)$ can be done in NL.

Let $\Pi^\dagger$ be the *linear stratified* datalog program with the following rules:

$$P(x) \leftarrow R(x,y) \wedge T(y) \wedge R(y,z) \wedge R(z,v) \wedge T(v), \qquad (r1)$$
$$P(x) \leftarrow R(x,y) \wedge T(y) \wedge R(y,z) \wedge R(z,v) \wedge P(v), \qquad (r1')$$
$$Q(x) \leftarrow R(x,y) \wedge P(y) \wedge R(y,z) \wedge R(z,v) \wedge T(v), \qquad (r2)$$
$$Q(x) \leftarrow R(x,y) \wedge P(y) \wedge R(y,z) \wedge R(z,v) \wedge P(v), \qquad (r2')$$
$$Q(x) \leftarrow R(x,y) \wedge Q(y), \qquad (r3)$$
$$G(x) \leftarrow T(x), \qquad (r4)$$
$$G(x) \leftarrow P(x), \qquad (r5)$$
$$G(x) \leftarrow Q(x). \qquad (r6)$$

Checking whether $\Pi^\dagger, \mathcal{A} \models G(a)$ can be done in NL. We claim that $\Pi^\dagger, \mathcal{A} \models G(a)$ iff $\Pi, \mathcal{A} \models T(a)$, for any ABox $\mathcal{A}$ and any $a \in \mathsf{ind}(\mathcal{A})$.

($\Rightarrow$) Suppose $\Pi^\dagger, \mathcal{A} \models G(a)$. By $(r4)$–$(r6)$, we have one of the following cases:

Case 1: $\Pi^\dagger, \mathcal{A} \models T(a)$. Then trivially $\Pi, \mathcal{A} \models T(a)$.
Case 2: $\Pi^\dagger, \mathcal{A} \models P(a)$. Then $\Pi, \mathcal{A} \models T(a)$ by $(r1)$ and $(r2)$.
Case 3: $\Pi^\dagger, \mathcal{A} \models Q(a)$. Then, by $(r3)$–$(r4)$, there are $a_0, a_1, \ldots, a_n, a_{n+1}$ such that

- $a = a_0$;
- $R(a_i, a_{i+1}) \in \mathcal{A}$, for $0 \le i \le n$;
- $\Pi^\dagger, \mathcal{A} \models P(a_{n+1})$;
- there are $z', v' \in \mathsf{ind}(\mathcal{A})$ with $R(a_{n+1}, z'), R(z', v') \in \mathcal{A}$ and $\Pi^\dagger, \mathcal{A} \models P(v')$.

As in case 2, $\Pi, \mathcal{A} \models T(a_{n+1})$ and $\Pi, \mathcal{A} \models T(v')$, from which $\Pi, \mathcal{A} \models T(a_n)$. As $\Pi^\dagger, \mathcal{A} \models P(a_{n+1})$, there is an R-successor a_{n+2} of a_{n+1} with $\Pi, \mathcal{A} \models T(a_{n+2})$. But then (r) is applicable at a_{n-1} (with y being a_n, z being a_{n+1} and v being a_{n+2}). By iteratively applying (r) for $i = n - 1, n - 2, \ldots, 0$, we conclude that $\Pi, \mathcal{A} \models T(a_0)$.

($\Leftarrow$) Suppose $\Pi, \mathcal{A} \models T(r)$. Then there is a finite 2-ary (derivation) tree $\mathfrak{T}$ such that

- the vertices v of $\mathfrak{T}$ are some elements from $\mathsf{ind}(\mathcal{A})$;
- r is the root of $\mathfrak{T}$;
- any vertex v of $\mathfrak{T}$ either is a leaf or has 2 successors: 'left' v_1 and 'right' v_2 such that $\mathcal{A} \models R(v, v_1) \wedge R(v_1, w) \wedge R(w, v_2)$, for some $w \in \mathsf{ind}(\mathcal{A})$;
- if v is a leaf, then $T(v) \in \mathcal{A}$.

We prove that $\Pi^\dagger, \mathcal{A} \models G(r)$ by induction on the depth of $\mathfrak{T}$. The basis of induction ($\mathfrak{T}$ of depth 0) is trivial. For the induction step, we define inductively a finite sequence $u_0, d_0, u_1, d_1, \ldots, d_{n-1}, u_n$, where the u_i are vertices of $\mathfrak{T}$ and $d_i \in \{\mathfrak{r}, \mathfrak{l}\}$. First, we set $u_0 = r$. Now, suppose u_i has been defined. If u_i is a leaf of $\mathfrak{T}$, we stop and set $n = i$. Otherwise, let v_1 and v_2 be, respectively, the left and right successors of u_i in $\mathfrak{T}$. If v_1 is not a leaf, we set $d_i = \mathfrak{l}$ and $u_{i+1} = v_1$. Otherwise, we set $d_i = \mathfrak{r}$ and $u_{i+1} = v_2$. Note that

- $d_{n-1} = \mathfrak{r}$, if $n \ge 1$;
- if $d_i = \mathfrak{r}$, there are $y, w \in \mathsf{ind}(\mathcal{A})$ with $R(u_i, y), T(y), R(y, w), R(w, u_{i+1}) \in \mathcal{A}$.

Now, we have two cases depending on the sequence $\mathfrak{dir} = d_0, d_1, \ldots, d_{n-1}$.

Case 1: $\mathfrak{dir}$ does not contain $\mathfrak{l}$. Then we can show by induction on i from $n - 1$ to 0 using (ii) that $\Pi^\dagger, \mathcal{A} \models P(u_i)$, for $0 \le i \le n - 1$. It follows that $\Pi^\dagger, \mathcal{A} \models G(r)$.

Case 2: $\mathfrak{dir}$ contains at least one $\mathfrak{l}$. Let k be such that the last occurrence of $\mathfrak{l}$ in $\mathfrak{dir}$ is between u_k and u_{k+1}. By (i), $k + 1 < n$, and so u_{k+2} is well defined. The argument from case 1 shows that $\Pi^\dagger, \mathcal{A} \models P(u_{k+1})$. By IH, $\Pi^\dagger, \mathcal{A} \models G(y)$ for the right successor y of u_k. This means that either $\Pi^\dagger, \mathcal{A} \models Q(y)$ or $\Pi^\dagger, \mathcal{A} \models P(y)$ or $\Pi^\dagger, \mathcal{A} \models T(y)$. In the first case, we obtain $\Pi^\dagger, \mathcal{A} \models Q(r)$ using $(r3)$ and the fact that y is accessible from r via R in $\mathcal{A}$. In last two cases (using $(r2)$ or $(r2')$), we have $\Pi^\dagger, \mathcal{A} \models Q(u_k)$. By construction, u_k is accessible from r via R in $\mathcal{A}$, and so $\Pi^\dagger, \mathcal{A} \models Q(r)$. It follows that $\Pi^\dagger, \mathcal{A} \models G(r)$.

Theorem 3. *Answering any OMQ* $(\mathcal{D}is_T, q_{TnT})$, *for* $n \geq 2$, *is* P-*complete.*

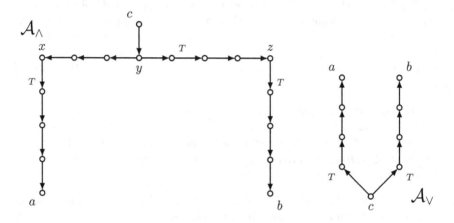

Proof. We sketch a proof for q_{T2T} shown in the picture above and leave the general case to the reader. Let $\varPi$ be the program with the single rule

$$T(x) \leftarrow R(x,y) \wedge T(y) \wedge R(y,z) \wedge R(z,u) \wedge R(z,v) \wedge T(v).$$

It suffices to show that checking whether $\varPi, \mathcal{A} \models T(a)$, for $\mathcal{A}$ and $a \in \mathrm{ind}(\mathcal{A})$, is P-hard. Consider the following two ABoxes:

It is routine to verify the following properties of these ABoxes:

$\wedge$ **-gadget**

$$\varPi, \mathcal{A}_\wedge \cup \{T(a), T(b)\} \models T(c),$$
$$\varPi, \mathcal{A}_\wedge \cup \{T(a)\} \not\models T(c),$$
$$\varPi, \mathcal{A}_\wedge \cup \{T(b)\} \not\models T(c),$$
$$\varPi, \mathcal{A}_\wedge \cup \{T(a), T(b), R(c',c)\} \not\models T(c');$$

$\vee$ **-gadget**

$$\varPi, \mathcal{A}_\vee \cup \{T(a)\} \models T(c),$$
$$\varPi, \mathcal{A}_\vee \cup \{T(b)\} \models T(c),$$
$$\varPi, \mathcal{A}_\vee \not\models T(c),$$
$$\varPi, \mathcal{A}_\vee \cup \{T(a), T(b), R(c',c)\} \not\models T(c').$$

Now, with any monotone Boolean circuit $\mathcal{C}$ with an output o and all gates having exactly two inputs, we associate an ABox $\mathcal{A}_\mathcal{C}$ by replacing every AND-gate in $\mathcal{C}$ with inputs a and b and output c by a fresh copy of $\mathcal{A}_\wedge$, and every

OR-gate with inputs a and b and output c by a fresh copy of $\mathcal{A}_\vee$. Given an input α for $\mathcal{C}$, we place atoms $T(a)$ on the input gates a (which are also individuals of $\mathcal{A}_\mathcal{C}$) with $\alpha(a) = 1$, and denote the resulting ABox by $\mathcal{A}_\mathcal{C}^\alpha$. We claim that $\mathcal{C}$ outputs 1 under α iff $\mathcal{A}_\mathcal{C}^\alpha, \Pi \models T(o)$.

The implication ($\Rightarrow$) is proved by induction, using the properties of $\mathcal{A}_\wedge$ and $\mathcal{A}_\vee$, that if a gate g of $\mathcal{C}$ outputs 1 under α, then $\Pi, \mathcal{A}_\mathcal{C}^\alpha \models T(g)$.

($\Leftarrow$) Suppose $\mathcal{C}$ outputs 0. Define an ABox $\mathcal{A}$ by extending $\mathcal{A}_\mathcal{C}^\alpha$ as follows. We add atoms $T(c)$ for all gates g that output 1 under α, atoms $T(x)$ for those copies of $\mathcal{A}_\wedge$ that correspond to an AND-gate having 1 as its left input, and atoms $T(y)$ and $T(z)$ for those copies of $\mathcal{A}_\wedge$ that correspond to an AND-gate having 1 as its right input. It is readily checked that no rule in Π can be applied to $\mathcal{A}$. Since $\mathcal{C}$ outputs 0, it follows that $\Pi, \mathcal{A}_\mathcal{C}^\alpha \not\models T(o)$.

(The reader may want to figure out which part of the proof goes wrong for $n = 1$.) On the other hand we have:

Proposition 3. *Answering the OMQ* $(\mathcal{D}is_A, q_{T1T})$ *is P-complete.*

The proof is similar to that of Theorem 3 and uses the following gadgets $\mathcal{A}_\wedge$, $\mathcal{A}_\vee$:

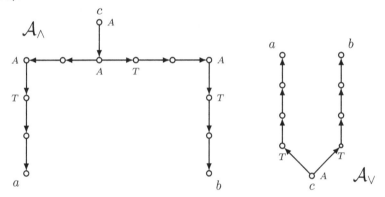

So far in this section we have considered OMQs with F-path CQs, thus excluding path CQs such as q in the picture below

$$\overset{T}{\underset{}{\circ}}\!\!\longrightarrow\!\!\overset{F}{\underset{}{\circ}}\!\!\longrightarrow\!\!\overset{T}{\underset{}{\circ}}$$

As shown in [10], answering the OMQ $(\mathcal{D}is_A, q)$ with this q is P-complete; in fact, it follows from the proof that $(\mathcal{D}is_T, q)$ is P-complete, too. This tempted us to conjecture that having a solitary F in the middle of a path CQ with solitary T's on both sides ensures P-hardness. To our surprise, there is a family of path CQs of this shape that are NL-complete.

A path CQ q_{TF} is called a TF-*path CQ* if it is of the form

$$q_{TF} = \underset{y_0}{\overset{T}{\circ\!\!\longrightarrow\!\!\circ}} \cdots \circ \underset{x}{\overset{F}{\longrightarrow\!\!\circ\!\!\longrightarrow\!\!\circ}} \cdots \underset{y_1}{\overset{T}{\circ\!\!\longrightarrow\!\!\circ}}\!\!\longrightarrow\!\!\circ \cdots \underset{y_m}{\overset{T}{\circ\!\!\longrightarrow\!\!\circ}}\!\!\longrightarrow\!\!\circ \cdots \underset{y_{m+1}}{\circ\!\!\longrightarrow\!\!\circ}$$

where the $T(y_i)$ and $F(x)$ are all the solitary occurrences of T and F in q_{TF}. We represent this CQ as

$$q_{TF} = \{T(y_0)\} \cup q_0 \cup q,$$

where q_0 is the sub-CQ of q_{TF} between y_0 and x with $T(y_0)$ removed and q is the same as in Theorem 2 (and q^*_{m+1} is q without $F(x)$).

Theorem 4. *If q satisfies the condition of Theorem 2 and there is a homomorphism $h\colon q^*_{m+1} \to q_0$ such that $h(x) = y_0$, then answering $(\mathcal{D}is_A, q_{TF})$ is NL-complete.*

Proof. We use the notations introduced for Theorem 2. Let Π be the following linear-stratified datalog program:

$$
\begin{aligned}
G &\leftarrow F(x) \wedge P(x) \wedge Q(x), & (r1) \\
G &\leftarrow F(x) \wedge Q(x) \wedge \tilde{q}^*_{m+1}, & (r2) \\
P(x) &\leftarrow A(x) \wedge \tilde{q}^*_{m+1}, & (r3) \\
P(x) &\leftarrow A(x) \wedge \tilde{q}^*_i \wedge P(y_i) \wedge Q(y_i), & (r4) \\
Q(x) &\leftarrow T(y_0) \wedge \tilde{q}_0(y_0, x), & (r5) \\
Q(x) &\leftarrow A(y_0) \wedge Q(y_0) \wedge \tilde{q}_0(y_0, x). & (r6)
\end{aligned}
$$

It suffices to prove that $\Pi, \mathcal{A} \models G$ iff $\mathcal{D}is_A, \mathcal{A} \models q_{TF}$, for all ABoxes $\mathcal{A}$.

($\Leftarrow$) Suppose $\mathcal{D}is_A, \mathcal{A} \models q_{TF}$. Let $V_P = \{a \in \mathrm{ind}(\mathcal{A}) \mid \Pi, \mathcal{A} \models P(a)\}$ and $V_Q = \{a \in \mathrm{ind}(\mathcal{A}) \mid \Pi, \mathcal{A} \models Q(a)\}$. Define an interpretation $\mathcal{I}$ with domain $\mathrm{ind}(\mathcal{A})$ by taking

$$
\begin{aligned}
T^{\mathcal{I}} &= \{a \mid T(a) \in \mathcal{A}\} \cup \{a \in V_P \text{ or } a \in V_Q \mid F(a) \notin \mathcal{A}\}, \\
F^{\mathcal{I}} &= \{a \mid F(a) \in \mathcal{A}\} \cup \{a \notin V_P \text{ and } a \notin V_Q \mid T(a) \notin \mathcal{A}\}.
\end{aligned}
$$

Note that we have both $a \in F^{\mathcal{I}}$ and $a \in T^{\mathcal{I}}$ only if $F(a), T(a) \in \mathcal{A}$. Clearly, $\mathcal{I}$ is a model of $(\mathcal{D}is_A, \mathcal{A})$, and so there is a homomorphism $f\colon q_{TF} \to \mathcal{I}$. We show now that $\Pi, \mathcal{A} \models G$. First, we have $\Pi, \mathcal{A} \models Q(f(x))$. Indeed, if $T(f(y_0)) \in \mathcal{A}$, then we can use $(r5)$. If $A(f(y_0)) \in \mathcal{A}$, then $f(y_0) \in V_Q$ (using $r6$) and $f(y_0) \in T^{\mathcal{I}}$ follows from the definition of $\mathcal{I}$. So, $f(x) \in V_Q$ is again obtained by $(r5)$. Second, there are two similar cases. If $T(f(y_i)) \in \mathcal{A}$, for $1 \le i \le m$, then $\Pi, \mathcal{A} \models G$ by $(r2)$. Otherwise, we take the smallest i such that $T(f(y_i)) \notin \mathcal{A}$. Then $A(f(y_i)) \in \mathcal{A}$ and, by the definition of $\mathcal{I}$, we have $f(y_i) \in V_P$ (using $(r3)$ or $(r4)$) and $f(y_i) \in T^{\mathcal{I}}$, and so again $\Pi, \mathcal{A} \models G$ by $(r1)$.

($\Rightarrow$) Suppose there is a derivation of G from Π and $\mathcal{A}$. Then $\mathcal{D}is_A, \mathcal{A} \models q$ and there exists a sequence $v^0, v^1, \ldots, v^n \in \mathrm{ind}(\mathcal{A})$ such that:

- $F(v^0) \in \mathcal{A}$;
- $A(v^i) \in \mathcal{A}$ and $v^i \in V_P$, for $1 \le i < n$;
- for each i $(0 \le i < n)$, we have $q^*_j(v^i, v^{i+1})$, for some $j \in \{1, \ldots, m\}$;
- $q^*_{m+1}(v^{n-1}, v^n) \wedge T(v^n) \in \mathcal{A}$.

Moreover, there are also paths $s^i_0, s^i_1, \ldots, s^i_{k_i}$, where $v^i = s^i_{k_i}$ and $0 \le i \le n$, such that

- $T(s^i_0) \in \mathcal{A}$;

- $A(s_j^i) \in \mathcal{A}$ and $s_j^i \in V_Q$, for $1 \leq j \leq k_i$;
- for each j $(0 \leq j < k_i)$, we have $q_0(s_j^i, s_{j+1}^i) \in \mathcal{A}$;
- $A(s_{k_i}^i) \in \mathcal{A}$ and $s_{k_i}^i \in V_P$ or, if $i = 0$, then $F(s_{k_0}^0) \in \mathcal{A}$.

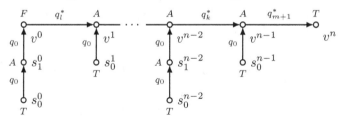

Let $\mathcal{I}$ be any interpretation based on $\mathcal{A}$. Let i be the maximal number such that $v_i \in F^{\mathcal{I}}$.

Case 1: $s_l^i \in T^{\mathcal{I}}$, for $0 \leq l < k_i$. In this case, there exists a homomorphism h_1 from $\boldsymbol{q}_{TF}$ to $\mathcal{I}$ such that $h_1(y_0) = s_{k_i-1}^i$, $h_1(x) = v_i$ and $h_1(y_j) = v_{i+1}$, where j is maximal with $T(y_j) \notin \mathcal{A}$. Then $\mathcal{D}is_A, \mathcal{A} \models \boldsymbol{q}_{TF}$, because $\boldsymbol{q}$ satisfies Theorem 2.

Case 2: otherwise. Let j be minimal with $s_j^i \in V_Q$ and $s_j^i \in F^{\mathcal{I}}$. Then there is a homomorphism h_2 from $\boldsymbol{q}_{TF}$ to $\mathcal{I}$ such that $h_2(y_0) = s_{j-1}^i$ and $h_2(x) = s_j^i$. We obtain $\mathcal{D}is_A, \mathcal{A} \models \boldsymbol{q}_{TF}$ using the homomorphism h.

Example 2. By Theorem 4, the following CQs $\boldsymbol{q}$ give NL-complete OMQs $(\mathcal{D}is_A, \boldsymbol{q})$:

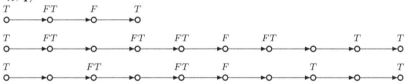

6   coNP

As shown in [10] answering $(\mathcal{D}is_A, \boldsymbol{q})$ with the CQ $\boldsymbol{q}$

T T F F
○———▸○———▸○———▸○

is coNP-complete. Here, we generalise this observation. We say that a path CQ $\boldsymbol{q}$ is a *2-2-CQ* if it has at least two solitary T, at least two solitary F all of which are located after all the T, and every occurrence of T or F in $\boldsymbol{q}$ is solitary. We represent any given 2-2-CQ $\boldsymbol{q}$ as shown below

T T F F
○—▸○—▸○—▸○—▸○—▸○
  p x r y s z u w v

where $\boldsymbol{p}$, $\boldsymbol{r}$, $\boldsymbol{u}$ and $\boldsymbol{v}$ do not contain F and T, while $\boldsymbol{s}$ may contain solitary occurrences of both T and F (in other words, the T shown in the picture are the first two occurrences of T in $\boldsymbol{q}$ and the F are the last two occurrences of F in $\boldsymbol{q}$). Denote by $\boldsymbol{q_r}$ the suffix of $\boldsymbol{q}$ that starts from x but without $T(x)$; similarly, $\boldsymbol{q_u}$

is the suffix of q starting from z but without $F(z)$. Denote by q_r^- the prefix of q that ends at y but without $T(y)$; similarly, q_u^- is the prefix of q ending at w but without $F(w)$.

Theorem 5. *Answering any OMQ (Dis_A, q) with a 2-2-CQ q is* CONP-*complete provided the following conditions are satisfied:*

- *there is no homomorphism $h_1 \colon q_u \to q_r$ with $h_1(z) = x$;*
- *there is no homomorphism $h_2 \colon q_r^- \to q_u^-$ with $h_2(y) = w$.*

Proof. We prove CONP-hardness by reduction of the NP-complete 3SAT. Given a 3CNF ψ, we construct an ABox $\mathcal{A}_\psi$ as follows. First, for every literal ℓ whose propositional variable is present in ψ, we take the following ℓ-gadget that contains sufficiently many occurrences of A:

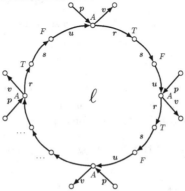

One can show that, for every model of Dis_A extending this ℓ-gadget, we have $\mathcal{I} \not\models q$ iff the A-points in the gadget are all in $T^{\mathcal{I}}$ or are all in $F^{\mathcal{I}}$.

Next, for every pair ℓ and $\neg\ell$ of literals as above, we connect the corresponding gadgets following the pattern in the picture below:

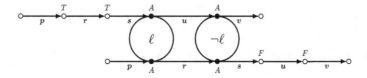

Now, one can show that, for every model of Dis_A extending this new gadget, we have $\mathcal{I} \not\models q$ iff either all A-points in the ℓ-gadget are in $T^{\mathcal{I}}$ and all A-points in the $\neg\ell$-gadget are in $F^{\mathcal{I}}$ or the other way round.

Finally, for every clause $c = (\ell_1 \vee \ell_2 \vee \ell_3)$ in ψ, we connect the $\neg\ell_1$-, ℓ_2- and ℓ_3-gadgets as shown below, always taking fresh A-points (by the construction, we have a sufficient supply of them):

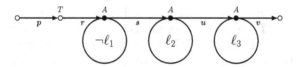

Denote the resulting structure by $\mathcal{A}_\psi$. We leave it to the reader to verify, using the properties of the gadgets mentioned above, that ψ is satisfiable iff $Dis_A, \mathcal{A}_\psi \not\models q$.

We do not know yet whether this theorem holds for $Dis_\top$ in place of Dis_A.

7   Conclusion

In this paper, we have obtained a few new results on the data complexity of answering a given ontology-mediated query (OMQ) that consists of a conjunctive query (CQ) and a covering axiom similar to the one used in the variant [16, Example 7] of the well-known 'Andrea example' [21]. We have observed that answering such OMQs is often tractable, with the respective OMQs being rewritable into standard datalog queries over the data. Sometimes we can even achieve rewritability into linear datalog, which guarantees OMQ answering in NL. We have given a few necessary and sufficient conditions for these phenomena. We have also discovered a few interesting counterexamples, in particular, a minimal CQ with solitary occurrences of both T and F that is FO-rewritable, a path CQ that is NL-complete for $Dis_\top$ but P-complete for Dis_A, and a path CQ with a solitary F in the middle and solitary Ts on either side of it that is NL-complete.

Acknowledgements. The work of O. Gerasimova and M. Zakharyaschev was carried out at the National Research University Higher School of Economics and supported by the Russian Science Foundation under grant 17-11-01294; the work of V. Podolskii was supported by the Russian Academic Excellence Project '5-100' and by grant MK-7312.2016.1.

References

1. Abiteboul, S., Hull, R., Vianu, V.: Foundations of Databases. Addison-Wesley, Menlo Park (1995)
2. Afrati, F.N., Gergatsoulis, M., Toni, F.: Linearisability on datalog programs. Theor. Comput. Sci. **308**(1—-3), 199–226 (2003). https://doi.org/10.1016/S0304-3975(02)00730-2
3. Artale, A., Calvanese, D., Kontchakov, R., Zakharyaschev, M.: The DL-Lite family and relations. J. Artif. Intell. Res. (JAIR) **36**, 1–69 (2009)
4. Baader, F., Horrocks, I., Lutz, C., Sattler, U.: An Introduction to Description Logic. Cambridge University Press, Cambridge (2017)
5. Bienvenu, M., ten Cate, B., Lutz, C., Wolter, F.: Ontology-based data access: a study through disjunctive datalog, CSP, and MMSNP. ACM Trans. Database Syst. **39**(4), 33:1–44 (2014)
6. Bienvenu, M., Ortiz, M.: Ontology-mediated query answering with data-tractable description logics. In: Faber, W., Paschke, A. (eds.) Reasoning Web 2015. LNCS, vol. 9203, pp. 218–307. Springer, Cham (2015). doi:10.1007/978-3-319-21768-0_9
7. Calvanese, D., De Giacomo, G., Lembo, D., Lenzerini, M., Rosati, R.: Tractable reasoning and efficient query answering in description logics: the DL-Lite family. J. Autom. Reasoning **39**(3), 385–429 (2007)

8. Calvanese, D., De Giacomo, G., Lembo, D., Lenzerini, M., Rosati, R.: Data complexity of query answering in description logics. Artif. Intell. **195**, 335–360 (2013). https://doi.org/10.1016/j.artint.2012.10.003

9. Feier, C., Kuusisto, A., Lutz, C.: Rewritability in monadic disjunctive datalog, MMSNP, and expressive description logics. CoRR abs/1701.02231 (2017). http://arxiv.org/abs/1701.02231

10. Gerasimova, O., Kikot, S., Podolskii, V., Zakharyaschev, M.: On the data complexity of ontology-mediated queries with a covering axiom. In: Proceedings of the 30th International Workshop on Description Logics (2017)

11. Gottlob, G., Papadimitriou, C.H.: On the complexity of single-rule datalog queries. Inf. Comput. **183**(1), 104–122 (2003). http://dx.doi.org/10.1016/S0890-5401(03)00012-9

12. Grau, B.C., Motik, B., Stoilos, G., Horrocks, I.: Computing datalog rewritings beyond horn ontologies. In: IJCAI 2013, Proceedings of the 23rd International Joint Conference on Artificial Intelligence, Beijing, China, August 3–9, 2013, pp. 832–838 (2013). http://www.aaai.org/ocs/index.php/IJCAI/IJCAI13/paper/view/6318

13. Hernich, A., Lutz, C., Ozaki, A., Wolter, F.: Schema.org as a description logic. In: Calvanese, D., Konev, B. (eds.) Proceedings of the 28th International Workshop on Description Logics, Athens, Greece, June 7–10, 2015, CEUR Workshop Proceedings, vol. 1350. CEUR-WS.org (2015). http://ceur-ws.org/Vol-1350/paper-24.pdf

14. Kaminski, M., Nenov, Y., Grau, B.C.: Datalog rewritability of disjunctive datalog programs and non-Horn ontologies. Artif. Intell. **236**, 90–118 (2016). http://dx.doi.org/10.1016/j.artint.2016.03.006

15. Kontchakov, R., Rodríguez-Muro, M., Zakharyaschev, M.: Ontology-based data access with databases: a short course. In: Rudolph, S., Gottlob, G., Horrocks, I., van Harmelen, F. (eds.) Reasoning Web 2013. LNCS, vol. 8067, pp. 194–229. Springer, Heidelberg (2013). doi:10.1007/978-3-642-39784-4_5

16. Kontchakov, R., Zakharyaschev, M.: An introduction to description logics and query rewriting. In: Koubarakis, M., Stamou, G., Stoilos, G., Horrocks, I., Kolaitis, P., Lausen, G., Weikum, G. (eds.) Reasoning Web 2014. LNCS, vol. 8714, pp. 195–244. Springer, Cham (2014). doi:10.1007/978-3-319-10587-1_5

17. Krisnadhi, A., Lutz, C.: Data complexity in the $\mathcal{EL}$ family of description logics. In: Dershowitz, N., Voronkov, A. (eds.) LPAR 2007. LNCS, vol. 4790, pp. 333–347. Springer, Heidelberg (2007). doi:10.1007/978-3-540-75560-9_25

18. Lutz, C., Sabellek, L.: Ontology-mediated querying with $\mathcal{EL}$: Trichotomy and linear datalog rewritabvility. In: Proceedings of the 30th International Workshop on Description Logics (2017)

19. Poggi, A., Lembo, D., Calvanese, D., De Giacomo, G., Lenzerini, M., Rosati, R.: Linking data to ontologies. J. Data Semant. X **4900**, 133–173 (2008)

20. Rosati, R.: The limits of querying ontologies. In: Schwentick, T., Suciu, D. (eds.) ICDT 2007. LNCS, vol. 4353, pp. 164–178. Springer, Heidelberg (2006). doi:10.1007/11965893_12

21. Schaerf, A.: On the complexity of the instance checking problem in concept languages with existential quantification. J. Intell. Inf. Syst. **2**, 265–278 (1993)

QSMat: Query-Based Materialization for Efficient RDF Stream Processing

Christian Mathieu[1(✉)], Matthias Klusch[2], and Birte Glimm[3]

[1] Computer Science Department, Saarland University, 66123 Saarbruecken, Germany
ChristianMathieu@gmx.net
[2] German Research Center for Artificial Intelligence, 66123 Saarbruecken, Germany
[3] Institute of Artificial Intelligence, University of Ulm, 89069 Ulm, Germany

Abstract. This paper presents a novel approach, QSMat, for efficient RDF data stream querying with flexible query-based materialization. Previous work accelerates either the maintenance of a stream window materialization or the evaluation of a query over the stream. QSMat exploits knowledge of a given query and entailment rule-set to accelerate window materialization by avoiding inferences that provably do not affect the evaluation of the query. We prove that stream querying over the resulting *partial* window materializations with QSMat is sound and complete with regard to the query. A comparative experimental performance evaluation based on the Berlin SPARQL benchmark and with selected representative systems for stream reasoning shows that QSMat can significantly reduce window materialization size, reasoning overhead, and thus stream query evaluation time.

1 Introduction

In many applications, such as machinery maintenance or social media analysis, large volumes of continuously arriving data must be processed in near-realtime. These data streams stem from sources such as thermometers, humidity or flow sensors, social media messages, price updates, news feeds and many others. Often, it is not only necessary to select and filter information from these streams, but also to infer implicit information with the aid of additional domain knowledge. This process of deriving implicit information (*reasoning*) and filtering (*querying*) is called *stream reasoning*. Inferences are drawn using a set of *inference rules*. While there exist increasingly mature solutions for reasoning in static ontologies, stream reasoning poses additional challenges due to the large volume and frequently unreliable and noisy nature of stream data and its transient nature. Stream data is often considered relevant only during a small time interval, called the *stream window*. Data leaving the window and inferences drawn from said data may thus become invalid over time, and newly arriving data may entail new inferences. A standard query language in this context is Continuous SPARQL (C-SPARQL [1]), an extension of SPARQL for continuous queries over streams of RDF data.

© Springer International Publishing AG 2017
P. Różewski and C. Lange (Eds.): KESW 2017, CCIS 786, pp. 159–174, 2017.
https://doi.org/10.1007/978-3-319-69548-8_12

Both reasoning and querying of the resulting *materialized stream window* can become significant runtime bottlenecks. Due to the practical relevance of the problem, there are many existing approaches to speed up either of these two components.

Incremental materialization approaches such as IMaRS [2] or SparkWave [7] attempt to speed up materialization by eliminating redundant recomputation between subsequent stream windows. As a result, they spend less time materializing each stream window, but they still materialize the window completely. The materialization might depend on the stream window interval defined by the query (or queries) being posed, but usually not on the specific query itself. This materialized window still contains *all* inferences that follow from window content (and static assertions) under the given entailment scheme. Many of these inferences might never be necessary to answer a concrete query, yet they are derived to guarantee completeness with regard to any query.

Query rewriting techniques such as StreamQR [4] exploit this, and compile ontological information into one or more rewritten queries. This has the advantage that window materialization is entirely avoided by encoding entailments directly into the query. To allow static rewriting (i.e. the rewriting happens upfront and not while stream data is processed), these approaches usually require that stream data does not contain schema knowledge.

Similar to query rewriting, our approach QSMat attempts to speed up reasoning by exploiting knowledge of the specific query, thus bridging querying and reasoning. In contrast to query rewriting, we retain the added flexibility and expressivity of explicit reasoning and we support custom, user-definable inference rules. To reduce window materialization overhead, we analyze background knowledge and the aforementioned provided inference rule-set during a static preprocessing step per query, and discard ontological information and rules that are provably never needed to satisfy a given query. This promises to reduce the problem size for subsequent materialization of each stream window. Query evaluation over each window then only requires the creation of a partial materialization, for which we prove completeness with regard to the query.

The remainder of this paper is structured as follows. Section 2 introduces QSMat, a novel approach to query based stream materialization. A comparative experimental evaluation is then presented in Sect. 3. Section 4 discusses related work, before we conclude in Sect. 5.

2   Query-Based Stream Materialization

In the following we give a broad overview of the idea, followed by the formal presentation of the algorithm itself in *Sect.* 2.2. Noteworthy properties of the algorithm are discussed in *Sect.* 2.3.

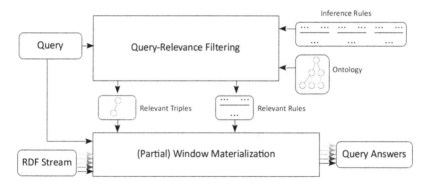

Fig. 1. The QSMat architecture

2.1 Overview

QSMat[1] is an approach for continuous query evaluation over RDF streams under reasoning. Its goal is accelerating reasoning for each stream window by suppressing inferences during window materialization that are not necessary to answer the query. In other words, it generates incomplete *partial window materializations* that are still sufficient to maintain completeness with respect to the query. Figure 1 illustrates the QSMat architecture. QSMat is given a query, a static ontology (in the form of triples) expressing background knowledge, and an inference rule-set. The algorithm extracts all triple patterns from the query's where clause, to analyze which triples of a future window materialization are potentially relevant. For each triple pattern, the algorithm guarantees that if a complete window materialization would have generated a triple matching the pattern, then the partial window materialization still generates that triple to maintain completeness. In a one-time per query preprocessing step, QSMat performs a backward search over inference rules to find all rules and all triples of the materialized static ontology (also called *static materialization*) that could be needed during window materialization to deduce query-relevant conclusion triples. Since schematic information is usually provided by background knowledge, but not from stream data (e.g. the stream usually does not define new subclasses), the user can tag rule premises that can only match schematic information. Such premises are called *cut-premises*. Using this cut-information, QSMat can restrict the relevance search further, since it can exclude inference trees which are not possible with the given background knowledge. This restricted search results in a more aggressive filtering that can exclude more unneeded inferences during subsequent partial window materializations (Fig. 1).

Afterwards, the algorithm has a triple and rule subset sufficient to materialize the triples needed for a triple pattern. By merging all of these per-pattern subsets, a set of triples and rules sufficient to answer the entire query is formed.

---

[1] Github: https://github.com/cmth/qsmat.

Algorithm 1. The main procedure of QSMat

1: **procedure** QSMAT
2:　　**Input:** Query string $\mathcal{Q}_{str}$, rule-set R, static ontology O, stream S
3:
4:　　Query $\mathcal{Q} \leftarrow \text{parse}(\mathcal{Q}_{str})$
5:　　$(M', R') \leftarrow \text{FILTERFORQUERY}(\mathcal{Q}, R, O)$
6:
7:　　**for each** stream window $w \in S$ **do**
8:　　　　$M_w \leftarrow$ materialization of $(M' \cup w)$ under R'
9:　　　　$\mathcal{Q}.\text{eval}(M_w)$
10:　　**end for**
11: **end procedure**

Subsequently, the stream window is materialized partially using this combined triple and rule-set and the query is evaluated over the result.

2.2　The QSMat Technique

The main procedure QSMAT is shown in Algorithm 1. At the highest level, a query string ($\mathcal{Q}_{str}$), a static ontology (O, interpreted as a set of triples), and a set of inference rules (R) are passed to QSMAT. Each rule $r \in R$ may have premise patterns tagged as a *cut-premise*, $r.P_{\text{cut}}$; other patterns are referred to as *live-premises*, denoted $r.P_{\text{live}}$. Both $r.P_{\text{cut}}$ and $r.P_{\text{live}}$ are treated as lists of patterns. The stream S is also treated as a parameter to ease notation. The procedure first parses the query string, yielding $\mathcal{Q}$, a representation of the abstract query tree. It then calls FILTERFORQUERY (cf. Algorithm 2). This function handles the relevance filtering and extracts a set M' of relevant triples from the materialization of the static ontology O, and a subset R' of relevant rules from the rule-set R needed for the query Q. The algorithm then listens to the stream and partially materializes each stream window w with M' and R', yielding M_w. Afterwards, M_w contains all triples potentially necessary to answer the query. The procedure then evaluates $\mathcal{Q}$ over M_w, handling the result as appropriate for the given query type.

The function FILTERFORQUERY (Algorithm 2) first computes the materialized schema M using the static ontology O and rule-set R. It then traverses the abstract query tree Q to find all triple patterns of the query. For each such triple pattern p, the function FILTERFORPATTERN (Algorithm 3) then extracts all triples from M and all rules from R that are (transitively) necessary to satisfy p, i.e. to create a partial materialization that still contains all triples of a complete materialization which match p. After FILTERFORPATTERN returns, FILTERFORQUERY aggregates the newly extracted relevant triples (in M') and rules (in R'). After all query triple patterns are processed, FILTERFORQUERY returns M' and R', which now contain all triples and rules necessary to satisfy any pattern of the query.

The function FILTERFORPATTERN (Algorithm 3) finds all triples in M and all rules in R necessary during window materialization to fulfill a given

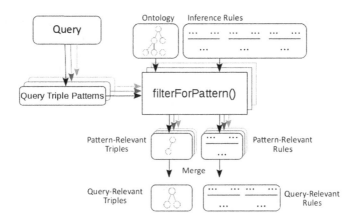

Fig. 2. FILTERFORQUERY concept

Algorithm 2. Extracts static ontology triples and rules relevant for a query

---

 1: **function** FILTERFORQUERY
 2:    **Input:** Query $\mathcal{Q}$, rule-set R, static ontology O
 3:    **Output:** Filtered static materialization M', filtered rule-set R'
 4:
 5:    $M' \leftarrow \emptyset, R' \leftarrow \emptyset$
 6:    $M \leftarrow$ materialization of O under R
 7:    **for each** triple pattern $p \in \mathcal{Q}$ **do**
 8:        $(M'', R'') \leftarrow$ FILTERFORPATTERN(R, M, p)
 9:        $M' \leftarrow M' \cup M''$
10:        $R' \leftarrow R' \cup R''$
11:    **end for**
12:    **return** (M', R')
13: **end function**

---

pattern p_{goal}. It does this by recursively backtracking over rules and finding the sets of triples that could match a rule premise in such a way that the rule can produce relevant output. Note that a pattern can be visualized as the (possibly infinite) set of all triples that match this pattern. By restricting a rule premise with a variable binding consistent with the rule conclusion, this restricted premise then subsumes all triples that potentially fulfill it during a rule application yielding a conclusion that matches the goal pattern (i.e. a potentially relevant pattern). A cut-premise can be thought of as the set of schema triples that may lead to relevant rule firings given the provided ontology, which in turn allows restricting live-premises to match only those triples from a stream that can possibly become relevant given the schema information. Since the latter is statically derivable, it is already explicitly contained in M and can be exhaustively enumerated (Fig. 2).

The function maintains a set P of patterns already processed *for the current query triple pattern* to avoid infinite recursion. If the current pattern p_{next} was not already processed before, then the function iterates over all rules that might yield triples matching p_{next}. Rules with no live-premises are skipped since they cannot fire due to triples derived during window materialization by definition. It then binds all variables in cut-premises of the rule r ($r.P_{cut}$) to matching ground terms in p_{next} and calls the function GROUNDRULE (Algoritm 4), which exhaustively backtracks over all restrictions of this rule given the triples in M matching the rule's cut-premises. In other words, it finds all ways this rule can be relevant during window materialization to produce a triple matching p_{next}, and thus transitively p_{goal}.

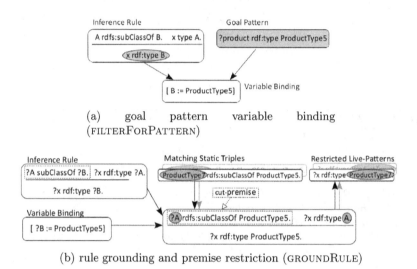

(a) goal pattern variable binding (FILTERFORPATTERN)

(b) rule grounding and premise restriction (GROUNDRULE)

Fig. 3. FILTERFORPATTERN rule predecessor search

The function GROUNDRULE (Algorithm 4) is responsible for finding all ways a rule r can fire yielding a relevant conclusion consistent with a supplied variable binding b. It then returns both the relevant triples from the static materialization and the transitive predecessor patterns needed in the derivation. In other words, it guarantees that all triples matching the goal pattern are still derivable during partial window materialization. It iterates over all combinations of static triples that consistently match all cut-premises, i.e. all ways the rule can fire given the static schema. To do so, it first finds all triples t in M matching the first cut-premise. It then extends the variable binding to be consistent with this triple and then subsequently processes the following cut-patterns in the same manner. Due to the refined variable binding, each following cut-pattern only matches static triples consistent with all previous patterns. For each consistent grounding, i.e. each way to satisfy *all* cut-patterns without disagreeing on variable bindings, a restricted live-pattern p' is generated for each live-premise by applying the variable binding. The pattern p' is then added to Q'. Each t involved in a consistent grounding is added to T'. Afterwards, T' and Q' contain the predecessors needed to guarantee relevant rule firings of r. The set T' contains all

Algorithm 3. Finds triples and rules relevant for a triple pattern

```
1: function FILTERFORPATTERN
2:     Input: Static materialization M, rule-set R, goal pattern p_goal
3:     Output: Relevant triples M'', relevant rules R''
4:
5:     P ← ∅, Q ← ∅, M'' ← ∅, R'' ← ∅
6:     Q.append(p_goal)
7:     while Q not empty do
8:         p_next ← Q.pop()
9:         if p_next ∉ P then
10:            P ← P ∪ {p_next}
11:            M'' ← M'' ∪ {t ∈ M | t matches p_goal}
12:            for each rule r ∈ R, |r.p_live| > 0 do          ▷ Skips static-only rules
13:                Binding b ← bind(r.p_c, p_next)
14:                if b ≠ ⊥ then  ▷ otherwise r cannot produce triples matching p_next
15:                    (T', Q') ← GROUNDRULE(M, r, b, 0)
16:                    if Q' ≠ ∅ then  ▷ Q' empty ⇒ r cannot yield t ∈ p_next given M
17:                        Q ← Q ∪ Q'
18:                        M'' ← M'' ∪ T'
19:                        R'' ← R'' ∪ {r}
20:                    end if
21:                end if
22:            end for
23:        end if
24:    end while
25:    return (M'', R'')
26: end function
```

static triples needed for cut-premises of r, while Q' contains patterns subsuming all triples that may fulfill a live-premise of r after being newly derived during window materialization. Both T' and Q' are then returned.

2.3 Correctness of QSMat

We next show that the partial materialization computed by QSMat is indeed sufficient to answer the given query. Since DESCRIBE queries are implementation dependent, we do not consider them here.

Theorem 1. *Let Q be any (non-DESCRIBE) C-SPARQL [1] query, R a monotonic inference rule-set, O a static ontology, and S a stream. Then O and S entail an answer to Q under R iff QSMat computes this answer given Q, R, O, and S as input.*

Proof (Sketch). The *if direction* corresponds to the *soundness* of QSMat: Note that the complete window materialization is the transitive closure of an inference rule-set on the union of static and window triples. Since the inference rules are monotonic and the filtered ontology and rule-set created by QSMat are subsets of O and R, respectively, soundness is trivial.

Algorithm 4. Finds predecessors necessary for a rule to fire and create relevant conclusions during partial window materialization

1: **function** GROUNDRULE
2:     **Input:** static materialization M, rule r, variable binding b, premise index i
3:     **Output:** relevant static triples T', relevant predecessor patterns Q'
4:
5:     **if** $i < |r.P_{\text{cut}}|$ **then**
6:         $p_{cut} \leftarrow$ apply b to $r.P_{\text{cut}}[i]$
7:         $T \leftarrow \{t \in M \mid t \text{ matches } p_{\text{cut}}\}$
8:         **for** $t \in T$ **do**
9:             $b' \leftarrow b \cup \text{bind}(p_{\text{cut}}, t)$
10:            $(T'', Q'') \leftarrow$ GROUNDRULE$(M, r, b', i+1)$
11:            **if** $Q'' \neq \emptyset$ **then**   ▷ Q'' empty $\Rightarrow$ grounding failed
12:                $T' \leftarrow T' \cup T'' \cup \{t\}$
13:                $Q' \leftarrow Q' \cup Q''$
14:            **end if**
15:        **end for**
16:    **else**   ▷ consistent grounding
17:        **for each** $p_{\text{live}} \in r.P_{\text{live}}$ **do**
18:            $p' \leftarrow$ apply b to p_{live}
19:            $Q' \leftarrow Q' \cup \{p'\}$
20:        **end for**
21:    **end if**
22:    **return** (T', Q') ▷ Note that Q' is empty if r cannot yield relevant output
23: **end function**

The *only if direction* corresponds to the *completeness* of QSMat: Since all C-SPARQL query types depend solely on triples matching query triple patterns (QTPs), we focus, w.l.o.g., on single QTPs. A QTP can be visualized as the set of all triples that would match this pattern. A triple of the complete materialization matching a QTP can only come from one or more of the following three sources: (i) It can be contained in the stream window directly, (ii) it can be entailed purely from static ontology triples, or (iii) it can be derived transitively from at least one stream triple and zero or more static triples. For (i), the QTP matches the triple since it is contained in the partial materialization because of Line 8 of QSMat. For (ii), the triple is marked by QSMat in Line 11 of FILTERFORPATTERN. Case (iii) covers newly derived query-relevant triples. If one such triple was missing from the partial window materialization, then there must have been at least one rule application that fired during complete materialization, but not during partial materialization. Hence, either (a) a rule was not marked or (b) a triple matching a rule premise was not contained in the partial materialization. For (a), the rule can only have fired during complete window materialization if it had at least one live-premise, by definition of live-premises.[2] Since the rule must have fulfilled all cut-premises to fire during complete window materialization, it has at least one

---

[2] Otherwise the stream would have to contain or imply schematic information that was specified as cut in the rule-set, which is a design error.

consistent grounding using the complete static materialization, thus it created at least one live-pattern in GROUNDRULE[3]. This means FILTERFORPATTERN added it to the set of relevant rules in Line 19. For (b), the only way to have a query-relevant triple missing from the partial materialization is if a rule entailed it, and was missing a premise triple. This premise triple matched either a cut-premise or a live-premise (or both, which is subsumed by the former two cases). If it was a cut-premise, the triple must have been part of a consistent grounding, and was contained in the complete static materialization. Thus it was added to the set of relevant triples by GROUNDRULE in Line 12. If it was a live-premise, it was subsumed by the live-premise created by GROUNDRULE in Line 18 and added to the set of relevant patterns in Line 19. In this case, the premise triple can either be a static triple, a window triple, or newly derived in the window materialization. The same argument above can be applied to the created live-pattern instead of the query triple pattern, which leads to a proof by structural induction over all possible inferences trees. This proof must terminate, since an inference tree that completed during complete materialization can only have finite depth and width. Either way, we have a contradiction with the assumption that the partial materialization employed by QSMat was not complete with regard to the query triple pattern. Since this holds for all query triple patterns, QSMat is complete with regard to the query.

3   Comparative Performance Evaluation

In this section we present a comparative experimental evaluation between QSMat and the selected state-of-the-art approaches C-SPARQL using Jena for materialization, SparkWave and StreamQR.

3.1   Experimental Setting

The core idea behind QSMat is avoiding unnecessary inferences by calculating a partial materialization that is sufficient to answer a given query. This can only yield noteworthy performance benefits if the query is independent of many inferences generated by a complete materialization[4] and if this is statically provable by QSMat. Conceptually, QSMat is similar both to query rewriting approaches, which attempt to speed up evaluation by analysis of the query, and to optimized reasoners, which attempt to reduce runtime overhead of the reasoning step.

Competitive approaches. To evaluate the feasibility of QSMat's query relevance filtering, we compare its performance to several state-of-the-art approaches with C-SPARQL [1,3] serving as a baseline. Since the C-SPARQL engine does not offer reasoning support as of the time of this writing, we extended it with a complete materialization step using Jena as a back-end. As a competitive reasoning

---

[3] That is why rules that never produced any restricted live-pattern during static search are not needed to satisfy the query, hence are safely removed by QSMat.

[4] As a pathological counterexample take a query with the where clause (?s ?p ?o): since the query matches all derivable triples, none can be excluded.

approach, `SparkWave` [7] was chosen, a fast stream reasoning engine using RETE-based inference [5] (the so-called *epsilon network*) and query pattern matching. Query rewriting approaches are represented by `StreamQR` [4], a stream reasoning engine that uses ontological background knowledge to translate the original query into a union of conjunctive queries, which explicitly express all ways query triple patterns can be satisfied under reasoning.[5]

Testing environment. The dataset used for testing is derived from the *Berlin SPARQL Benchmark* (BSBM). While BSBM is a SPARQL benchmark, and thus not stream oriented, it features a flexible data generator that allows for creating variable-size problem instances simulating e-commerce use-cases in the context of vendors offering products. To adapt these datasets to a suitable stream setting, offers and the underlying product data are treated as stream data, while the concept hierarchy of product types and all information not directly related to products and offers are used as background knowledge. To allow for comparable measurements between approaches, a test framework using `Jena` was created, where the triples contained in each window are first aggregated, and the complete window is then passed to a stream-enabled reimplementation of each approach. While this does effectively process data in a one-shot fashion per window, instead of an admittedly more natural streaming scheme, it allows much more transparent performance measurements on equal footing.

The BSBM benchmark defines parametric query templates, which are used to create concrete, randomized benchmark queries by sampling template parameters (e.g. %ProductFeature1%) over the corresponding input data (e.g. product features). For each such query instance, all algorithms are evaluated on identical background knowledge and identical stream window content, and for an equivalent query in the respective query language. This avoids sampling noise due to differing problem sizes or real-time-dependent window semantics. It further allows a direct measurement of the elapsed wall-clock time spent during query evaluation, is independent of triple arrival rate, and also offers scaling information for problem sizes where e.g. a throughput metric might reach the point of saturation first for given window sizes. Before the actual evaluation, a warm-up phase is performed. This gives the system time to stabilize, and reduces sampling noise e.g. due to initialization effects or at-runtime optimizations from the Java Just-In-Time compiler. The result times for individual query instances are averaged per query template. All tests were run on a system using an Intel Xeon W55990 3.33GHz CPU with 32GB of DDR3-1333 RAM.

3.2   Evaluation Results

Scalability with regard to ontology size and query. As mentioned above, performance of `QSMat` depends on the reasoning demands of the specific query. Product information generated by BSBM is classified in a product type hierarchy.

---

[5] To give an example: Assume a query matches (?x rdf:type A), and A has subclasses B and C. If no other way to derive membership in A, B or C exists, then it is sufficient to find all (?x rdf:type A), (?x rdf:type B) and (?x rdf:type C).

Leaf products in this hierarchy are classes without subclasses, and class membership of instances cannot be derived in any way except by explicit assertion, while product types further toward the root of the product type hierarchy have increasing numbers of subclasses and larger reasoning demands. As end-points of this, the root product type offers the least, while leaf product types offer the most opportunity for optimization, both for QSMat and for StreamQR. Deeper product hierarchies further increase the reasoning demands for product types near the root, while the opportunity for optimization in more shallow hierarchies is expected to be lower. A query template chosen to verify this assumption is shown in Listing 3.1.

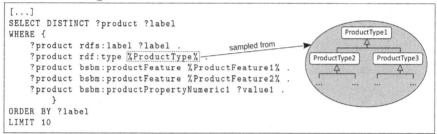

Listing 3.1: Query Template

The results of the static scaling test based on the query template in Listing 3.1 for the Jena-based implementations is shown in Fig. 4a and Fig. 4b for leaf and for root product types respectively. Both QSMat and StreamQR result in a substantial speedup compared to the completely materializing reference implemenation C-SPARQL, especially for leaf types. This is because for leaves, the complete materialization needs to derive all superclasses of each type in the product type hierarchy, although none of them are ever relevant to the query. Noteworthy is the extremely similar performance of QSMat and StreamQR. This is no coincidence. For this type of query, both QSMat and StreamQR converge to the same behavior: QSMat excludes all inference rules for this type of query, avoids materialization entirely, and then evaluates the query over the raw stream window content. StreamQR creates a trivially rewritten query that is identical to the original query, since there are no relevant schematic dependencies to be compiled into the query. While they arrive at this result differently, both algorithms evaluate the original query over the raw stream window, resulting in almost identical performance.

For root types, the performance of C-SPARQL is comparable to the leaf case, which is to be expected, since the cost of the complete materialization does not depend on the specific product type queried. QSMat still outperforms C-SPARQL, but to a lesser degree than for the leaf case, especially for large type hierarchies and thus higher reasoning demand per stream triple. While QSMat can still exclude most rules not needed for the query, all product types are relevant for this type of query. This means that QSMat cannot exclude any part of the product type hierarchy and needs to derive all superclass memberships. As a

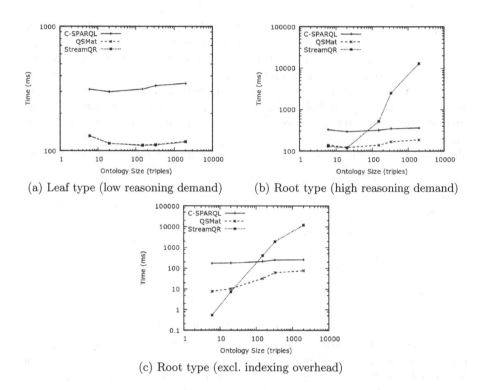

(a) Leaf type (low reasoning demand)   (b) Root type (high reasoning demand)

(c) Root type (excl. indexing overhead)

Fig. 4. Query evaluation time with regard to hierarchy size (Jena-based implementations, 100 k stream triples/window)

consequence, a higher fraction of inferences in C-SPARQL's complete materialization is relevant to the query, which is why QSMat's advantage compared to the complete materialization is less dramatic.

It is also apparent that the performance of StreamQR is far more sensitive to ontology size for this type of query instance. Since all product types are relevant when querying instances of the root type, a larger type hierarchy implies that StreamQR must create a larger number of sub-queries to encode subclass relations, which results in increasingly costly query evaluation. These sub-queries need to replicate triple patterns even if they are not related to the rewritten rdf:type pattern, which becomes quite costly for very large and deep hierarchies. As a consequence, its runtime increases with ontology size to a much larger degree than for the leaf case.

The indexing of stream triples employed by our Jena-based query processor dominated query evaluation time for small ontologies, which masked differences between the evaluated algorithms for small ontology sizes and thus low reasoning demands. To emphasize these differences, Fig. 4c shows the results for the more interesting root product type while excluding indexing overhead. This reveals that low hierarchy sizes are most beneficial to StreamQR, which then needs to create only a small number of sub-queries, leading to a performance advantage over the algorithms employing explicit reasoning. Results of the same static scaling test for a SparkWave-based reasoning backend with and without query-relevance filtering of background knowledge with QSMat are shown in Fig. 5a and b for leaf and root type products respectively. Since inference rules are fixed for SparkWave, the rule-set was not subjected to relevance-filtering. As above, leaf type queries allow QSMat to exclude more of the type hierarchy than root type queries, which translates into a larger performance benefit compared to regular SparkWave for leaf types. The fairly small performance benefit for root types can be explained by QSMat discarding background knowledge that is irrelevant to the query even for root types.

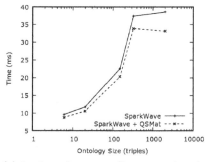
(a) Leaf product type (low reasoning demand)

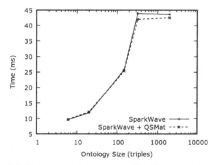
(b) Root product type (high reasoning demand)

Fig. 5. Query evaluation time with regard to hierarchy size (Sparkwave-based Implementations, 100 k stream triples/window)

Scalability with regard to stream window size. Another important consideration is scaling with regard to window size. While a larger ontology and more complex type hierarchy can require more reasoning overhead per stream triple, a larger window contains more stream triples triggering said reasoning. Since this benchmark solely requires independently classifying streamed instances in a static schema, reasoning demands per window triple do not depend on window size. The *query template* from above was reused for this test, but product types were sampled uniformly both among hierarchy levels and among types in each level, instead of the more pathological root and leaf types above. Figure 6a and b show the results for a small and a large static ontology and thus hierarchy size. The cost per stream triple varies between algorithms, and is dependent on the static ontology size, but all evaluated algorithms scale fairly consistently with growing window size, leading to the conclusion that per-triple reasoning demands are the primary factor regarding scaling.

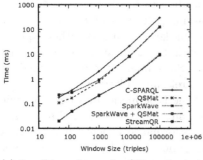
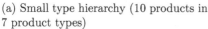

(a) Small type hierarchy (10 products in 7 product types)

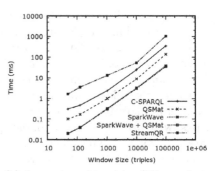

(b) Large type hierarchy (10k products in 329 product types)

Fig. 6. Query evaluation time with regard to hierarchy size (Sparkwave-based Implementations, 100 k stream triples/window)

4   Related Work

There are several existing approaches to accelerate stream reasoning. For our purposes, they can be broadly classified into approaches that focus either on faster query evaluation or reasoning.

The C-SPARQL [1,3] engine is a query execution engine for C-SPARQL queries. Its query semantics distinguish between *logical windows*, which are time-based, and *physical windows*, which contain a given number of triples. Logical windows are advanced with a given time-step and can either overlap (*sliding*), or perfectly tile (*tumbling*). Unlike QSMat, the currently available implementation does not include materialization.

The CQELS [8] engine is a query processor employing window semantics and query operators similar to C-SPARQL, including tumbling and sliding windows. Query evaluation is triggered by the arrival of new triples (*streaming*) and not at the end of each window as with C-SPARQL. The engine further supports reordering of query operators during query runtime to yield a less costly execution order. However, unlike QSMat, it provides no reasoning support.

SparkWave [7] uses a query language closely related to C-SPARQL and supports time-based windows. In contrast to C-SPARQL and CQELS, it supports reasoning for a subset of RDFS entailment plus inverse and symmetric properties. The RETE-based algorithm combines a reasoning layer with a pattern matching layer. Both reasoning and query matching is streaming, i.e. triggered by newly arriving triples. As a tradeoff for increased performance, its query format is more restrictive than that of C-SPARQL (and thus QSMat) and CQELS. Unlike QSMat, the fixed subset of RDFS entailment rule cannot be customized.

The Hoeksema S4 Reasoner [6] is a C-SPARQL engine that supports a fixed subset of the C-SPARQL query language and RDFS entailment. Unlike QSMat, it is implemented as a distributed system of processing elements that are realized in S4. The approach only supports time-based windows, since count-based windows would require more synchronization between individual processing elements.

IMaRS [2] is a C-SPARQL engine with reasoning support for generic rule systems and incrementally maintains a full materialization. The approach supports only logical windows, which allows it to exploit the specific window semantics of C-SPARQL to precompute expiration times for newly derived triples. However, QSMat only needs to perform a partial materialization depending on the needs to satisfy the given stream query.

StreamQR [4] is a query rewriting approach that aims to circumvent the need for explicit reasoning by encoding schematic knowledge in a set of conjunctive sub-queries. That is possible, since this conversion is syntactical and independent of the specific stream triples arriving during query runtime. The sub-queries are then evaluated over the raw stream without the need for reasoning. This is both an advantage and a disadvantage compared to materialization like in QSMat: The number of sub-queries generated depends on the complexity of the relevant background knowledge, hence it is most beneficial if this number is low. Both approaches perform analysis on a syntactic level, where StreamQR preprocesses the query with the help of relevant static information, while QSMat preprocesses static information with the help of the query.

5   Conclusions

We presented a flexible approach, QSMat, to accelerate stream reasoning by exploiting knowledge of the query and a configurable inference rule-set. It works backwards from the query, and extracts parts of the static materialization and the rule-set that are not provably irrelevant to the query, which allows it to maintain completeness with regard to the query. The performance evaluation of QSMat based on the Berlin SPARQL Benchmark (BSBM) revealed that it can reduce reasoning overhead significantly both on its own or as a preprocessing step for another state-of-the-art reasoner. Future work is concerned with cross-dependencies between multiple query triple patterns, which could yield a more restrictive filtering, thus exclude unneeded inferences that cannot provably be discarded on a per-pattern basis yet.

Acknowledgments. This research was partially supported by the German Federal Ministry for Education and Research (BMB+F) in the project INVERSIV and the European Commission in the project CREMA.

References

1. Barbieri, D.F., Braga, D., Ceri, S., Della Valle, E., Grossniklaus, M.: C-SPARQL: SPARQL for Continuous Querying. In: Proceedings of 18th International Conference on World Wide Web (WWW). ACM (2009)

2. Barbieri, D.F., Braga, D., Ceri, S., Della Valle, E., Grossniklaus, M.: Incremental reasoning on streams and rich background knowledge. In: Aroyo, L., Antoniou, G., Hyvönen, E., ten Teije, A., Stuckenschmidt, H., Cabral, L., Tudorache, T. (eds.) ESWC 2010. LNCS, vol. 6088, pp. 1–15. Springer, Heidelberg (2010). doi:10.1007/978-3-642-13486-9_1

3. Barbieri, D.F., Braga, D., Ceri, S., Della Valle, E., Grossniklaus, M.: C-SPARQL: a continuous query language for RDF data streams. Semant. Comput. **4**(1), 3–25 (2010)

4. Calbimonte, J.-P., Mora, J., Corcho, O.: Query rewriting in RDF stream processing. In: Sack, H., Blomqvist, E., d'Aquin, M., Ghidini, C., Ponzetto, S.P., Lange, C. (eds.) ESWC 2016. LNCS, vol. 9678, pp. 486–502. Springer, Cham (2016). doi:10.1007/978-3-319-34129-3_30

5. Forgy, C.L.: Rete: a fast algorithm for the many pattern/many object pattern match problem. Artif. Intell. **19**, 17–37 (1982)

6. Hoeksema, J., Kotoulas, S.: High-performance distributed stream reasoning using S4. In: Proceedings of Workshop OrdRing at International Semantic Web Conference (2011)

7. Komazec, S., Cerri, D., Fensel, D.: Sparkwave: continuous schema-enhanced pattern matching over RDF data streams. In: Proceedings of the 6th ACM International Conference on Distributed Event-Based Systems. ACM (2012)

8. Le-Phuoc, D., Dao-Tran, M., Xavier Parreira, J., Hauswirth, M.: A native and adaptive approach for unified processing of linked streams and linked data. In: Aroyo, L., Welty, C., Alani, H., Taylor, J., Bernstein, A., Kagal, L., Noy, N., Blomqvist, E. (eds.) ISWC 2011. LNCS, vol. 7031, pp. 370–388. Springer, Heidelberg (2011). doi:10.1007/978-3-642-25073-6_24

Employing Link Differentiation in Linked Data Semantic Distance

Sultan Alfarhood[1(✉)], Susan Gauch[2], and Kevin Labille[2]

[1] King Saud University, Riyadh, Saudi Arabia
sultanf@ksu.edu.sa
[2] University of Arkansas, Fayetteville, AR, USA
{sgauch,kclabill}@uark.edu

Abstract. The use of Linked Open Data (LOD) has been explored in recommender systems in different ways, primarily through its graphical representation. The graph structure of LOD is utilized to measure inter-resource relatedness via their semantic distance in the graph. The intuition behind this approach is that the more connected resources are to each other, the more related they are. The drawback of this approach is that it treats all inter-resource connections identically rather than prioritizing links that may be more important in semantic relatedness calculations. In this paper, we show that different properties of inter-resource links hold different values for relatedness calculations between resources, and we exploit this observation to introduce improved resource semantic relatedness measures, *Weighted Linked Data Semantic Distance (WLDSD)* and *Weighted Resource Similarity (WResim)*, which are more accurate than the current state of the art approaches. Exploiting these proposed weighted approaches, we also present two different ways to calculate links weights: *Resource-Specific Link Awareness Weights (RSLAW)* and *Information Theoretic Weights (ITW)*. To validate the effectiveness of our approaches, we conducted an experiment to identify the relatedness between musical artists in *DBpedia*, and it demonstrated that approaches that prioritize link properties resulted in more accurate recommendation results.

1 Introduction

Due to the massive amount of digital information available in recent years, it has become necessary to tailor the right information to the right user at the right time. Accordingly, new techniques and approaches have started to emerge that focus on matching information to users in order to help them make proper decisions. Some systems actively alert users to information or items that might be of interest to them; these methods are called recommender systems. Such systems have been embraced widely in various online platforms including commerce, news, and entertainment. There are numerous research works in this field that attempt to improve different aspects of recommender systems some of which are their recommendation accuracy, diversity, and novelty [1]. Researchers developing these systems continue to face several challenges, particularly a lack of a priori data needed in order for these systems to work appropriately. Several systems also lack sufficient semantic information about

© Springer International Publishing AG 2017
P. Różewski and C. Lange (Eds.): KESW 2017, CCIS 786, pp. 175–191, 2017.
https://doi.org/10.1007/978-3-319-69548-8_13

items, and semantic information about the relationships between items, so that related items can be accurately identified and recommended.

Because of its extensive offering of structured data in different domains, Linked Open Data (LOD) has been investigated in the field of recommender systems. In particular, LOD provides extensive open datasets containing multi-domain concepts with their relationships to each other, and these relations enable recommender systems to infer related concepts across collections [2]. Additionally, LOD standards and technologies ease the task of recommender systems by providing standard interfaces to retrieve the required data, eliminating the need for additional computing processing of raw data in addition to providing ontological knowledge of the data [3].

The use of LOD has been explored in recommender systems in different ways, primarily through exploiting its graph representation or through statistical approaches [4]. One approach that utilizes the graph structure of LOD in recommender system is to measure resources relatedness through their semantic distance in the graph [5–7]. The intuition behind this semantic distance approach is that the more connected resources are to each other in the LOD graph, the more related they are. This concept is the core of a resource relatedness measure, the Linked Data Semantic Distance (*LDSD*) [5], as well as a more recent measure based on it, Resource Similarity (*Resim*) [6]. The drawback of these approaches is that they treat all links between resources equally rather than prioritizing inter-resource links that may deliver additional value in semantic relatedness calculations. However, we argue in this paper that different properties of resources links hold different values for relatedness calculations. Furthermore, we exploit this observation to introduce improved resource semantic relatedness measures, *Weighted Linked Data Semantic Distance (WLDSD)* and *Weighted Resource Similarity (WResim)*, which are more accurate than the current state of the art approaches. Exploiting these proposed weighted approaches, we also present two different ways to calculate links weights: Resource-Specific Link Awareness Weights (*RSLAW*) and information theoretic weights (*ITW*). To validate the effectiveness of our approaches, we conducted an experiment to identify the relatedness between musical artists in *DBpedia*[1], and it demonstrated that approaches that prioritize link properties resulted in more accurate recommendation results.

The remainder of this paper is organized as follows: Sect. 2 presents related concepts and works followed by background information that this document adopts and improves in Sect. 3. Next, Sect. 4 presents our approaches. Afterward, how our proposed approaches are evaluated against current state of art approaches in Sect. 5. Finally, Sect. 6 offers a summary of this paper and presents some future work.

2 Related Work

Passant [5] suggests an approach to exploiting LOD in recommender systems by calculating the semantic distance between resources in the LOD. His approach, called Linked Data Semantic Distance (*LDSD*), uses direct links between resources in

---

[1] www.dbpedia.org.

addition to indirect resources through an intermediate resource to calculate a semantic distance between these resources. Utilizing this approach, Passant in [8] has created a music recommender system, called *dbrec*, that is built on top of the popular LOD provider, *DBpedia*, in order to recommend musical artists and bands. This system starts by reducing the LOD dataset to a compact one that enables efficient semantic distance computations. It then calculates this semantic distance for each pair of musical artists or bands. Finally, utilizing these distances, related artists are generated for the user. Utilizing the aforementioned concept, Piao et al. [6] introduced an improved linked data semantic distance approach called Resource Similarity (*Resim*) that revised the original *LDSD* approach, overcoming some its weaknesses such as equal self-similarity, symmetry, and minimality issues. They also improved their approach in [7] by applying different normalization methods that depend on the path appearances in the data set. The drawback of these approaches is that they handle all the resources connections equally and do not prioritize resources links that hold additional value in semantic relatedness calculations. Alfarhood et al. [9] suggests an approach that expands the coverage of LDSD-based approaches to include additional resources. They employ an all-pair shortest path algorithm, namely, the well-known Floyd-Warshall algorithm, to compute semantic distances based on resources beyond one or two links away.

Di Noia et al. [10, 11] show that content-based recommender systems can effectively use LOD to overcome issues of items that are described by limited content. They describe a content-based recommender system that employs LOD datasets, for instance, *DBpedia, Freebase,* and *LinkedMDB* to recommend movies. They utilize these LOD datasets to collect contextual information about movies such as actors, directors, and genres and then apply a content-based recommendation approach to generate recommendation results.

Using different tactics, Nguyen et al. [12] investigate the usage of two structural context similarity approaches of graphs in the field of LOD recommender systems. They found that these two metrics *SimRank* and *PageRank*, are promising in this field and can produce some novel recommendations, but with a high performance cost. Furthermore, Damljanovic et al. [2] present a concept recommender system based on LOD that assists users choosing proper concept tags and topics to improve their web search experience. They introduced a similarity-based approach relying on the relationship between concepts in the LOD graph. They also present another statistical-based method to calculate concept similarities and a comparison of both approaches to the *Google Adwords Keyword Tool*. They conclude that the graph-based method outperforms their baseline in relatedness measures while the statistical method came up with better-unexpected results. Correspondingly, Fernández-Tobías et al. [13] have developed a cross-domain recommender system that relies on LOD to link concepts from two different domains. They extract information about the two domains from the LOD sources and then link concepts using a graph-based distance between these concepts. Based on this approach, they developed in [14] a recommender system for the domains of architecture and music to suggest musical artists based on a selected location built on top of *DBpedia*.

Meymandpour and Davis [15] describe a LOD-based recommender system that combines semantic analysis of items with collaborative filtering approaches to overcome the item cold-start problem. They found that their semantic approach works well when combined with collaborative filtering methods to improve recommendations. The collaborative filtering is particularly helpful when there is limited information available in the user profile. Likewise, Heitmann and Hayes in [16] exploit LOD to overcome common collaborative-filtering challenges such as the new-user, new-item, and sparsity problems. Heitmann [17] has also developed an open framework for cross-domain personalization relying on the data representation in LOD. The LOD representation is used to model the user profile in addition to the system catalog of items which results in an open framework for recommender systems.

In another work, Peska and Vojtas in [18] show that LOD can be used effectively to enhance recommender systems in current e-commerce sites. They rely on LOD sources to fetch additional information about items in current systems in order for content-based recommender systems to work properly.

Clearly, there is a very active research community focusing on applying LOD sources to recommender systems. Our work builds on these projects but differs in that it takes the advantage of the LOD nature to improve current relatedness measures approaches through prioritizing some links that hold more relatedness value between the LOD resources.

3 Background

Linked Open Data (LOD) is designed as resources (nodes) connected semantically to each other via links (edges) as in a graph. This graph-based nature is essential to formally define LOD instances. This document adopts the same definition for LOD datasets as the one described in [5]:

A Linked Open Data dataset is a graph G such as $G = (R, L, I)$ in which:

$R = \{r_1, r_2, ..., r_x\}$ is a set of resources identified by their URI (Unique universal identifier)

$L = \{l_1, l_2, ..., l_y\}$ is a set of typed links identified by their URI

$I = \{i_1, i_2, ..., i_z\}$ is a set of instances of these links between resources, such as $i_i = <l_j, r_a, r_b>$

As connectivity between resources in the graph can show relatedness, the more connected the resources the more related they are. In this context, a direct connection between two resources exists when there is a direct distinct link (directional edge) between them. Therefore, we define a direct distance (C_d) between two resources r_a and r_b through a link with a property l_i is equal to one if there a link with a property l_i exists that connects the resource r_a to the resource r_b as follows:

$$C_d(l_i, r_a, r_b) = \begin{cases} 1 & \text{if the link } \langle l_i, r_a, r_b \rangle \text{ exists} \\ 0 & \text{otherwise} \end{cases} \tag{1}$$

Resources can also be indirectly connected through other resources. Indirect connectivity between two resources occurs when they are connected through another resource, and these connections can be either incoming or outgoing from the intermediate resource. Thus, there are two types of indirect connections: incoming and outgoing. An incoming indirect distance (C_{ii}) between two resources r_a and r_b is equal to one if there is a resource r_c such that r_c is directly connected to both r_a and r_b via links of property l_i as follows:

$$C_{ii}(l_i, r_c, r_a, r_b) = \begin{cases} 1 & \{\exists r_c | \langle l_i, r_c, r_a \rangle \& \langle l_i, r_c, r_b \rangle\} \\ 0 & \text{otherwise} \end{cases} \tag{2}$$

Similarly, an outgoing indirect distance (C_{io}) between two resources r_a and r_b is equal to one if there is a resource r_c such that both r_a and r_b are directly connected to r_c via links of property l_i as follows:

$$C_{io}(l_i, r_c, r_a, r_b) = \begin{cases} 1 & \{\exists r_c | \langle l_i, r_a, r_c \rangle \& \langle l_i, r_b, r_c \rangle\} \\ 0 & \text{otherwise} \end{cases} \tag{3}$$

The indirect distance notation can be generalized for all intermediate resources as follows[2]:

$$C_{ii}(l_i, r_a, r_b) = \sum_n C_{ii}(l_i, r_n, r_a, r_b)$$

$$C_{io}(l_i, r_a, r_b) = \sum_n C_{io}(l_i, r_n, r_a, r_b)$$

3.1   Linked Data Semantic Distance (LDSD)

Based on the previous definitions, [5] defines a measurement for relatedness between two resources in the LOD using the concepts of direct and indirect connections. This approach is called Linked Data Semantic Distance (*LDSD*), and it is calculated as follows:

$$LDSD(r_a, r_b) = \frac{1}{1 + DC(r_a, r_b) + DC(r_b, r_a) + IC_i(r_a, r_b) + IC_o(r_a, r_b)} \tag{4}$$

where $DC(r_a, r_b)$ is merely the direct distance (C_d) between the resources r_a and r_b normalized by the log of all outgoing links from r_a as follows:

$$DC(r_a, r_b) = \sum_i \frac{C_d(l_i, r_a, r_b)}{1 + log\left(\sum_n C_d(l_i, r_a, r_n)\right)}$$

$DC(r_b, r_a)$ is the direct distance (C_d) between the resources r_b and r_a normalized by the log of all outgoing links from r_b as follows:

---

[2] These versions of C_i accept three inputs instead of four as in the regular C_i.

$$DC(r_b, r_a) = \sum_i \frac{C_d(l_i, r_b, r_a)}{1 + log\left(\sum_n C_d(l_i, r_b, r_n)\right)}$$

$IC_i(r_a, r_b)$ is the incoming indirect distance (C_{ii}) between the resources r_a and r_b normalized by the log of all incoming indirect links from r_a as follows:

$$IC_i(r_a, r_b) = \sum_i \frac{C_{ii}(l_i, r_a, r_b)}{1 + log\left(\sum_n C_{ii}(l_i, r_a, r_n)\right)}$$

$IC_o(r_a, r_b)$ is the outgoing indirect distance (C_{io}) between the resources r_a and r_b normalized by the log of all outgoing indirect links from r_a as follows:

$$IC_o(r_a, r_b) = \sum_i \frac{C_{io}(l_i, r_a, r_b)}{1 + log\left(\sum_n C_{io}(l_i, r_a, r_n)\right)}$$

This *LDSD* approach is our first baseline in this document in which we will introduce link weights. We also discuss a second baseline in the next section.

3.2   Resource Similarity (Resim)

Resource Similarity (*Resim*) [6] is an improved linked data semantic distance approach that enhances the original *LDSD* approach by overcoming some of its weaknesses. *Resim* can be calculated as follows:

$$Resim(r_a, r_b) = \begin{cases} 0 & \text{if } URI(r_a) = URI(r_b) \text{ or } r_a \text{ owl : same As } r_b \\ LDSD_\gamma(r_a, r_b) & \text{if } LDSD_\gamma(r_a, r_b) \neq 1 \\ Property_{sim}(r_a, r_b) & \text{otherwise} \end{cases} \tag{5}$$

The linked data semantic distance $(LDSD_\gamma)$ component is calculated as follows:

$$LDSD_\gamma(r_a, r_b) = \frac{1}{1 + RDC(r_a, r_b) + RDC(r_b, r_a) + RIC_i(r_a, r_b) + RIC_o(r_a, r_b)} \tag{6}$$

where $RDC(r_a, r_b)$ is simply the direct distance (C_d) between the resources r_a and r_b normalized by the log of number of instances of link l_i as follows:

$$RDC(r_a, r_b) = \sum_i \frac{C_d(l_i, r_a, r_b)}{1 + log\left(\sum_m \sum_n C_d(l_i, r_m, r_n)\right)}$$

$RDC(r_b, r_a)$ is the direct distance (C_d) between the resources r_b and r_a normalized by the log of number of instances of link l_i as follows:

$$RDC(r_b, r_a) = \sum_i \frac{C_d(l_i, r_b, r_a)}{1 + log\left(\sum_m \sum_n C_d(l_i, r_m, r_n)\right)}$$

$RIC_i(r_a, r_b)$ is the incoming indirect distance (C_{ii}) between the resources r_a and r_b through a resource r_j normalized by the log of all incoming indirect links from r_j with a link property of l_i as follows:

$$RIC_i(r_a, r_b) = \sum_i \sum_j \frac{C_{ii}(l_i, r_j, r_a, r_b)}{1 + log\left(\sum_m \sum_n C_{ii}(l_i, r_j, r_m, r_n)\right)}$$

$RIC_o(r_a, r_b)$ is the outgoing indirect distance (C_{io}) between the resources r_a and r_b through a resource r_j normalized by the log of all outgoing indirect links from r_j with a link property of l_i as follows:

$$RIC_o(r_a, r_b) = \sum_i \sum_j \frac{C_{io}(l_i, r_j, r_a, r_b)}{1 + log\left(\sum_m \sum_n C_{io}(l_i, r_j, r_m, r_n)\right)}$$

In addition, $Property_{sim}$ calculates the similarity of shared links properties between the resources r_a and r_b if they are not directly or indirectly connected as follows:

$$Property_{sim}(r_a, r_b) = 1 - \left(\frac{\sum_i \left(\frac{C_{sip}(l_i, r_a, r_b)}{\sum_m \sum_n C_d(l_i, r_m, r_n)} \right)}{C_{ip}(r_a) + C_{ip}(r_b)} + \frac{\sum_i \left(\frac{C_{sop}(l_i, r_a, r_b)}{\sum_m \sum_n C_d(l_i, r_m, r_n)} \right)}{C_{op}(r_a) + C_{op}(r_b)} \right) \quad (7)$$

where $C_{sip}(l_i, r_a, r_b)$ is the number of shared incoming links of property l_i between the resources r_a and r_b. $C_{sop}(l_i, r_a, r_b)$ is the number of shared outgoing links of property l_i between the resources r_a and r_b. $\sum_m \sum_n C_d(l_i, r_m, r_n)$ represents the number of instances of the link property l_i. $C_{ip}(r_a)$ is the total number of incoming links to a resource r_a. $C_{op}(r_a)$ is the total number of outgoing links from a resource r_a.

The *Resim* measure is the second baseline we enhance in this document by introducing link weights. Our proposed enhancements are discussed in the following sections.

4 Approach

Links between different resources in LOD can be of different importance for recommender systems. Recognizing these differences could be vital in order to produce better recommendation results. For instance, singers who create a joint work (duet) together are likely more related to each other than singers who just share the same birth city since performing on the same work implies more similarities between these two artists. Therefore, the "collaboration" link based on the shared work is likely to have a higher impact on relatedness than a "born in" link and it should carry more weight for recommendation purposes. Therefore, it is our belief that links should be distinguished based on the level of relatedness between resources indicated by the link. However, both approaches *LDSD* and *Resim* treat all link properties equally, and they do not distinct between links that have no significant impact on recommendations and those that should influence recommendations. As a result, enhancements to *LDSD* and *Resim* are introduced by including a weighting factor that modifies the semantic distance value based on individual link importance for relatedness calculations.

4.1 Weighted Linked Data Semantic Distance (WLDSD)

The weighting factor is introduced to the first baseline, *LDSD*, as follows:

$$WLDSD(r_a, r_b) = \frac{1}{1 + WDC(r_a, r_b) + WDC(r_b, r_a) + WIC_i(r_a, r_b) + WIC_o(r_a, r_b)} \quad (8)$$

where $WDC(r_a, r_b)$ is the direct distance (C_d) weighted by W_{l_i} for each link with a property l_i as follows:

$$WDC(r_a, r_b) = \sum_i \left(\frac{C_d(l_i, r_a, r_b)}{1 + \log\left(\sum_n C_d(l_i, r_a, r_n)\right)} \times W_{l_i} \right)$$

$WDC(r_b, r_a)$ is the direct distance (C_d) weighted by W_{l_i} for each link with a property l_i as follows:

$$WDC(r_b, r_a) = \sum_i \left(\frac{C_d(l_i, r_b, r_a)}{1 + \log\left(\sum_n C_d(l_i, r_b, r_n)\right)} \times W_{l_i} \right)$$

$WIC_i(r_a, r_b)$ is the incoming indirect distance (C_{ii}) weighted by W_{l_i} for each link with a property l_i as follows:

$$WIC_i(r_a, r_b) = \sum_i \left(\frac{C_{ii}(l_i, r_a, r_b)}{1 + \log\left(\sum_n C_{ii}(l_i, r_a, r_n)\right)} \times W_{l_i} \right)$$

$IC_o(r_a, r_b)$ is the outgoing indirect distance (C_{io}) weighted by W_{l_i} for each link with a property l_i as follows:

$$WIC_o(r_a, r_b) = \sum_i \left(\frac{C_{io}(l_i, r_a, r_b)}{1 + \log\left(\sum_n C_{io}(l_i, r_a, r_n)\right)} \times W_{l_i} \right)$$

such that the value of every weight W_{l_i} is a positive rational number between zero and one ($0 \leq W_{l_i} \leq 1$).

The weighting factor W_{l_i} is introduced in the original *LDSD* method for every link-based operation. Therefore, higher direct and indirect distance values are generated for those links with a high weight (W_{l_i}); conversely, less emphasis is resulted on these links when their weight is low while some link properties are cancelled if their corresponding weight is zero.

4.2 Weighted Resource Similarity (WResim)

The weighting factor is also introduced to the second baseline in this document, *Resim*, as follows:

$$
WResim(r_a, r_b) = \begin{cases} 0 & if\ URI(r_a) = URI(r_b) or r_a owl: sameAs r_b \\ WLDSD_\gamma(r_a, r_b) & if\ WLDSD_\gamma(r_a, r_b) \neq 1 \\ Property_{sim}(r_a, r_b) & otherwise \end{cases}
$$

$$(9)$$

The weighted linked data semantic distance ($WLDSD_\gamma$) component is calculated as follows:

$$
WLDSD_\gamma(r_a, r_b) = \frac{1}{1 + WRDC(r_a, r_b) + WRDC(r_b, r_a) + WRIC_i(r_a, r_b) + WRIC_o(r_a, r_b)}
$$

$$(10)$$

where $WRDC(r_a, r_b)$ is simply the direct distance (C_d) between the resources r_a and r_b normalized by the log of number of instances of link l_i weighted by W_{l_i} for each link with a property l_i as follows:

$$
WRDC(r_a, r_b) = \sum_i \left(\frac{C_d(l_i, r_a, r_b)}{1 + \log\left(\sum_m \sum_n C_d(l_i, r_m, r_n)\right)} \times W_{l_i} \right)
$$

$WRDC(r_b, r_a)$ is the direct distance (C_d) between the resources r_b and r_a normalized by the log of number of instances of link l_i weighted by W_{l_i} for each link with a property l_i as follows:

$$
WRDC(r_b, r_a) = \sum_i \left(\frac{C_d(l_i, r_b, r_a)}{1 + \log\left(\sum_m \sum_n C_d(l_i, r_m, r_n)\right)} \times W_{l_i} \right)
$$

$WRIC_i(r_a, r_b)$ is the incoming indirect distance (C_{ii}) between the resources r_a and r_b through a resource r_j normalized by the log of all incoming indirect links from r_j with a link property of l_i weighted by W_{l_i} for each link with a property l_i as follows:

$$
WRIC_i(r_a, r_b) = \sum_i \sum_j \left(\frac{C_{ii}(l_i, r_j, r_a, r_b)}{1 + \log\left(\sum_m \sum_n C_{ii}(l_i, r_j, r_m, r_n)\right)} \times W_{l_i} \right)
$$

$WRIC_o(r_a, r_b)$ is the outgoing indirect distance (C_{io}) between the resources r_a and r_b through a resource r_j normalized by the log of all outgoing indirect links from r_j with a link property of l_i weighted by W_{l_i} for each link with a property l_i as follows:

$$WRIC_o(r_a, r_b) = \sum_i \sum_j \left(\frac{C_{io}(l_i, r_j, r_a, r_b)}{1 + \log\left(\sum_m \sum_n C_{io}(l_i, r_j, r_m, r_n)\right)} \times W_{l_i} \right)$$

such that the value of every weight W_{l_i} is a positive rational number between zero and one ($0 \le W_{l_i} \le 1$).

Similar to *WLDSD*, the weighting factor W_{l_i} is introduced in the original *Resim* method for every link-based operation and, therefore, higher direct and indirect distance values are generated for those links with a high weight (W_{l_i}).

5   Link Weights Computation

The previous section raises a critical question: how to measure the weight of each link (W_{l_i})? This section introduces two approaches to this calculation: Resource-Specific Link Awareness Weights (*RSLAW*) and Information Theoretic Weights (*ITW*).

5.1   Resource-Specific Link Awareness Weights (RSLAW)

LOD resources are connected using different link properties, and most of these properties are used to connect different classes of resources. For example, the link property "genre" can be used to connect artists to their corresponding genre, and it can be used to link a song or a movie to its corresponding genre too. On the other hand, some link properties tend to be very specific in only connecting resources within similar classes such as the link property "*associatedMusicalArtist*" that is mostly used to connect musical artists to each other. It is our belief that link properties that are typically used to link specific resource classes together indicate more relatedness than links used to connect a wide variety of resource classes. Based on this intuition, LOD link weights can be generated to emphasize those links that are specific to particular resources.

As R is already defined as the set of all resources in the linked data dataset, a recommender system can define a subset of R to indicate those resources that the recommender system is interested to include in the recommendation process. Formally, γ is a set of resources with a resource class intended for recommendation specified by the recommender system ($\gamma \subseteq R$).

In this approach, the weight of a link l_x is the probability that this link is associated with γ. In particular, the weight of a link l_x is the total number of instances of the link l_x between resources r_i and r_j that belong to a specific resource class set (γ) divided by the total number of instances of the link l_x between all resources regardless of their resource class as follows:

$$W_{l_x} = \frac{\sum_i \sum_j C_d(l_x, r_i, r_j)}{\sum_m \sum_n C_d(l_x, r_m, r_n)}, \left\{ \forall r_i, r_j | r_i \in \gamma \text{ and } r_j \in \gamma \right\} \tag{11}$$

To illustrate this approach, Table 1 shows the number of link property instances in a LOD dataset. The total number of instances of each link property in the whole dataset is shown in the left portion of the table whereas the right portion of the table shows the number of instances of each link property between specific resource classes only ("*dbpedia:MusicalArtist*" or "*dbpedia:MusicalBand*"), therefore γ is the set of all resources with class of ("*dbpedia:MusicalArtist*" or "*dbpedia:MusicalBand*"). For illustration, we classify link properties in this example into three categories: highly resource-specific link properties such as *associatedMusicalArtist* (95/100) and *associatedBand* (70/75), poorly resource-specific link properties such as *influencedBy* (10/50) and *relative* (5/25), in addition to generic link properties *occupation* (0/10), *hometown* (0/6). This approach prioritizes highly resource-specific link properties as they carry more value between those resources whereas generic link properties tend to describe general information about all classes of resources.

Table 1. RSLAW example

(a) Total link property instances		(b) Resource-specific link property instances	
Link Property	Count	Link Property	Count
associatedMusicalArtist	100	*associatedMusicalArtist*	95
associatedBand	75	*associatedBand*	70
influencedBy	50	*influencedBy*	10
relative	25	*relative*	5
occupation	10	*occupation*	0
hometown	6	*hometown*	0

5.2 Information Theoretic Weights (ITW)

The second approach to calculating the link weights is inspired from the well-known method from the information retrieval field, *TF-IDF* (Term Frequency–Inverse Document Frequency) [19], which is used to weight the importance of a term in a document within a pool of documents. In our scope, the importance of a link to a resource is assessed based on the entire collection of resources and links in the LOD dataset. Unlike the *RSLAW* approach which considers only the links distribution in the dataset, the weights in this approach are dynamically calculated and take into consideration the relationship between the link and other links in the dataset in addition to the relationship between the link and the resources linked to. In contrast, the disadvantage of this approach is that the whole LOD dataset must be traversed in order to compute the weights for each link which can be heavy on computing resources. Yet, this value can be computed once and stored in a preprocessing step, then integrated into the LOD engine to use the weights on the fly as needed.

Since the weights are calculated dynamically, they are referred here as $W(l_x, r_a, r_b)$ instead of W_{l_i} because they require all the link information for their calculation. Additionally, this approach results in weights that do not meet our proposed constraints of the link weights range ([0–1]); therefore, rescaling these values back into this range is required as discussed later in this section. Initially, the non-scaled information theoretic weights $W_{ns}(l_x, r_a, r_b)$ are calculated as follows:

$$W_{ns}(l_x, r_a, r_b) = LF(l_x, r_a, r_b) \times IRF(l_x, r_a, r_b) \qquad (12)$$

In this formula[3], the link frequency $LF(l_x, r_a, r_b)$ is the average normalized frequency of the link l_x that connects either r_a or r_b to others. This normalized frequency is calculated as the total number of both incoming and outgoing links of a property l_x to either the resource r_a or r_b normalized by the total number of both incoming and outgoing links to either the resource r_a or r_b as follows:

$$LF(l_x, r_a, r_b) = \frac{\left(\dfrac{\sum_j C_d(l_x, r_a, r_j) + \sum_j C_d(l_x, r_j, r_a)}{\sum_i \sum_j C_d(l_i, r_a, r_j) + \sum_i \sum_j C_d(l_i, r_j, r_a)}\right) + \left(\dfrac{\sum_j C_d(l_x, r_b, r_j) + \sum_j C_d(l_x, r_j, r_b)}{\sum_i \sum_j C_d(l_i, r_b, r_j) + \sum_i \sum_j C_d(l_i, r_j, r_b)}\right)}{2}$$

The inverse resource frequency $IRF(l_x, r_a, r_b)$ is the total number of resources in the LOD dataset intended for recommendation ($|\gamma|$) divided by total instances of the link l_x as follows:

$$IRF(l_x, r_a, r_b) = \log \frac{\sum_i r_i}{\sum_i \sum_j C_d(l_x, r_i, r_j)}, \{\forall r_i | r_i \in \gamma\}$$

Finally, weights calculated using this approach must be rescaled back in the range [0–1] as follows:

$$W(l_x, r_a, r_b) = \frac{W_{ns}(l_x, r_a, r_b) - min}{max - min} \qquad (13)$$

where *min* is the value of the minimum calculated weight, and *max* is the maximum value.

6   Evaluation

In order to assess whether or not our proposed approaches are effective, we conducted several experiments to measure their effectiveness against three baselines *LDSD*, *Resim*, and *Jaccard Index*. The *Jaccard Index* [20], also called the *Jaccard Similarity Coefficient*, is a statistical measure to estimate the similarity between two sets. It is calculated as the number of items shared by the sets divided by the number of items in either set as follows:

$$\text{Jaccard}(r_a, r_b) = \frac{|N(r_a) \cap N(r_b)|}{|N(r_a) \cup N(r_b)|} \qquad (14)$$

such that $N(r_a)$ is the set of neighbor resources to a resource r_a which is directly linked to each member of the set.

---

[3] We have tried different variations of this approach, and we report the best performing version only.

Like several related works in this field [5, 8, 12], we applied these experiments in the music domain to measure the relatedness between musical artists and bands. The following sections detail the dataset of the experiments followed by their methodology.

6.1 Dataset

We conducted the experiment using a dataset from the second Linked Open Data-enabled recommender systems challenge[4]. This dataset was built using Facebook profiles by collecting personal preferences (likes) for items in several domains. It contains the preferences of 52,069 users and it includes 5751 distinct resources in the music domain mapped to their corresponding resources in *DBpedia* (with the resource class "*dbpedia:MusicalArtist*" or "*dbpedia:MusicalBand*"). The total number of users' musical preferences in this dataset was 1,013,973 with an average of 19.47 preferences per user. The semantic distance calculation between resources in the previously mentioned dataset was conducted on a live *DBpedia* server (version 2015-10)[5].

6.2 Methodology

Similar to the approach taken by previous studies [6, 7], we randomly selected 500 users who have at least 10 preferences from the aforementioned dataset. Five preferences per user were reserved for testing purposes while the other preferences were used to build a user profile for each user. Next, we generated a list of recommended resources for each user based on the similarity of the user and every resource in the dataset. The similarity score between each user and resource is calculated based on the semantic distance generated by every approach as in the following:

$$similarity(u_i, r_a) = \frac{\sum_{r_b \in Profile(u_i)} (1 - SemanticDistance(r_a, r_b))}{|Profile(u_i)|} \quad (15)$$

where *SemanticDistance*(r_a, r_b) is the semantic distance approach used for the evaluation (*LDSD, WLDSD, Resim,* or *WResim*). *Profile*(u_i) is the user profile of the user u_i which contains all resources that the user u_i has liked minus five resources (those used for testing purposes).

The resulting list of resources was sorted in a descending order per user, and then testing resources were used to measure the effectiveness of each semantic distance approach using the standard metrics F_1 Score and the Mean Reciprocal Rank (MRR). The F1 score, the harmonic mean of precision and recall, is calculated as follows:

$$F_1 = 2 \times \frac{precision \times recall}{precision + recall} \quad (16)$$

---

[4] http://sisinflab.poliba.it/events/lod-recsys-challenge-2015/dataset/.
[5] http://wiki.dbpedia.org/Downloads2015-10.

Also, the Mean Reciprocal Rank (MRR) that takes into account how early a relevant result appears within ranked results is calculated as follows:

$$MRR = \frac{\sum_{i=1}^{|Q|} \frac{1}{rank_i}}{|Q|} \tag{17}$$

where $rank_i$ is the highest rank of relevant results in a query Q_i.

6.3 Results

We ran the experiment on both baselines *LDSD* (formula 4) and *Resim* (formula 5) in addition to their weighted versions (formulas 8 and 9) using three weighting approaches *RSLAW* (formula 11), *ITW* weights (formula 13), and random weights. The latter is selected as a baseline for the weighting approaches and to analyze their effects on the original baselines. Table 2 shows the results of the experiment using the F_1 score and MRR metrics on all approaches. The F_1 score values are presented at different ranked results cutoffs, i.e., 5, 10, and 20. In this table, the best results are shown in bold.

Table 2. Evaluation results for all LDSD-based and Resim-based approaches

	Jaccard	LDSD-based approaches				Resim-based approaches			
		LDSD	Weighted (*WLDSD*)			Resim	Weighted (*WResim*)		
			RSLAW	*ITW*	*Random*		*RSLAW*	*ITW*	*Random*
MRR	0.010	0.028	0.036	0.029	0.026	0.037	**0.040**	0.038	0.035
F_1@5	0.009	0.031	0.041	0.033	0.016	0.049	**0.052**	0.049	0.046
F_1@10	0.011	0.044	0.048	0.045	0.043	0.053	**0.054**	**0.054**	0.048
F_1@20	0.012	0.046	0.053	0.052	0.051	0.051	**0.054**	0.051	0.046

The first conclusion we can draw from this experiment is that our basic baseline, *Jaccard Index*, indeed scored the lowest among all the metrics (F_1 and MRR). The MRR of the *Jaccard Index* was 0.010 whereas it was 0.028 and 0.037 for *LDSD* and *Resim* respectively. *LDSD* and *Resim* approaches which take resource connections in the LOD graph into considerations performs better in LOD resource similarity approaches. Also, the *Resim* baseline outperforms the *LDSD* baseline among all metrics (MRR of 0.037 for *Resim* versus 0.028 for *LDSD*), confirming results reported by [7].

Second, our weighted approaches using either *RSLAW* or *ITW* weights outperform all baselines *Jaccard Index*, *LDSD* and *Resim* in all metrics. These improvements were statistically significant ($p < 0.05$) based on a paired student t-test. The MRR was 0.036 for our *WLDSD-RSLAW* approach versus 0.028 for the non-weighted *LDSD* approach, an improvement of 29%, whereas it was 0.040 for *WResim-RSLAW* versus 0.037 for the original *Resim*, an 8% improvement.

Confirming the MRR metric, the F_1 score of *WLDSD-RSLAW* was 0.041 for the top five results versus a score of 0.031 for *LDSD*, an improvement of 32% while it was 0.052 for *WResim-RSLAW* versus 0.049 versus the unweighted *Resim*, an improvement of 6%. These results also hold at other results cutoff points as displayed in Fig. 1. Even

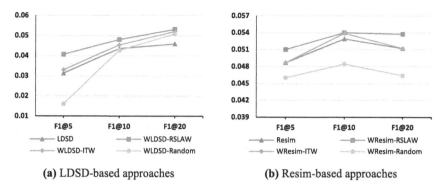

(a) LDSD-based approaches   **(b)** Resim-based approaches

Fig. 1. F_1 scores for all approaches

though the improvement rate in *WResim-RSLAW* is not as large as it is in *WLDSD-RSLAW*, it achieved the most accurate recommendation results among all approaches in the experiment.

The third conclusion we can draw from this experiment is that the *RSLAW* weights produced a bigger improvement than the *ITW* weights in both weighted approaches *WLDSD* and *WResim*. The information theoretic weights approach performed slightly better than the baselines, *LDSD* (MRR of 0.029 vs 0.028) and *Resim* (MRR of 0.038 vs 0.037) but worse than the *RSLAW* weighting approach in general. Even though it was not as accurate as the *RSLAW* approach, it still confirms the importance of exploiting the link properties in order to achieve better recommendation results.

Lastly, the experiment results also demonstrate that using random weights in both *WLDSD-Random* and *WResim-Random* results in reduced accuracy against both baselines *LDSD* (MRR of 0.026 vs 0.028) and *Resim* (MRR of 0.035 vs 0.037). This observation also holds at all the F_1 results cutoffs (@5, @10, and @20), and it confirms that the higher accuracy achieved by the *RSLAW* weights was not due to chance.

Overall, the results demonstrate that, although both baselines (*LDSD* & *Resim*) and their weighted variations (*WLDSD* & *WResim*) calculate the semantic distance between resources using the same underlying techniques, our approaches that weight links differentially provide increased accuracy. Also, weighting links using the *RSLAW* approach based on their association with specific classes of resources enables us to identify, and incorporate, latent semantic correlations between links and entities. *WLDSD* and *WResim* demonstrate that links play different roles and should be exploited in any semantic relatedness process for further accurate results.

7   Conclusion and Future Work

In this paper, we showed that different properties of resources links hold different values for relatedness calculations. We exploited this observation to introduce improved resource semantic relatedness measures, *WLDSD* and *WResim*, which are more accurate than the current state of the art approaches with two different ways to calculate links weights *RSLAW* and *ITW*. To verify our observations, we conducted an experiment in the music domain, and its results showed that the accuracy of our proposed weighted approaches (*WLDSD* and *WResim*) outperformed their original non-weighted versions. Furthermore, our *WResim* with *RSLAW* weights achieved the most accurate recommendation results among other approaches in the experiment.

In the future, we will explore different ways to calculate the links weights. One possible approach is to combine link property nature with path-based normalization to achieve higher relatedness accuracy. Additionally, we will explore expanding current semantic distance techniques to more than one hub away resources to include resources that are not reachable by the current radius of semantic distance approaches.

References

1. Cremonesi, P., Tripodi, A., Turrin, R.: Cross-domain recommender systems. In: Proceedings of the 2011 IEEE 11th International Conference on Data Mining Workshops, Washington, DC, USA, pp. 496–503 (2011)
2. Damljanovic, D., Stankovic, M., Laublet, P.: Linked data-based concept recommendation: comparison of different methods in open innovation scenario. In: Simperl, E., Cimiano, P., Polleres, A., Corcho, O., Presutti, V. (eds.) ESWC 2012. LNCS, vol. 7295, pp. 24–38. Springer, Heidelberg (2012). doi:10.1007/978-3-642-30284-8_9
3. Di Noia, T., Ostuni, V.C.: Recommender Systems and Linked Open Data. In: Faber, W., Paschke, A. (eds.) Reasoning Web 2015. LNCS, vol. 9203, pp. 88–113. Springer, Cham (2015). doi:10.1007/978-3-319-21768-0_4
4. Figueroa, C., Vagliano, I., Rocha, O., Morisio, M.: A systematic literature review of Linked Data-based recommender systems. Concurrency Comput. Pract. Exper. **27**(17), 4659–4684 (2015)
5. Passant, A.: Measuring semantic distance on linking data and using it for resources recommendations. In: AAAI Spring Symposium: Linked Data Meets Artificial Intelligence, vol. 77, p. 123 (2010)
6. Piao, G., showkat Ara, S., Breslin, J.: Computing the semantic similarity of resources in DBpedia for recommendation purposes. In: Joint International Semantic Technology Conference, pp. 185–200 (2015)
7. Piao, G., Breslin, J.: Measuring semantic distance for linked open data-enabled recommender systems. In: Proceedings of the 31st Annual ACM Symposium on Applied Computing, pp. 315–320 (2016)
8. Passant, A.: dbrec — music recommendations using DBpedia. In: Patel-Schneider, P.F., Pan, Y., Hitzler, P., Mika, P., Zhang, L., Pan, J.Z., Horrocks, I., Glimm, B. (eds.) ISWC 2010. LNCS, vol. 6497, pp. 209–224. Springer, Heidelberg (2010). doi:10.1007/978-3-642-17749-1_14
9. Alfarhood, S., Labille, K., Gauch, S.: PLDSD: propagated linked data semantic distance. In: 2017 IEEE 26th International Conference on Enabling Technologies: Infrastructure for Collaborative Enterprises (WETICE), Poznan, Poland, pp. 278–283 (2017)
10. Di Noia, T., Mirizzi, R., Ostuni, V., Romito, D.: Exploiting the web of data in model-based recommender systems. In: Proceedings of the Sixth ACM Conference on Recommender Systems, pp. 253–256 (2012)
11. Di Noia, T., Mirizzi, R., Ostuni, V., Romito, D., Zanker, M.: Linked open data to support content-based recommender systems. In: Proceedings of the 8th International Conference on Semantic Systems, pp. 1–8 (2012)

12. Nguyen, P., Tomeo, P., Di Noia, T., Di Sciascio, E.: An evaluation of SimRank and personalized PageRank to build a recommender system for the web of data. In: Proceedings of the 24th International Conference on World Wide Web, pp. 1477–1482 (2015)

13. Fernández-Tobías, I., Cantador, I., Kaminskas, M., Ricci, F.: A generic semantic-based framework for cross-domain recommendation. In: Proceedings of the 2nd International Workshop on Information Heterogeneity and Fusion in Recommender Systems, pp. 25–32 (2011)

14. Kaminskas, M., Fernández-Tobías, I., Ricci, F., Cantador, I.: Knowledge-based music retrieval for places of interest. In: Proceedings of the Second International ACM Workshop on Music Information Retrieval with User-Centered and Multimodal Strategies, pp. 19–24 (2012)

15. Meymandpour, R., Davis, J.G.: Enhancing recommender systems using linked open data-based semantic analysis of items. In: Davis, J. (ed.) 3rd Australasian Web Conference (AWC 2015), pp. 11–17 (2015)

16. Heitmann, B., Hayes, C.: Using linked data to build open, collaborative recommender systems. In: AAAI Spring Symposium: Linked Data Meets Artificial Intelligence, pp. 76–81 (2010)

17. Heitmann, B.: An open framework for multi-source, cross-domain personalisation with semantic interest graphs. In: Proceedings of the Sixth ACM Conference on Recommender Systems, pp. 313–316 (2012)

18. Peska, L., Vojtas, P.: Using linked open data in recommender systems. In: Proceedings of the 5th International Conference on Web Intelligence, Mining and Semantics, pp. 17:1–17:6 (2015)

19. Salton, G., McGill, M.: Introduction to Modern Information Retrieval. McGraw-Hill Inc., New York (1983)

20. Jaccard, P.: Etude de la distribution florale dans une portion des Alpes et du Jura. Bulletin de la Societe Vaudoise des Sciences Naturelles **37**(142), 547–579 (1901)

Ontologies and Controlled Vocabularies

Ontology for Representing Human Needs

Soheil Human[1,2(✉)], Florian Fahrenbach[1], Florian Kragulj[1],
and Vadim Savenkov[1]

[1] Institute for Information Business,
Vienna University of Economics and Business (WU Wien), Vienna, Austria
`soheil.human@univie.ac.at,`
`{soheil.human,florian.fahrenbach,florian.kragulj,vadim.savenkov}@wu.ac.at`
[2] Department of Philosophy, University of Vienna, Vienna, Austria

Abstract. Need satisfaction plays a fundamental role in human well-being. Hence understanding citizens' needs is crucial for developing a successful social and economic policy. This notwithstanding, the concept of need has not yet found its place in information systems and online tools. Furthermore, assessing needs itself remains a labor-intensive, mostly offline activity, where only a limited support by computational tools is available. In this paper, we make the first step towards employing need management in the design of information systems supporting participation and participatory innovation by proposing OPENEED, a family of ontologies for representing human needs data. As a proof of concept, OPENEED has been used to represent, enrich and query the results of a needs assessment study in a local citizen community in one of the Vienna districts. The proposed ontology will facilitate such studies and enable the representation of citizens' needs as Linked Data, fostering its co-creation and incentivizing the use of Open Data and services based on it.

Keywords: Human needs ontology · OPENEED · Needs · Satisfiers · Need studies · Representation · Linked data

1 Introduction

Needs form an essential basis for human well-being [38]. It has been argued that addressing needs is an effective approach to guide organizational change [40], increase employees' well-being [18] and support decision making [34]. No less important is understanding of citizens' needs for developing a successful social and economic policy [6,7]. Due to the importance of need satisfaction, several need theories have been developed in humanities and social sciences, such as psychology, economics, philosophy, sociology, anthropology and social policy over the last century. However, the concept of need has not yet been fully taken into account in information systems and online tools as it deserves.

Furthermore, because of the existing narrow computational support, needs assessment methodologies themselves have remained labor-intensive, mostly

© Springer International Publishing AG 2017
P. Różewski and C. Lange (Eds.): KESW 2017, CCIS 786, pp. 195–210, 2017.
https://doi.org/10.1007/978-3-319-69548-8_14

offline activities. Specifically, while only a few methodologies for assessing and systematically studying needs exist to date (e.g. [3,21]), tasks such as acquisition, representation, analysis, and visualization of citizens' needs still belong to a largely offline area, in which support by dedicated computational tools is either limited or not existing[1].

The contribution of our paper towards bridging this gap is twofold. Firstly, based on selected publications from the vast body of research studying human needs in various contexts, we develop a modular ontology consisting of a small robust core capturing central theoretical concepts present in most need theories, and a number of extensions reflecting specific need theories and need assessment methodologies. Secondly, we report on a need study experiment performed in a local community in one Viennese district. The ultimate goal of the project was to intensify online participation and narrow the divide between offline and online urban communities through co-creation of citizen-centric Open Data.

We envisage that OPENEED will open up the vast body of existing interdisciplinary research for use in the Semantic Web context, encouraging the sharing and analyzing the data about human needs across fields and communities. Done properly, such data can have numerous applications, from strengthening online participation and fueling open science to better decision making at the local level and improved open government processes. Thus, we see open needs datasets as important for both citizens and policy-makers. With numerous need theories underpinning the need data, semantically enriched RDF format is a natural choice for data publishers wishing to facilitate the interpretation and enable re-use of the data.

The paper is organized as follows: in Sect. 2, we provide a necessary background, namely an overview of relevant need theories (Sect. 2.1), and a survey of existing relevant ontologies (Sect. 2.2). In Sect. 3, we present our family of ontologies for representing human need data. In Sect. 4, we first present a concrete method for creating and inferring explicit knowledge of needs and then show how the results of a specific study can be represented using OPENEED in Sect. 4.3. Section 5 offers concluding remarks and discusses some further potential positive impacts of the presented ontology.

2   Background and Related Work

2.1   Theoretical Background

Since need satisfaction is one of the most fundamental aspects of humans, several need theories have been proposed over the last century. Although no consistent usage of the term "need" can be found across or within various disciplines [13],

---

[1] Due to differences in researchers' epistemological positions, some researches prefer to use other terms instead of *need assessment*, such as *creating and inferring explicit knowledge of needs* (see e.g. [21]). In this paper, we try to use the term *need assessment* and other similar terms from a neutral perspective, without following or supporting a particular epistemological position.

it can be said in many cases the identification and categorizing human needs constitutes the main focus of many studies operating with the term need, thus resulting in numerous alternative systematics. Table 1, summarizes some of the main need categorizations that have been proposed in the literature. As a non-normative supporting tool, the ontology that is presented in this paper, makes it possible for experts to use any need-categorizations that they prefer to represent human need data.

Besides categorizations of needs, it has been proposed that needs should be differentiated from desires and satisfiers. While needs are the most fundamental requirements and the basis for one's desires and satisfiers, desires are personal and intentional. Satisfiers are either objects or states in which needs and desires are fulfilled. Figure 1 shows a hierarchy of needs, desires, and satisfiers [19,21]. Needs, desires, and satisfies have been defined as distinct classes in the OPENEED ontology. This is an important difference of OPENEED and other few existing similar approaches that enables many new innovative applications, which will be discussed later.

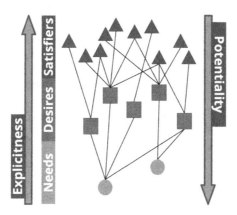

Fig. 1. The hierarchy of needs, desires, and satisfiers according to [21]

In addition to general proposed categorizations, a vast sets of adjectives or tags may be used to define needs (and also desires and satisfiers). These adjectives have provided a virtually inexhaustible binary distinctions between different kinds or levels of needs [6]. Table 2 shows some of the most used adjectives and binary distinctions that can be used for defining needs (see [6] for a detailed definition of these adjectives). OPENEED makes it possible to assign these adjectives to need entities.

Ontologies in the OPENEED family are provided with a set of optional rules in the Semantic Web Rule Language (SWRL) [17] to be used, modified and extended by experts. Our current application of rules is to enable automatic inference of appropriate adjectives based on the provided data. E.g. an expert with a non-constructivist epistemological position can define the following rule:

Table 1. Over the last century, many categorizations of human needs have been proposed. This table summarizes some of such categorisations.

Author	Concise summary/Categorization
Aristotle [36]	Necessity is closely related to needs. Two types of necessities or needs: 1. Absolute needs, 2. Relative needs. Three types of goods: 1. Goods of the soul, 2. Goods of the body, 3. External goods
Murray [33]	Psychogenic needs: 1. Ambition needs, 2. Materialistic needs, 3. Power needs, 4. Status defence needs, 5. Affection needs, 6. Information needs
Alderfer [2]	1. Growth, 2. Relatedness, 3. Existence
Kano et al. [23]	1. Basic needs, 2. Delights, 3. Performance needs
Deci and Ryan [8]	Psychological needs: 1. Competence, 2. Relatedness, 3. Autonomy
Maslow [28,29]	1. Physiological needs, 2. Safety needs, 3. Love needs, 4. Esteem needs, 5. Cognitive needs, 6. Aesthetic needs, 7. Self-actualization, 8. Self-transcendence
Max-Neef [31]	A 36 cell matrix of needs; First dimension: 1. Subsistence, 2. Protection, 3. Affection, 4. Understanding, 5. Participation, 6. Leisure, 7. Creation, 8. Identity, 9. Freedom; Second dimension (existential categories): 1. Being (qualities), 2. Having (things), 3. Doing (actions), 4. Interacting (settings)
Doyal and Gough [9]	1. Health needs, 2. Intermediate needs, 3. Autonomous needs
Price [35]	Children Needs: 1. Physical, 2. Physiological, 3. Psychological, 4. Social, 5. Emotional, 6. Intellectual, 7. Educational, 8. Spiritual
Glasser [14]	1. Survival (food, clothing, shelter, breathing, personal safety, security and sex, having children), 2. Belonging/connecting/love, 3. Power/significance/competence, 4. Freedom/autonomy, 5. Fun/learning
Thomson [39]	Fundamental versus instrumental needs
McLeod [32]	Absolute versus relative needs; Universal versus particular needs; Existence versus welfare needs

if a need (also a desire or a satisfier) has been declared by a citizen, it can be automatically identified as a *felt* or *subjective* need. Similarly, if a need has been assessed by an expert, it can be automatically marked as *interpreted* or *objective* need. The rules have limited applicability scope and need to be applied with care. For instance, the distinction between felt/subjective and interpreted/objective needs does not hold for constructivist approaches like Bewextra [21].

Table 2. Some of the most used adjectives and binary distinctions that can be used for defining needs [6]

Some of the adjectives that may be used to define different kinds of needs:
e.g. absolute + need = absolute need

absolute, basic, circumstantial, common, comparative, derivative, discursive, experiential, expressed, false, felt, higher, inherent, instrumental, intermediate, interpreted, normative, objective, ontological, particular, real, relative, social, subjective, substantive, technical, thick, thin, true, universal

---

Some of the binary distinctions between different kinds or levels of human need:

absolute ⇔ relative | objective ⇔ subjective | basic ⇔ higher | material ⇔ non-material | positive ⇔ negative | non-instrumental ⇔ instrumental | non-derivative ⇔ derivative | physical/somatic ⇔ mental/spiritual | physiological ⇔ cultural | viscerogenic ⇔ sociogenic | intrinsic ⇔ procedural | natural ⇔ artificial | true ⇔ false | inherent ⇔ interpreted | constitutional ⇔ circumstantial | thin ⇔ thick | hedonic ⇔ eudaimonic

---

Based on these perspectives some requirements considered to be needs if they are declared as necessities by the very individual. Although experts (analysts) make claims about them, the final judgment about need or non-need is done by the individuals. Thus, when using OPENEED, researchers standing on certain epistemological positions should choose or extend the rules according to their perspective.

Finally, it has been argued that people are not aware of their hidden needs [15] and tend to express their satisfiers when they are asked to declare their needs. Hence, need assessment is not an easy and straightforward task [3,21]. Accordingly such studies normally include different steps of interpretation and evaluation (see Sect. 4 for an example of such studies). Therefore, it is very important to be able to keep track of the source of an information and to know whether it is a citizen's direct declaration or an expert's interpretation based on a particular method in a particular step of a need assessment study. The OPENEED ontology, that is presented in this paper, makes such kind of source-tracking possible.

2.2   The Concept of Need in the Semantic Web Context

The hypothesis that explication of needs as a driving force of human activity can bring positive impact on the economy has been articulated by [42]. There are several projects (mostly in the research or prototype phase) that bring this concept in the Semantic Web context and even make it a cornerstone block of large architectures. One of the most notable examples of this is the project *Web of Needs* [24,25] that seeks to build a distributed social network where matchings between human participants are established automatically based on the needs they publish or want to help satisfy [12,26]. The Web of Needs consists of agents (nodes), who exchange semantic information, which is captured in the RDF format [26]. The ontology used in the system[2] is rather operational, with the focus on matching users based on needs they publish or intend to fulfill, and enabling the software agents to communicate in order to enabling the matching process. In fact, the ontology does not distinguish between needs and satisfiers and is insufficient to describe studies of needs.

An approach with a similar scope to ours is described in the Master's thesis by S. Dsouza [10] developing the Fundamental Human Needs Ontology (FHN) which does distinguish between satisfiers and needs, it also distinguishes particular needs identified by Max-Neef [30] including *affection, creation, freedom, identity, leisure, participation, protection, subsistence* and *understanding*. The ontology itself is described in the thesis but unfortunately not currently available in a machine readable form. As can be inferred from the description, FHN lacks means of describing both the *provenance* of needs (e.g., how these needs have been obtained) and their assessments and evaluations, which is crucial for applications. The ontology [10] is evaluated using a prototype mobile application by the ontology author rather than any study with human subjects.

As indirectly follows from the recent survey [1], there are to the best of our knowledge no other attempts to formalize the concept of human need along the lines of literature sources mentioned in Sect. 2.1. It follows that an ontology capable of representing human needs resulting from some principled assessment methods hardly exist yet. We address this shortcoming with the OPENEED ontology described in the next section.

3   The OpeNeeD Ontology

This section presents OPENEED, a modular OWL ontology for publishing human need profiles. The following general requirements motivate the design of the ontology:

(i) The ontology must be modular, allowing for expressing relations between different need theories while allowing to capture the semantics of needs, desires (wants, wishes) and satisfiers in different approaches.

(ii) It should capture the categorizations and adjectives commonly used to describe and specify needs, desires and satisfiers (see Sect. 2.1 for a sample of such common categorizations and adjectives).

(iii) It should support principled rigorous studies, like Bewextra (see Sect. 4.1). It should be possible to clearly identify the input data created by the study participants, along with the relevant context information in which the data has been obtained.

(iv) It should be possible to track the lineage of measurements associated with the derived concepts, including the input data used, and the methodology underpinning the computation.

---

[2] http://purl.org/webofneeds/model.

(v) The ontology must allow for (and possibly encourage) publishing of the input data and intermediate steps, to facilitate verification and reuse of the raw collected material.

As demonstrated by Fig. 2, besides the core need theory concepts (subclasses of HumanRequirement in Fig. 2), the focus of OPENEED-CORE ontology is on specifying the context of the study, in particular experts performing the study, community of study participants (human subjects), methodology used, identified human requirements: satisfiers, desires and needs, and metrics assessing of identified requirements.

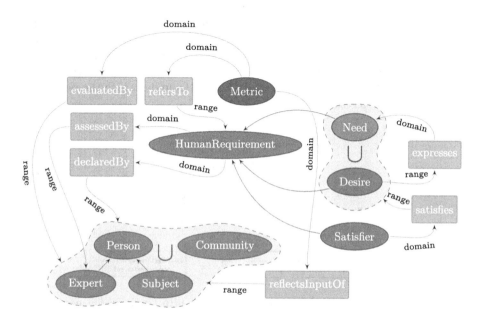

Fig. 2. An outline of the OPENEED-CORE ontology

The following external ontologies are referenced by OPENEED-CORE (Table 3) using the owl:imports instruction.

Table 3. External ontologies used in the TBox of OPENEED-CORE

OpeNeeD concept	Equivalent external concept	Reference
Person	http://www.w3.org/2006/vcard/ns#Individual	[16]
Metric	http://purl.obolibrary.org/obo/IAO_0000027	[4]

While no consistent usage of the terms and concepts can be found across various need theories [13], we tried to make OPENEED-CORE as lightweight and robust as possible. Hence, it only includes basic classes and properties that commonly occur in need theories. Considering that it is not obligatory to use all classes (e.g. *Desire*) and properties of the OPENEED-CORE ontology, researchers are free to apply the ontology based on their own interpretation. To enable concrete applications, OPENEED-CORE can be extended by a number of further ontologies as depicted in Fig. 3: OPENEED-ADJECTIVES, an ontology for representing needs adjectives and binary distinctions between different kinds or levels of human needs (see Table 2), OPENEED-MASLOW, an ontology for representing Maslow's hierarchy of human needs (see Table 1), OPENEED-MAX-NEEF an ontology for representing Max-Neef's classification of human needs (see Table 1), and OPENEED-BEWEXTRA, an ontology for representing human need data derived by the Bewextra needs assessment methodology (see Sect. 4.1). We plan to extend the range of formalized theories in the course of future work. As it was mentioned before, experts can develop their own external ontologies and connect them to the OPENEED-CORE. Figure 3 illustrates the current structure of the OPENEED family and specifies the URLs under which the respective ontologies can be accessed.

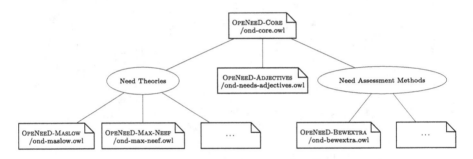

Fig. 3. OPENEED ontology family, available at http://purl.org/openeed

4   A Use Case for OpeNeeD

We have applied the OPENEED ontology to create an open dataset capturing the need profile of a local community in one of Viennese districts (Stuwerviertel, part of the second district of Vienna, also known as Leopoldstadt). In this section, we provide an outline of the need study and the underlying methodology used to perform it.

4.1   Bewextra: A Method for Investigation of Needs

Among a few other methods to identify needs (e.g. [3,15]), Bewextra is an action research approach to investigate substantial needs of people. This methodology aims at creating need knowledge in social systems that should inform and enhance decision making [19,21,22]. The approach is radically bottom-up which means that it starts with the perspective of the very individual and tries to find common patterns among the members of the social system. It uses qualitative methods for data acquisition and has been applied in organizational learning projects with up to 300 participants [20].

The methodological framework consists of three consecutive steps. The first step covers data acquisition based on the approach of learning from the future. The output of this step is a number of satisfiers, articulated by the members of an organization in a process where facilitators ask questions about an ideal future scenario. In the second step, analysts propose hypotheses about substantial needs underlying the satisfiers, stimulated by the observations of the first step and enabled by different views on these observations. Finally, the third step covers the validation of the generated hypotheses by communicative validation and quantitative analysis. In the following, we describe the three steps in more detail. Figure 4 shows the whole Bewextra framework at a glance.

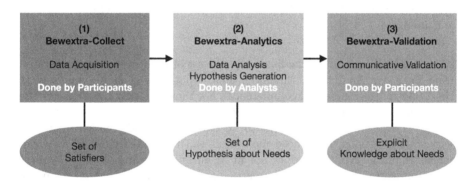

Fig. 4. Overview of the Bewextra process

Bewextra-Collect. The purpose of the first step is to create a trustful work-shop environment that enables the participants to explicate a large number of wishes, dreams, visions and ideas of an ideal future (in the context of their social system). The workshop interventions are designed to facilitate the detachment from the current situation in a way that participants can fantasize about their ideal future scenarios. A facilitator makes the participants imagine that they were actually present in a scenario taking place in the future (5 to 10 years from now); the narrative time journey takes up to several minutes and the imagined time leap is illustrated with appropriate music. Whilst being engaged in these scenar-ios, the participants are asked to answer two questions: "What has emerged and is new?" and "What has come to an end?". In this setting, people should shift their thinking, i.e. to detach from today's circumstances, including restrictions and impossibilities and to come up with visionary and creative results tran-scending current boundaries. Following the Stakeholder Theory [11], we involve all stakeholder groups concerned, i.e. learn from the future from different points of view. This allows for investigating the overall social system holistically.

Bewextra-Analytic. For the data analysis and the generation of hypotheses about needs we follow a hermeneutic approach [5] and use generative listening [37,41]. The method of generative listening aims at hearing the essence of what the participants say and thus, trying to hermeneutically understand which need(s) they try to express by the satisfier they mention. It is about capturing the essence by not letting prejudice take over, trying to see the world with the eyes of the participant.

Using the method of generative listening on the satisfiers which were generated in step-1, we are coding the articulated ideas, wishes and answers. For this purpose, we regularly use the software suite ATLAS.ti to organize codes (and groundedness) and to illustrate hierarchies. The unit of the analysis (defined as a quotation in ATLAS.ti) is each participant. Finally, we utilize a haptic approach and put the codes (often several hundreds) on the floor. We then organize and cluster them so that patterns emerge and main categories of hypotheses about possible hidden needs can be generated. In short, Bewextra-Analytic enables the emergence of hidden needs of the participants and results in a set of hypotheses about needs.

Bewextra-Validation. In the final step, the set of need hypotheses generated during Bewextra-Analytic is validated. The hypotheses shall be validated in terms of both correctness and completeness. For the validation of correctness we use an online questionnaire containing the hypotheses generated in Bewextra-Analytic. This questionnaire is sent to all participants and consists mainly of Likert scale questions. Each need hypothesis can be rated from 1 to 4, where 1 means that the hypothesis does not fit at all and 4 means that the hypothesis fits perfectly. Additionally, the participants are asked to comment on the completeness of the proposed need hypotheses in case that relevant needs or need aspects are missing. This communicative validation can either be done in a workshop setting or as part of the online questionnaire. The simultaneous use of completeness (qualitative) and correctness (quantitative) validation allows us to accept or reject the generated hypotheses about needs in order to finally create a catalogue containing explicit knowledge about substantial needs.

4.2   Needs Assessment for a Viennese Quarter

Originally conceived as a means of inferring implicit need knowledge in organizations, the Bewextra method has been applied successfully by public administrations, for instance to define the urban development strategy in the German city of Andernach[3]. Here, we outline a results of another recent study in an urban context, specifically a study of the residents' needs of a Viennese quarter *Stuwerviertel*, belonging to a city's second district Leopoldstadt [27]. The context of the study was to explore the means of incentivising online participation of a currently mostly offline local community.

---

[3] "Leitbild   Andernach   2030"   http://www.andernach.de/de/leben_in_andernach/leitbildstadt.html.

To this end, the finding of the need study will be used as an input for online discussion towards establishing a development agenda of the quarter. A needs ontology derived from the study using OPENEED will furthermore exemplify co-creation in the Open Data context. Below, we sum up the results of the study and then in Sect. 4.3, we show how these results can be represented with the help of OPENEED.

Data Acquisition and Process. In the first Bewextra-Collect part, 1503 satisfiers from 80 participants in three workshops have been collected. This was achieved by "learning from an envisioned future" [21]: the workshop participants tried to imagine themselves in the year 2030 and provided a short account of their wishes, dreams, thoughts that came to their minds during the thought experiment. The subsequent data analysis step based on a method called *generative listening*, resulted in 355 codes — hypotheses about underlying needs of this large number of satisfiers (see Table 4).

Table 4. Overview of the phases Bewextra-Collect and Bewextra-Analytic

Workshop	Participants	Satisfier	Codes	Needs
School 1	39	599	156	**12**
School 2	27	581	135	
Adult residents	14	323	64	
Sum	**80**	**1503**	**355**	

Table 5. Identified needs and their acceptance rate by 122 survey participants

Need for	Acceptance rate
opportunities to spend spare and leisure-time	85%
cleanliness	84%
local supply	83%
security	81%
education	80%
aesthetics and beauty	79%
good human relations	79%
modernity and continuous development	77%
opportunities to meet and feel connected	77%
a supportive political frame	77%
positive recognition of the district	76%
highly ecologically compatible mobility	76%

In the Bewextra-Analytic phase a catalogue of 12 hypotheses about shared and contextualized needs in the quarter has been produced via *semantic clustering* performed by several experts. These hypothesized needs were then validated via communicative validation by a subset of individuals living, working, and going to school in the respective local area (122 online survey participants). The acceptance rate reflects the degree to which participants of the survey share the identified needs on a four point scale ranging from "I agree" and "I rather agree" to "I rather not agree" and "I do not agree". The results of the study can be found in Table 5, and their OPENEED encoding is discussed next.

4.3   Representing Study Results in OpeNeeD

To capture the result of a study outlined in the previous section, we instantiated the classes of the OPENEED-CORE and OPENEED-BEWEXTRA ontologies. In particular, from the latter one, the community approval metric, a subclass of ond:Metric (or, equivalently, of OBI_0000027 "data item" of [4], with the class ond:CommunityApproval defined in OPENEED-BEWEXTRA (since community approval is part of Bewextra methodology but not necessarily relevant for other theories and methods). An excerpt of the ontology instance representing a row from Table 5 is graphically depicted in Fig. 5.

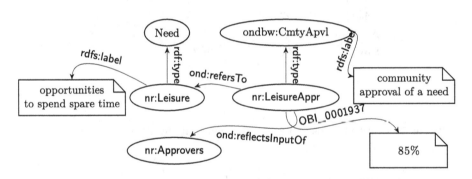

Fig. 5. Example of the instance data in OPENEED

With more than thousand random satisfiers mentioned by the participants of the experiment, we do not instantiate them all in the ontology. Raw data remains available in an accompanying datasets as questionnaire scans. Extracted are selected *codes* (created during the workshop for adult residents – cf. Table 4) which in OPENEED terminology correspond to *desires*, and the *needs*. We believe that such a flexibility of reflecting the raw data in the ontology is desirable, and designed OPENEED with a possibility of describing the origin of satisfiers, desires and needs as individuals (class odn:Subject) or community as a whole (class odn:Community). This allows to publish the data with varying degrees of granularity, as permitted by resources.

4.4   Expected Applications of the Ontology

The OPENEED ontology introduced in Sect. 3 has many potential applications. Firstly, it serves for publishing the results of studies as a self-contained dataset, for instance as Open Data. Secondly, it facilitates querying the results of a single or multiple studies thus enabling comparative analysis. For instance, satisfiers mentioned by the study participants within the Bewextra-Collect phase can be analyzed according to common social science criteria such as age or gender. Regarding age, it is typically desirable to distinguish between the needs of adolescents and elderly, in order to specify policy interventions. Queries would allow for drawing a connection between satisfiers and needs, enabling one, e.g., to identify most representative needs. Thirdly, it enables data enrichment, e.g., using SWRL rules as described in Sect. 2.1. Fourthly, the design of OPENEED encourages tracking provenance of declarations, assessments, and evaluations, be it individual experts who assess needs, user communities who declare or evaluate them, or any other person or entity. Last but not least, the modular structure of OPENEED allows both the requirements of larger communities, as well as of subcommunities and even individuals within them, to be represented. It is a matter of available resources if the results of need studies are captured with a high degree of granularity. Investing effort into preserving the inputs and requirements of as small groups as possible is a way towards embracing pluralism and properly reflecting the needs of minorities even if these are not preserved in an aggregated assessment.

5   Conclusion and Outlook

In this paper, we presented OPENEED, a family of ontologies for representing human need data. As a non-normative support tool, the ontology can be used to create human need profiles derived based on different epistemological or methodological standpoints. Hence, it consists of a small core ontology and an extendable collection of ontologies covering specific need theories and methodologies.

As an evaluation, we used OPENEED to represent, enrich and query the results of a recent needs assessment study. This allowed us to convert the identified need profiles of the local community into a Linked Open Dataset, making the results of the needs assessment study available for citizens, policy makers and researchers. At the same time, equipped with the ontology and semantic representation of the study results, one can query, compare, enrich, analyze, evaluate and archive the data obtained in the past, current and future studies, helping to scale up needs assessment methods and broaden the range of their applications.

Beyond the presented use case, OPENEED will facilitate declaration, assessment, processing, representation and visualization of individual citizens' needs or communities' collective needs. Clearly, not all such data can be openly published: individual or sensitive needs should be only processed while fully respecting citizens' privacy and security. In particular, (i) in personal use cases, such data can be processed by private secure artificial agents, who are under the control of the individual citizens (data owners); (ii) in public use cases, such data can be used as anonymized or pseudonymized information by the permission of the individual citizens. A detailed discussion of these concerns falls outside the scope of this paper. We believe that creation of people's need profiles, when done correctly, can have wide-reaching applications, e.g., supporting citizen participation, direct democracy, open government, open science, and ultimately pluralism in the society.

Acknowledgments. This work is funded through the project 855407 by the Austrian Federal Ministry of Transport, Innovation and Technology (BMVIT) under the program "ICT of the Future" between Nov. 2016 and Apr. 2019. Soheil Human is supported by the Jubilee Fund of the City of Vienna.

References

1. Abaalkhail, R., Guthier, B., Alharthi, R., Saddik, A.E.: Survey on ontologies for affective states and their influences. Semant. Web Preprint, 1–18 (2017)
2. Alderfer, C.P.: Existence, Relatedness, and Growth: Human Needs in Organizational Settings. Free Press, New York (1972)
3. Altschuld, J.W., Watkins, R.: A primer on needs assessment: more than 40 years of research and practice. New Dir. Eval. **2014**(144), 5–18 (2014)
4. Bandrowski, A., Brinkman, R., Brochhausen, M., Brush, M.H., Bug, B., Chibucos, M.C., Clancy, K., Courtot, M., Derom, D., Dumontier, M., et al.: The ontology for biomedical investigations. PLoS ONE **11**(4), e0154556 (2016)
5. Davis, B.: Listening for differences: an evolving conception of mathematics teaching. J. Res. Math. Educ. **28**(3), 355–376 (1997)
6. Dean, H.: Understanding Human Need. Policy Press, Bristol (2010)
7. Dean, H.: Welfare Rights and Social Policy. Routledge, London (2014)
8. Deci, E.L., Ryan, R.M.: Cognitive evaluation theory. Intrinsic Motivation and Self-determination in Human Behavior. Perspectives in Social Psychology, pp. 43–85. Springer, Boston (1985). doi:10.1007/978-1-4899-2271-7_3
9. Doyal, L., Gough, I.: Intermediate needs. A Theory of Human Need, pp. 191–221. Palgrave, London (1991). doi:10.1007/978-1-349-21500-3_11
10. Dsouza, S.D.: Cloud-based Ontology Solution for Conceptualizing Human Needs. Master's thesis, University of Ottawa (2015)
11. Freeman, R.E., Wicks, A.C., Parmar, B.: Stakeholder theory and "the corporate objective revisited". Organ. Sci. **15**(3), 364–369 (2004)
12. Friedrich, H., Kleedorfer, F., Human, S., Huemer, C.: Integrating matching services into the web of needs. In: SEMANTiCS (2016)
13. Gasper, D.: Conceptualising human needs and wellbeing. In: Gough, I., McGregor, J.A. (eds.) Wellbeing in Developing Countries: New Approaches and Research Strategies, pp. 47–70. Cambridge University Press (2007)
14. Glasser, W.: Choice Theory: A New Psychology of Personal Freedom. Harper Collins, New York (1998)
15. Goffin, K., Lemke, F., Koners, U.: Identifying Hidden Needs. Palgrave Macmillan, London (2010)
16. Halpin, H., Iannella, R., Suda, B., Walsh, N.: Representing vCard objects in RDF. W3C Member Submission 20 2010

17. Horrocks, I., Patel-Schneider, P.F., Boley, H., Tabet, S., Grosof, B., Dean, M., et al.: SWRL: a semantic web rule language combining OWL and RuleML. W3C Member Submission 21, 79 (2004)
18. Jost, P.J.: Individual differences between employees. In: The Economics of Motivation and Organization: An Introduction. Edward Elgar Publishing, Cheltenham (2014)
19. Kaiser, A., Fordinal, B., Kragulj, F.: Creation of need knowledge in organizations: an abductive framework. In: 2014 47th Hawaii International Conference on System Sciences, pp. 3499–3508 (2014)
20. Kaiser, A., Kragulj, F., Fahrenbach, F., Grisold, T.: Developing a knowledge based vision for a city. In: Proceedings of the 18th European Conference on Knowledge Management (ECKM 2017). Academic Conferences and Publishing International Limited, Reading (2017)
21. Kaiser, A., Kragulj, F.: Bewextra: creating and inferring explicit knowledge of needs in organizations. J. Futures Stud. **20**(4), 79–98 (2016)
22. Kaiser, A., Kragulj, F., Grisold, T.: Taking a knowledge perspective on needs: presenting two case studies within an educational environment in Austria. Electron. J. Knowl. Manage. **14**(3), 114–126 (2016)
23. Kano, N., Seraku, N., Takahashi, F., Tsuji, S.: Attractive quality and must-be quality. J. Jpn. Soc. Qual. Control **14**, 147–156 (1984)
24. Kleedorfer, F., Busch, C.M.: Beyond data: building a web of needs. In: Proceedings of the WWW2013 Workshop on Linked Data on the Web, Rio de Janeiro, Brazil, 14 May 2013
25. Kleedorfer, F., Busch, C.M., Grill, G., Khosravipour, S., Salcher, F., Tus, A., Gstrein, E.: Web of needs-a new paradigm for e-commerce. In: 2013 IEEE 15th Conference on Business Informatics (CBI), pp. 316–322. IEEE (2013)
26. Kleedorfer, F., Human, S., Friedrich, H., Huemer, C.: Web of needs: a process overview. In: SEMANTiCS (2016)
27. Kragulj, F., Fahrenbach, F.: From needs to satisfiers: how design thinking can inform organizational learning processes. In: Tomé, E., Gaby Neumann, B.K. (eds.) Proceedings of the International Conference Theory and Applications in the Knowledge Economy, pp. 470–482 (2017)
28. Maslow, A.: Motivation and Personality. Harper & Row, New York (1970)
29. Maslow, A.H.: A theory of human motivation. Psychol. Rev. **50**(4), 370–396 (1943)
30. Max-Neef, M., Elizalde, A., Hopenhayn, M.: Human Scale Development: Conception, Application and Further Reflections. No. Bd. 1 in Human Scale Development: Conception, Application and Further Reflections. Apex Press (1991)
31. Max-Neef, M., Elizalde, A., Hopenhayn, M.: Development and human needs. In: Real-life Economics: Understanding Wealth Creation, pp. 197–213 (1992)
32. McLeod, S.K.: Knowledge of need. Int. J. Philos. Stud. **19**(2), 211–230 (2011)
33. Murray, H.A.: Explorations in Personality. Oxford University Press, New York (1938)
34. Patnaik, D.: System logics: organizing your offerings to solve people's big needs. Design Manage. Rev. **15**(3), 50–57 (2004)
35. Price, S.: The special needs of children. J. Adv. Nurs. **20**(2), 227–232 (1994)
36. Reader, S.: Aristotle on necessities and needs. R. Inst. Philos. Suppl. **57**, 113–136 (2005)
37. Senge, P.M., Scharmer, C.O., Jaworski, J., Flowers, B.S.: Presence: An Exploration of Profound Change in People, Organizations, and Society. Crown Business, New York (2005)

38. Tay, L., Diener, E.: Needs and subjective well-being around the world. J. Pers. Soc. Psychol. **101**(2), 354–365 (2011)
39. Thomson, G.: Fundamental needs. R. Inst. Philos. Suppl. **57**, 175–186 (2005)
40. Watkins, R., Kavale, J.: Needs: defining what you are assessing. New Dir. Eval. **2014**(144), 19–31 (2014)
41. Yackel, E., Stephan, M., Rasmussen, C., Underwood, D.: Didactising: continuing the work of leen streefland. Educ. Stud. Math. **54**(1), 101–126 (2003)
42. Zhang, Y.-C.: The information economy. In: Johnson, J., Nowak, A., Ormerod, P., Rosewell, B., Zhang, Y.-C. (eds.) Non-Equilibrium Social Science and Policy. UCS, pp. 149–158. Springer, Cham (2017). doi:10.1007/978-3-319-42424-8_10

Knowledge Graph: Semantic Representation and Assessment of Innovation Ecosystems

Klaus Ulmschneider[(✉)] and Birte Glimm

Institute of Artificial Intelligence, Ulm University, Ulm, Germany
{klaus.ulmschneider,birte.glimm}@uni-ulm.de

Abstract. Innovative capacity is highly dependent upon knowledge and the possession of unique competences can be an important source of enduring strategic advantage. Hence, being able to identify, locate, measure, and assess competence occupants can be a decisive competitive edge. In this work, we introduce a framework that assists with performing such tasks. To achieve this, NLP-, rule-based, and machine learning techniques are employed to process raw data such as academic publications or patents. The framework gains normalized person and organization profiles and compiles identified entities (such as persons, organizations, or locations) into dedicated objects disambiguating and unifying where needed. The objects are then mapped with conceptual systems and stored along with identified semantic relations in a Knowledge Graph, which is constituted by RDF triples. An OWL reasoner allows for answering complex business queries, and in particular, to analyze and evaluate competences on multiple aggregation levels (i.e., single vs. collective) and dimensions (e.g., region, technological field of interest, time). In order to prove the general applicability of the framework and to illustrate how to solve concrete business cases from the automotive domain, it is evaluated with different datasets.

Keywords: Competence analysis · Competence detection · Competence assessment · Computational linguistics · Corporate strategy · Data mining · Decision making · Expert matching · Expert mining · Information extraction · Information retrieval · Innovation ecosystem · Knowledge graph · Knowledge representation · Machine learning · Name normalization · Name disambiguation · Natural language processing · Ontology · Patent analysis · Question-answering · Reasoning · Semantic technologies · Semantic analysis

1 Introduction

Continuous change is undeniably one of the main characteristics of our modern world. New technologies, techniques, business models, and processes are developed and evolve rapidly over time, primarily driven by the requirement to diversify from others and to cope with the pace of change. Individual and collective creativity, paired with existing knowledge and know-how, can be constituted as the backbone of this development and result in new knowledge or, consequently,

© Springer International Publishing AG 2017
P. Różewski and C. Lange (Eds.): KESW 2017, CCIS 786, pp. 211–226, 2017.
https://doi.org/10.1007/978-3-319-69548-8_15

in inventions. Effective R&D management, and, in particular, innovative capacity, is highly dependent upon knowledge and, above all, human individuals, and therefore inevitable linked with general business strategies [11]. Consequently, it is of increasing importance to be capable of analyzing knowledge occupants and specialists, who are driving the change. In principle, innovation is driven by mutual complementarities between individuals and organizations, i.e., the actual know-how carriers and problem-solving capacity. The interaction of knowledge occupants such as researchers, engineers, or organizations, as well as their created artifacts in consideration of time and location can be constituted as an innovation ecosystem. Thus, being able to capture, measure, locate, and assess such causal and contextual, local and global (competence) correlations, and making them explicit, can be a significant advantage from a strategic viewpoint (e.g., innovative impact, knowledge flows, knowledge gap identification, competitive assessments) and constitutes a competitive advantage when being capable of adapting rapidly to environmental transformation processes.

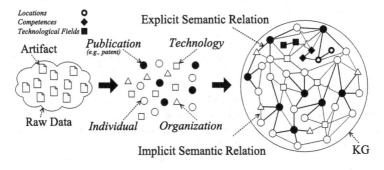

Fig. 1. Knowledge graph construction

In order to keep up with the process of continuous change, this paper introduces a novel methodology using several NLP, machine learning, and AI techniques combined with OWL reasoning. The aim is to gain multilevel competence information from publications on individual level (persons) and on several collective levels (e.g., department, institute, firm, university, industry, sector) as well as to capture their structural (e.g., institute belongs to university), temporal, and spatial (e.g., state, region, country, continent) arrangement in order to enable the analysis of an overall knowledge ecosystem (e.g., investigation of interactions and collaboration). The proposed framework allows for mutual exploitation of knowledge and know-how complementarities as well as knowledge flows and interactions between individuals and among organizations, or,

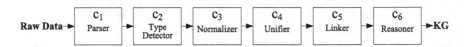

Fig. 2. Fundamental processing components

their analysis with regard to regional (e.g., a firm's branches) or geopolitical (e.g., a nation's) aspects with the fundament of a graph-based representation.

Specifically, knowledge occupants are identified, extracted, normalized, disambiguated, unified (if required), and brought into context with the help of the created *Knowledge Graph* (KG), which constitutes a (semantic) RDF graph and allows to store structure-lending entities as well as complex (semantic) analyses with regard to various dimensions. For example, temporal, spectral, spatial, topical features in the KG design allow to capture and react to weak signals (e.g., competence transitions) and to track evolutionary (knowledge) pathways as well as to generate innovation stimuli or support decision-making on future directions. To achieve this, a Semantic Pipeline (SP) [14], which processes raw data and stores the gathered information in the *Knowledge Graph*, is applied, primarily comprising the following tasks:

1. Parse raw data (e.g., patents, academic publications) into uniform objects
2. Identify and extract real-life entities which can occupy knowledge (candidates) and determine their type
3. Normalize the surface forms of names and store them together with extracted metadata in standardized properties
4. Disambiguate and, if referring to the same entity, unify competence occupant mentions
5. Semantically link (identified) competence occupant entities with each other, with related (business) artifacts (e.g., patents) as well as aligning them with, for example, topical, organizational, or spatial entities, and, on this basis, with entities which define a competence (an ontology)
6. Semantic analysis of knowledge interdependencies on multiple levels and in consideration of various dimensions (e.g., identification of competence clusters, knowledge flows, or competence paths by deducing implicit information by means of the semantic relations created in the previous step)

In particular, the presented framework allows for extracting and determining different types of entities from unstructured, semi-structured and structured textual content and transforming them into respective (structured) entities, and interlink them semantically by means of the KG. Further, the framework can distinguish between different types of competence possessing entities, which enables the detection of competence ownership and analysis of competence clusters on several aggregation levels based on technological fields of interest. By enriching traditional (text) processing with components from the field of Semantic Web and Machine Learning, (knowledge) interdependencies and inferences of (implicit) information on multiple (competence) levels are enabled.

The benefits of the presented framework and its underlying semantic representation are illustrated with scientific publications as well as with patents. Patents are employed since they represent, besides academic publications, a large proportion of technological and procedural knowledge and occupy interesting (implicit) relational knowledge. Rapidly increasing filed numbers of patents as well as diverse patent application strategies and complex ambiguous writing styles make patent analyses a difficult and time-consuming task (see e.g.,

[13–15]). In contrast, a significant amount of intellectual capital is reflected in patents which make their analysis a worthwhile exercise. Patents are applied for by physical (individual) or legal (organizational) entities, i.e., the actual carriers of knowledge (e.g., skills, qualifications, experiences) and creative potential. The sheer amount of filed patents and the increasingly growing number of patent applications consequently result in high technological impact. Hence, patents are indicators of research and development efforts and the analysis of patenting activity constitutes one way of measuring intellectual capital (e.g., technological competence owned by an organization) [2]. Therefore, patents are a valuable (large scale) source of scientific and technological information for R&D processes as well as for strategic decision making (see e.g., [4]).

It has be shown that patents can be analyzed across various dimensions, e.g., with regard to technological progress, technology planning, technological forecasting, R&D portfolio management, or infringement analyses (see e.g., [3,5,13]) and that they can be aligned with technological fields of interest [14]. However, what has been neglected so far is to extend their exploitation with regard to new methodologies, processes, technologies, or materials, to the experts as well as the think tanks who possess this know-how (i.e., the physical and legal entities) in order to create value for multiple application scenarios, which include knowledge identification, knowledge acquisition, knowledge localization as well as the management and early development of resources with regard to market changes (e.g., human resource development, knowledge transfer, recruiting, headhunting, R&D management, competitor foresight, supplier identification).

The contributions of this paper are as follows. A framework is introduced, which combines various interdisciplinary research areas from several fields of computer science and an algorithmic model to accurately derive competence information in order to identify competence occupants and their interdependencies as well as allowing their analysis along multiple (aggregated) dimensions. For this reason the identified individual and collective competence occupying entity types are integrated into the KG along with other types of entities (e.g., with the patents they applied for, patent classification systems as well as spatial, geopolitical, or topical entities) which, overall, formally represent a complex innovation ecosystem. On this basis, a reasoner is employed to exploit competences and their (semantic) interrelations and knowledge pathways being capable of deducing implicit information which provides more insights in depth and breadth. Various analysis techniques, such as text mining, network analysis, citation analysis, or index analysis are combined to discover meaningful implications. The results can be reused in combination with gathered temporal information. Besides increased general transparency of non-obvious valuable information, various use cases are enabled, and therefore, the proposed framework can serve multiple stakeholders and application scenarios.

The presented framework overcomes the limitations of existing approaches with regard to the following key aspects: improved normalization, unification, and integration algorithms which identify, organize, and interlink competence occupants based on structured, semi-structured, or unstructured data comprising syntactic, lexical, semantic, relational, and machine learning techniques as well as aligning the identified competence occupants with conceptual systems and adding reasoning capability to the underlying (processed) data, i.e., the KG. The KG embodies a complex ecosystem, which is capable of representing competence occupants and their peripheral (i.e., established, structural, or interacting) entities (e.g., scientific publications, technological fields of interest, employers) in a semantic manner as well as evolutionary knowledge pathways and clusters (i.e., knowledge creation and diffusion over time). Therefore, this work extends the focus from analyzing information from the actual raw artifacts to the layer of processing, deducing, and analyzing derivative information, which is not directly observable, i.e., obtaining derivative information from artifacts of other purpose. For example, the purpose of patents is not primarily analyzing knowledge occupants and knowledge flows.

The paper is organized as follows. Section 2 describes the central components of the presented framework. Section 3 demonstrates the applicability of the approach and presents the findings along with real-world business cases from the automotive domain. Finally, Sect. 4 discusses related work and Sect. 5 concludes with a summary and an outlook.

2 Competence Analysis

Identifying competence occupants and their mapping to actual competences and skills is a nontrivial task. In order to achieve high accuracy, multiple (pre-) processing steps need to be accurately performed. These processing steps include the determination of their (entity) type and structural level (e.g., individual vs. collective) as well as the normalization of occurring surface forms and their disambiguation. Then, the gathered competence occupants are semantically mapped with other entities (e.g., topical, spatial) to gain valuable insights from the *Knowledge Graph*. This section outlines inherent challenges of revealing competences, demonstrates the crucial processing steps to identify them accurately, and illustrates the capitalization of the gathered semantic graph for strategic purposes.

2.1 Entity Type Recognition and Disambiguation

In real-world, many types of entities exist. Examples include persons, organizations, locations, technologies, or materials. All of them can be helpful for knowledge analyses and are usually referenced within publications, such as patents, without being explicitly labeled with their (entity) type. In order to identify such named entities and to map them to their corresponding type for further processing, they have to be identified and labeled correctly (e.g., determination whether they can occupy knowledge, their structural level, or to semantically capture interdependencies). Depending on the data source, entities must be parsed from structured or semi-structured sections of the respective artifacts, or extracted from unstructured content (with help of NLP components, i.e., *Named Entity Recognition* (NER)). In the context of patents, important information regarding competences is available in semi-structured formats and therefore can be parsed with reasonable effort. However, entities do not occur separated by their type. In case of scientific publications, algorithms exist to extract authors and their affiliations, but nowadays such information is usually available in structured or semi-structured formats as well. Having raw names of potential competence occupants available, their type is determined by using specific patterns and indicators which are employed during the normalization process as illustrated in the next section.

2.2 Entity Name Normalization

Entity Name Normalization (ENN), also known as *Name Standardization*, refers to the standardization of name variants which can occur due to, for example, misspellings, abbreviations, or different naming conventions. Therefore, the process of name normalization attempts to transform surface forms, i.e., name variants, into a common format.

Competence occupants and their respective names, which are mentioned in artifacts such as scientific publications, frequently occur with several surface forms, might be incomplete, not formatted according to a common standard, extended with additional information not belonging to the actual name (e.g., titles, addresses, states), or acronyms are used. Hence, name normalization is a nontrivial task [1,8,9]. Table 1 illustrates some typical ENN challenges.

Table 1. Selected prevalent Entity Name Normalization challenges

Challenge	NameType	Group	Example #1	Example #2
Capitalization	Individual	Orthography	John Doe	JOHN DOE
Capitalization	Organization	Orthography	Nike Inc	NIKE INC
Punctuation	Individual	Orthography	Doe, John	John Doe
Punctuation	Organization	Orthography	Nike, Inc	Nike, Inc.
Diacritical Marks	Individual	Orthography	Ulf Lindström	Ulf Lindstroem (Ulf Lindstrom)
Diacritical Marks	Organization	Orthography	Telefónica S.A.	Telefonica
Compound Names	Individual	Orthography	Jean-Claude van Damme	van Damme, Jean-Claude
Compound Names	Organization	Orthography	Wal-Mart	Wal*mart
Transliteration	Individual	Orthography	Крассимир	Krassimir
Transliteration	Organization	Orthography	上海大学	Shànghâi Dàxué (Shanghai University)
Prefixes	Individual	Semantic	Mr. John Doe	Dr. John Doe
Prefixes	Organization	Semantic	Walt Disney	The Walt Disney Company
Suffixes	Individual	Semantic	John Doe Jr.	John Doe, MSc
Suffixes	Organization	Semantic	Exxon Mobil Corp.	Exxon Mobil Corporation
Acronyms	Individual	Semantic	John D. Doe	John Daniel Doe
Acronyms	Organization	Semantic	Univ. Beijing Technology	Beijing Tech Univ
(Other) Metadata	Individual	Semantic	John Doe, NY	John Doe, USA
(Other) Metadata	Organization	Semantic	Daimler AG, Stuttgart	Daimler 70567 Stuttgart DE

The proposed normalization component covers such challenges by means of rule-based, syntactical, orthographical, semantical, statistical, lexical, and relational aspects. It attempts to reduce a raw name to its core components, i.e., removing all information not belonging to the actual name itself while extracting as much metadata and pieces of evidence as possible for further processing (e.g., academic or honorific prefixes for individuals or legal forms for organizations). Note that this process is entity type dependent and respective entities with their dedicated properties are created. For example, in case of an individual name the normalizer separates first name, middle names (if any), and last name into dedicated fields and is capable of taking several combinations, orders, or punctuation (e.g., Doe, John D.) into account.

Additionally, names are normalized with regard to three normalization stages: the raw normalized name (core), a human display name (e.g., uniform casing, acronyms expanded) and one form for machine processing (e.g., punctuation, diacritical marks, special characters, stop words removed). Moreover, the structural (aggregation) level, in particular with regard to competence occupants, is determined as well (i.e., single competence vs. collective competence, e.g., institute vs. university) and respective associations are created. Note that additional information and (extracted) metadata is, if possible, normalized and filed in the same name object during this processing step as well (e.g., NY = New York, Jr. = Junior). Table 2 illustrates a successful normalization process for two individual and three organization names. Note that the overall entity type detection and normalization process is implemented with a feedforward and feedback loop to update potential incorrect assignments. For example, individuals are usually not mentioned together with legal forms, however person names can be the same as or part of a company name.

Table 2. Exemplary initial rudimentary Entity Name Normalization steps

Raw Name	Normalized Name Object (Extract)
Dr. John Francis D. Smith Jr.	[FirstName=John,middleNames[Francis,D],FamilyName=Smith,prefixes[Doctor],suffixes[Junior]]
Smith, Dr. John F. D.	[FirstName=John,middleNames[F,D],FamilyName=Smith,prefixes[Doctor],suffixes[]]
North Texas University	[Name=North Texas University,type=Academic,level=University]
University of North Texas	[Name=North Texas University,type=Academic,level=University]
Nike, Inc.	[Name=Nike,type=Business,legalForm=Incorporated]

2.3 Entity Unification

Entity Unification (EU), also known as *Entity Linking, Entity De-Duplication, Reference Normalization, Instance Unification, Record Linkage, Coreference Resolution,* or *Entity Resolution,* refers to the process of determining whether two entities (i.e., name mentions) refer to the same object (e.g., a person) in real-world, and, if referring to the same entity, mapping them to a canonical unambiguous referent (see e.g., [1,10]).

This processing step, which is building up on the *ENN* component, is essential, because competence occupants occur in various forms and the process of disambiguation and mapping therefore can have strong effect on the accuracy of single and collective competence assignments [7,9], and, in particular, on higher level deductions. In consequence, the (normalized) surface form of potential competence occupants, such as individual or organization names, need to be disambiguated and, in case of ambiguity, unified in preparation for further processing steps and, eventually, the population of the *Knowledge Graph* to accomplish the competence analysis task. Table 3 illustrates some typical EU challenges.

Table 3. Selected prevalent Entity Unification challenges

Challenge	NameType	Group	Example #1	Example #2
Spelling Mistakes/OCR Errors	Individual	Semantic	I.F. Kennedy	John Fitzgerald Kemedy
Spelling Mistakes/OCR Errors	Organization	Semantic	Tesla Inc.	Telsa Inc.
Similar Names	Individual	Semantic	Jonathan Meier	Jonathan Meyer
Similar Names	Organization	Semantic	TLG Immobilien	TAG Immobilien
Acronyms	Individual	Semantic	J.F. Kennedy	John Fitzgerald Kennedy
Acronyms	Organization	Semantic	IBM	International Business Machines Corporation
Translations	Individual	Multilinguality	Franz Lieber	Francis Lieber
Translations	Organization	Multilinguality	Universidad de Chile	University of Chile
Marriage/Name Changes	Individual	Time	Hillary Diane Rodham	Hillary Diane Rodham Clinton
Mergers/Splits/Acquisitions	Organization	Time	Mannesmann	Vodafone Group
One Name - Multiple Entities	Individual	Semantic/Time	John Doe (Dover)	John Doe (New York)
One Name - Multiple Entities	Organization	Semantic/Time	Merck & Co., Inc.	Merck KGaA
Multiple Names - Same Entity	Individual	Semantic/Time	President Trump	Donald Trump
Multiple Names - Same Entity	Organization	Semantic/Time	Daimler AG	Daimler-Benz AG
Ambiguities/Missing Information	Both	Semantic	Trump	Trump
Structural (level)	Organization	Semantic	Ulm University	Institute of Artificial Intelligence, Ulm University

Entity Unification is achieved using a fuzzy matching approach which combines several techniques and matching rules. For example, phonetic (distance-based), feature-based, and probabilistic similarity measures are employed. Furthermore, meta information as well as graph-based (i.e., implicit and explicit references) and statistical indicators, which can be derived from the source artifacts, are incorporated. In particular, the algorithms combine exact, partial and approximate matching (experiments were conducted with several similarity measures such as Cosine Similarity, Jaro-Winkler, Levenstein) on all normalization stages (RawProcessed, Display, Machine). Further, peripheral features, i.e., explicitly and implicitly (derived) references, are examined. Examples include topical and spatial associations, citations, references to (patent) classification systems, or the analysis of coreferences (e.g., co-authorship among authors in case of scientific publications or among inventors and assignees in case of patents as well as individual-organization associations). One important factor, which is often neglected, is time. Names can change due to marriage, mergers, splits, and acquisitions. If such information is available, it is reused for the unification process as well. Note that, in case of unification, all references are updated and every known surface form (variants, including the raw name and all normalized forms) is stored with the unified object (entity) and reused for further normalization and unification tasks. The best normalized name (in format and without metadata) is selected as reference name.

2.4   Multidimensional Competence Assessment

The main challenges of effective competence assessment are to accurately determine the possession and location of individual and collective competences as well as the capability to track their temporal evolution.

In order to cope with these challenges, the presented framework allows to identify, structure, (semantically) interlink, and therefore measure and analyze competence occupants with regard to multiple dimensions and in consideration of temporal aspects. The idea behind the approach is that all types of entities have causal relationships with each other (cf. Fig. 3), i.e., competences are mutually dependent or influenced from other entities and several inverse deductions can be made.

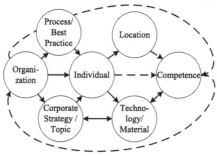

Fig. 3. Exemplary causal competence relationships in a business context

In particular, the framework semantically connects the gathered normalized and unified competence occupants with their established artifacts (i.e., their publications) as well as with other entities in the KG. These entities can represent other business-related artifacts, such as technology fact sheets, or conceptual systems, such as technological fields of interest, locations, or patent classification systems, which are also semantically interlinked (see [14] for more information). The semantic relations include, among other properties, the (relation) type (e.g., 'hasTopic', 'hasInventor', 'hasAssignee', 'belongsTo[Company]', 'isLocated') and respective temporal information (e.g., point in time of person-company association). On this basis, the competence occupants are aligned with entities which (formally) describe their actual competence(s) using respective 'hasCompetence' relations. Note that such (defining) skills and competences are hierarchically organized and also semantically interlinked with other knowledge-related entities in the KG (e.g., entities representing topical information). Hence, other knowledge-related entities in the KG bridge the gap between the competence occupying and the competence defining entities.

One important factor of the *Knowledge Graph* design is the concept of subsumption. In previous processing steps the (implicit) aggregation level of competence occupants is derived and transformed to respective (causal) relations, i.e., single competences are associated with collective competences (competence clusters). As example, consider research groups, institutes, R&D teams, laboratories, medical centers, think thanks, or organizational units, which can be further aggregated. Their associated competences can be attributed to their related firms, universities, or (joint) research projects. Moreover, aggregations in consideration of other dimensions are enabled, for example with regard to an industry's, a region's, or a nation's competence range. Note that the concepts of subsumption and transitivity also allow to respect several stages of competence defining entities (e.g., Semantic Web Technologies → Artificial Intelligence → Computer Science). Thus, further 'hasCompetence' relations, competence clusters, and knowledge pathways in the KG can be inferred (e.g., by applying transitivity rules) on several levels using a reasoner, thus, enabling competence assessments on multiple levels and along several dimensions. As example, consider multiple employees having associations with a competence such as Artificial Intelligence. If such a pattern is detected, the respective competence can be attributed to the overall company or corresponding academic institution. Hence, by analyzing the overall KG, such (collective) competences can be deduced and become explicit on higher level associations. The reasoner can, in combination with SPARQL queries, which consider the established semantic relations as well as the concepts of transitivity and subsumption, deal with queries regarding multiple dimensions (e.g., dedicated technological disciplines, localization of competence occupants) and temporal aspects (e.g., competence development paths), and, therefore, provide answers to concrete questions. For example, the reasoner is able to derive institutes which are associated with a university as well as their employees and therefore deduce the competence range of the university based on the employees' competence relationships. Specifically, with the gathered *Knowledge Graph*, the framework is further capable of identifying competences which can assist to solve a given problem, detecting potential knowledge gaps, tracking changing competence strategies of competitors or which nations are building up competences in a certain technological field of interest.

Summing up, dedicated entities from processed information are created and interlinked with other explicitly or implicitly gathered entities. This conceptual and representational difference allows the framework to assess competences on multiple levels, to drill up and down, and to answer complex (business-related) questions based on other related entities, which are also associatively and hierarchically organized. Moreover, individual and collective (competence) interactions can be captured by utilizing temporal features.

3   Validation

This section presents the research design and experimental results conducted with the proposed framework using several general datasets as well as patent datasets of interest within the automotive domain.

3.1   Research Design

The analysis framework is written in Java and extends the (semantic) processing pipeline used by Ulmschneider and Glimm [14] with normalization-, unification-, and competence-related components (see previous sections). The *Knowledge Graph* (KG), which is constituted by RDF triples and its defining ontology (OWL 2 RL profile), is prepopulated with several conceptual systems such as competences,[1] technological fields of interest, locations, and several patent classifications systems (IPC, CPC, USPC). All of them are hierarchically organized and semantically interlinked. In order to evaluate the presented framework, multiple datasets are used. The preprocessing components are evaluated with the following datasets:

– University names from the literature [6,10]

---

[1] Incorporates a domain-specific competence taxonomy combined with the ESCO ontology (European Skills, Competences, Qualifications and Occupations, see http://ec.europa.eu/esco/ for more information).

– World universities (all university names of the world)
– Large companies with high impact (companies listed on major stock market indices)
– Abstracts of scientific publications (KDDCup hep-th papers 1992–2003[2]), containing metadata such as titles, authors, affiliations, dates

Further, two multilingual patent datasets from two emerging technological fields of interest, which are relevant for the automotive domain, are used for evaluating the overall competence recognition and analysis framework with regard to business-related questions:

– Alternative Mobility Concepts (AMC)
  • Electro Mobility (EM)
  • Hydrogen Mobility (HM)
– Artificial Intelligence (AI)

The two integrated patent datasets contained more than 13,600 patents from the areas of Artificial Intelligence (28.38%), Electro Mobility (41.08%) and Hydrogen Mobility (30.54%). Most patents were applied for in the United States (51.4%), followed by Japan (25.35%), Germany (7.17%), and China (3.56%). All other patents were filed in other countries (12.52%), whereas the distribution of identified languages (textual content) was 71.13% (English), 16.01% (Japanese), 5.47% (German), 3.67% (Chinese), and 3.72% (other languages).

Note that all datasets are based on real-life data. The general datasets are used as baseline to evaluate entity type detection, name normalization, and entity unification whereas the integrated patent dataset is employed for detecting and evaluating expertise for the technological fields of AMC and AI on multiple dimensions. In order to analyze patents and align them with technological fields of interest the above-mentioned extended semantic processing pipeline (SP) was employed. Remember that the SP was enhanced by competence-specific components. Based on assignments to technological fields of interest as well as associations from other derived features competence occupant profiles were then semantically interlinked with actual competences.

As evaluation metrics accuracy A, which we define as the percentage of correct results R_c out of all non-null results R_w as well as coverage C, defined as the percentage of non-null results to total results, are used. Hence, coverage incorporates null results R_n as well.

$$A = \frac{R_c}{R_c + R_w} \qquad C = \frac{R_c + R_w}{R_c + R_w + R_n}$$

Additionally, the success rate ($SR = A \times C$), which indicates how likely the framework succeeds to generate a correct result, as well as the F-measure (F_1 score), which constitutes the harmonic mean of accuracy and coverage ($F_1 = 2 \times A \times C/(A + C)$), are calculated and measure the overall performance of the respective processing steps.

---

[2] http://www.cs.cornell.edu/projects/kddcup/datasets.html.

3.2　Findings

The most important evaluation step constitutes the detection of individual and collective competence occupants and determining their correct type. Since, to the best of our knowledge, no curated data is available for patents as a source of competence allocation and the creation of such a dataset is very labor-intensive, this step is evaluated with alternative datasets containing academic institution names, company names as well as person names. For academic institution names, a baseline dataset containing UK universities as used in Liu et al. [10] and Jacob et al. [6] is employed. Further, we extended this dataset with all other university names of the world, resulting in almost 12,000 instances. Note that both datasets exclusively include academic institution names. In order to evaluate the same with business organizations, we employed a dataset containing around 600 companies (i.e., their names) which are listed on major stock market indices. Table 4 lists the evaluation results.

Table 4. Evaluation results

Dataset	A_t	C_t	SR_t	F_{1t}	A_{n1}	C_{n1}	SR_{n1}	F_{1n1}	A_{n2}	C_{n2}	SR_{n2}	F_{1n2}
UK Universities	1.00	1.00	1.00	1.00	1.00	1.00	1.00	1.00	1.00	1.00	1.00	1.00
World Universities	1.00	0.99	0.99	0.99	0.78	0.99	0.77	0.87	0.85	0.99	0.84	0.91
Large Companies	1.00	0.93	0.93	0.96	0.97	0.93	0.90	0.95	1.00	0.93	0.93	0.96

A = Accuracy, C = Coverage, SR = Success Rate, F_1 = F-measure, (t) = type detection, (n1) = name normalization (standard), (n2) = name normalization (considering all normalization stages)

Overall, the performance was satisfying with regard to accuracy and coverage for the name type detection and the name normalization task. We further evaluated whether the type of organization was identified correctly. For both, the UK and the World University dataset, the accuracy and coverage was close to 100% with regard to the determination of their organization type (academic) and structural level (e.g., institute, faculty, university). For business organizations the coverage of the organization type was slightly lower (91%) but also with an accuracy reaching almost 100%. Only the accuracy of identified legal forms was relatively low with 86%. After evaluating datasets containing collective competences we shifted our attention to single competences. Therefore, we utilized the KDDCup dataset (scientific publications) and extracted the included author information. After parsing, we received 62,664 person names and further processed them to finally retrieve unambiguous 19,993 person names. We then evaluated whether the framework is able to correctly assign the authors to their respective type (individual) and achieved 89% accuracy. Additionally, more than 2,200 associated organizations could be derived from author affiliation strings. Finally, we processed the integrated patent dataset and obtained 75,715 potential competence occupant mentions. After processing the patents with the framework, 48,046 name mentions were annotated as individuals and 22,103 as organizations. For 5,566 mentions the name type could not be determined. Among the recognized organizations 20,625 were detected as business organizations and 858 as academic institutions, both with their respective kind (e.g., legal forms such as 'Incorporated', academic types such as 'University'). For 620 organizations the type could not be uniquely determined. Moreover, the conducted experiments revealed that accurately determining entity types and profound name normalization can increase the accuracy of unification tasks for more than 9%, which, in consequence, improves the overall quality of the analyses to be conducted on basis of the KG.

After integration of all processed profiles, we applied the reasoner on the KG and studied several business cases (cf. Sect. 2). For example, we found that Microsoft attempt to concentrate their core competences (including cooperation partners) regarding Artificial Intelligence in the US. Further, regarding their patent strategy, they massively reduced patent applications in this area beginning from 2009 which might be a weak signal that the company will focus on other technologies (and competences) in the future. In contrast, Daimler increased their efforts with regard to electric storage systems in the past view years and protect their corresponding inventions globally, which can be interpreted as that they are building up and protecting their (core) competences in this area worldwide.

Summing up, the experiments revealed that considering metadata, such as spatial, temporal, topical, or relational information as well as implementing self-improvement mechanisms can increase the accuracy of the overall gathered and computationally represented innovation ecosystem. Further, the capability of answering complex (business-related) queries on top of the KG makes the framework a considerably powerful tool. However, it must be noted that the overall processing and, in particular, the entity unification process with pairwise examinations is computationally extensive (i.e., the creation of the *Knowledge Graph*). In contrast, the upstream name normalizer turned out to be inexpensive and accurate. Nonetheless, we did not compare the results with (other) machine learning approaches so far and leave this as a future task.

4   Related Work

As illustrated in the previous sections, the presented competence analysis framework is interdisciplinary and combines as well as enhances techniques from several research areas. Therefore, related work is partitioned based on the fundamental components of the framework: Named Entity Detection and Name Normalization, Named Entity Disambiguation and Unification, and competence-, skill-, or expert-related analyses.

The task of entity name normalization and unification has been studied extensively. Solutions range from rule-based, dictionary-based, or string matching techniques to machine learning and hybrid approaches based on several types of data (e.g., (domain-specific) databases, websites) and application scenarios (newspaper articles, genes, diseases, employers, job postings, academic institutions) (e.g., see [1,5–10]). Some combine Named Entity Recognition (NER) with ENN, but few consider multilinguality, temporal aspects, or (hierarchical) dependencies (e.g., institute vs. university, Germany vs. Europe). Most approaches, however, have in common, that important information, such as metadata and (implicit) relationships, are neglected. Many authors examine the problem of normalization and disambiguation as one single, isolated task (e.g., normalized strings vs. objects with extracted and normalized meta information and their relations with each other). Moreover, the differentiation between structural (competence) levels and the obligatory process to use the gathered information with regard to higher level associations, including respective (business-related) analyses, receives almost no attention. In contrast, this work shifts the document-centric view (e.g., search engines, cross-document person name normalization) to the actual (multiple types of) entity mentions within documents, their inline references and cross-references among artifacts, which are, altogether, transformed into a respective graph-based representation.

Studies dealing with competences have several purposes. Some focus on creating a thesaurus or taxonomy, e.g., to improve search engines. Others create visualizations (e.g., competence maps), mostly based on quantitative techniques and for different purposes (e.g., see [2,11]). For example, Moehrle et al. [11] create inventor competence maps from patents with focus to HR management and Barirani et al. [2] create competence maps based on patent citations to assess national and firmlevel competences. They are able to identify and locate the largest invention communities in a given technological discipline. However, the approach requires patent citations as a prerequisite. While graphical presentations allow for deriving insights with regard to the big picture (e.g., to understand interdependencies on higher levels), concrete (qualitative) questions cannot be answered based on the underlying data (e.g., with regard to specific competences, competence occupant interactions, or competence developments). Zhao et al. [16] propose a system to recognize and normalize professional skills from resumés and matching them with a taxonomy created from Wikipedia categories and resumé sections. However, exact matches between taxonomy entries and extracted skills from textual content are required. Ronda-Popu and Guerras-Martín [12] analyze collaboration correlations by measuring scientific output and impact of institutions in the academic community by employing graph-based metrics (degree centrality) to derive insights about an institution's relevance within a collaboration network in the discipline of management.

5   Summary and Outlook

Continuous change and the requirement to diversify from others to remain competitive requires highly qualified specialists who possess cutting-edge intellectual capital and who are capable of transforming ideas, technological know-how, and constraining specifications into business value. However, accurately identifying and allocating such expertise is a challenging exercise.

Hence, this paper presents an integrated framework to detect competence occupants in publications such as patents, and to represent them, along with the actual publications, conceptual systems and other business-related artifacts, as a semantic graph (KG). The resulting KG allows for their topical, structural, and spatial analysis and supports inferences on multiple dimensions while considering individual as well as several collective competence levels.

Accordingly, the pro-active exploitation and management of competences can be achieved for multiple application scenarios and HR managers, procurement managers, technical engineers, innovation managers, patent analysts, researchers, existing think thanks, or business analysts are capable of utilizing competence intelligence according to their specific needs.

Controlled experiments with multilingual patents on emerging technological disciplines, which are emphasized along with real-world application scenarios and business cases from the automotive domain, demonstrate the feasibility of extracting, processing, and analyzing expertise on multiple dimensions. In particular, we have shown how to identify individuals and organizations referenced in (scientific) publications (e.g., patents) and how to map them with indicators of expertise (e.g., related topical information) on multiple aggregation levels.

The conducted controlled experiments emphasize that the illustrated improved processing techniques can indeed increase the accuracy of identification and disambiguation of competence occupants and their alignment with other entities on individual and organizational level. Moreover, the implemented techniques allow for accurately determining the type of collective competence occupants (e.g., academic institution, business organization) and their structural level (e.g., institute vs. university). Hence, additional implicit competence information can be deduced using a reasoner which is capable of analyzing the structure and pathways (e.g., by traversing the KG) as well as higher level associations based on the semantic graph-based representation.

With this work we demonstrate how to detect, normalize, unify, aggregate, and interrelate competences in a structured and analyzable form. In order to enhance the proposed framework and add additional value, we will integrate further complementary types of relevant (business) artifacts (e.g., technology fact sheets, invention reports) to the KG and extend the KG with further semantics.

Altogether, the framework and its analysis pipeline will be further developed and enhanced with focus on integrated interlinked views on competences and complementary entities as well as their (latent) interdependencies targeting a broader view and allowing additional predictive features based on the representation of the gathered innovation ecosystem (e.g., competence requirement foresight, (competitor) competence activity predictions).

References

1. Aswani, N., Bontcheva, K., Cunningham, H.: Mining information for instance unification. In: Cruz, I., Decker, S., Allemang, D., Preist, C., Schwabe, D., Mika, P., Uschold, M., Aroyo, L.M. (eds.) ISWC 2006. LNCS, vol. 4273, pp. 329–342. Springer, Heidelberg (2006). doi:10.1007/11926078_24
2. Barirani, A., Agard, B., Beaudry, C.: Competence maps using agglomerative hierarchical clustering. J. Intell. Manufact. **24**(2), 373–384 (2013). doi:10.1007/s10845-011-0600-y

3. Ernst, H.: Patent information for strategic technology management. World Pat. Inf. **25**(3), 233–242 (2003). doi:10.1016/S0172-2190(03)00077-2

4. Giereth, M., Stäbler, A., Brügmann, S., Rotard, M., Ertl, T.: Application of semantic technologies for representing patent metadata. In: Informatik 2006, vol. 1, pp. 297–304 (2006)

5. Huang, S., Yang, B., Yan, S., Rousseau, R.: Institution name disambiguation for research assessment. Scientometrics **99**(3), 823–838 (2014). doi:10.1007/s11192-013-1214-2

6. Jacob, F., Javed, F., Zhao, M., Mcnair, M.: sCooL: a system for academic institution name normalization. In: Proceedings of 15th International Conference on Collaboration Technologies and Systems (CTS 2014), pp. 86–93 (2014). doi:10.1109/CTS.2014.6867547

7. Jijkoun, V., Khalid, M.A., Marx, M., de Rijke, M.: Named entity normalization in user generated content. In: Proceedings of 2nd Workshop on Analytics for Noisy Unstructured Text Data (AND 2008), pp. 23–30 (2008). doi:10.1145/1390749.1390755

8. Jonnalagadda, S., Topham, P.: NEMO: extraction and normalization of organization names from PubMed affiliation strings. J. Biomed. Discov. Collab. **5**, 50–75 (2010)

9. Khalid, M.A., Jijkoun, V., de Rijke, M.: The impact of named entity normalization on information retrieval for question answering. In: Macdonald, C., Ounis, I., Plachouras, V., Ruthven, I., White, R.W. (eds.) ECIR 2008. LNCS, vol. 4956, pp. 705–710. Springer, Heidelberg (2008). doi:10.1007/978-3-540-78646-7_83

10. Liu, Q., Javed, F., Mcnair, M.: CompanyDepot: employer name normalization in the online recruitment industry. In: Proceedings of 22nd ACM SIGKDD International Conference on Knowledge Discovery and Data Mining (KDD 2016), pp. 521–530 (2016). doi:10.1145/2939672.2939727

11. Moehrle, M.G., Walter, L., Geritz, A., Müller, S.: Patent-based inventor profiles as a basis for human resource decisions in research and development. R&D Manag. **35**(5), 513–524 (2005). doi:10.1111/j.1467-9310.2005.00408.x

12. Ronda-Pupo, G.A., Guerras-Martín, L.Á.: Collaboration network of knowledge creation and dissemination on management research: ranking the leading institutions. Scientometrics **107**(3), 917–939 (2016). doi:10.1007/s11192-016-1924-3

13. Tseng, Y.H., Lin, C.J., Lin, Y.I.: Text mining techniques for patent analysis. Inf. Process. Manag. **43**(5), 1216–1247 (2007). doi:10.1016/j.ipm.2006.11.011

14. Ulmschneider, K., Glimm, B.: Semantic exploitation of implicit patent information. In: Proceedings of 7th IEEE Symposium Series on Computational Intelligence (SSCI 2016) (2016). doi:10.1109/SSCI.2016.7849943

15. Zhang, L., Li, L., Li, T.: Patent mining: a survey. SIGKDD Explor. **16**(2), 1–19 (2014)

16. Zhao, M., Javed, F., Jacob, F., Mcnair, M.: SKILL: a system for skill identification and normalization. In: Proceedings of 29th Conference on Innovative Applications of Artificial Intelligence (AAAI 2015), pp. 4012–4017 (2015)

Scalable Data Access and Storage Solutions

RDF Updates with Constraints

Mirian Halfeld-Ferrari[1], Carmem S. Hara[2], and Flavio R. Uber[2,3]([✉])

[1] Université d'Orléans, INSA CVL, LIFO EA 4022, 45067 Orléans, France
mirian@univ-orleans.fr
[2] Universidade Federal do Paraná, Curitiba, PR, Brazil
carmem@inf.ufpr.br, flavio.uber@gmail.com
[3] Universidade Estadual de Maringá, Maringá, PR, Brazil

Abstract. This paper deals with the problem of updating an RDF database, expected to satisfy user-defined constraints as well as RDF intrinsic semantic constraints. As updates may violate these constraints, side-effects are generated in order to preserve consistency. We investigate the use of nulls (blank nodes) as placeholders for unknown required data as a technique to provide this consistency and to reduce the number of side-effects. Experimental results validate our goals.

Keywords: RDF · RDFS · Constraints · Updates

1 Introduction

Due to the increasing number of distributed RDF datasets and their dynamic nature, the development of techniques for ensuring their consistency becomes a fundamental data quality issue. However, when analyzing the database and the web semantics worlds, a dichotomy on the notion of consistency can be observed. The web semantics world adopts the open world assumption (OWA) and ontological constraints are, in fact, inference rules. The database world usually adopts the closed world assumption (CWA) and constraints impose data restrictions. Let us consider the rule $r : Researcher(X) \rightarrow Professor(X)$ and a database storing the fact that Bob is a researcher ($D = \{Reseacher(Bob)\}$). When r is an inference rule, D is consistent because $Professor(Bob)$ is inferred from D and r. However, if r is a constraint, D is inconsistent because the fact $Professor(Bob)$ is not true in D (here, facts which are not stored in the database are considered false). Although inference rules and constraints can co-exist ([11,17]) their mechanisms are usually defined separately.

This paper adopts the database point of view and deals with the problem of updating an RDF database. Traditionally, whenever a database is updated, if constraint violations are detected, either the update is refused or compensation actions, which we call side-effects, must be executed in order to guarantee their satisfaction. Our work tackles the problem of "active rules" for RDF and computes the side-effects required by update operations. The originality of our approach is in the use of blank nodes as free nulls in the computation of side-effects. Although blank nodes have different capabilities [4], in this paper, we are

© Springer International Publishing AG 2017
P. Różewski and C. Lange (Eds.): KESW 2017, CCIS 786, pp. 229–245, 2017.
https://doi.org/10.1007/978-3-319-69548-8_16

only interested in their standard interpretation as existential variables (which can be replaced by free labeled nulls). To avoid confusion, we will refer to them as null nodes (or just nulls), used as placeholders for unknown required data. Notice however that in our approach user's update requirements have no nulls: nulls can only be generated automatically during side-effect computation.

We work with a logical formalism using special predicates to describe RDF data. For instance, we write: (i) $CI(Bob, Professor)$ to express that Bob is an instance of the class $Professor$ and (ii) $PI(Bob, DB, teaches)$ to indicate an instance of property $teaches$, assuming that $Professor$ and $Courses$ are, respectively, the property's domain and range. In this context, the following example illustrates our challenges and gives an overview of our approach.

(a)	(b)
CI(Bob, Researcher)	r_1:PI(X_1,X_2,coordinates)→CI(X_1,Researcher)
CI(Bob, Professor)	r_2:CI(X_1,Researcher)→PI(X_1,X_2,isMember)
PI(Bob, Jupiter, isMember)	r_3:PI(X_1,X_2,coordinates)→PI(X_1,X_2,isMember)
PI(Bob, DB, teaches)	r_4:CI(X_1,Researcher)→ CI(X_1,Professor)
PI(Bob, CNPq,grantFrom)	r_5:CI(X_1,Professor)→ ¬CI(X_1,Student)
CI(Ann, Student)	r_6:CI(X_1,Professor)→ PI(X_1,X_2,teaches)
PI(Tom, Java, teaches)	r_7:PI(X_1,X_2,grantFrom)→CI(X_1,Researcher)

Fig. 1. Database instance D (a) and Constraints $\mathcal{C}$ (b) for Example 1

Example 1. Let $\mathcal{C}$ (Fig. 1) be a set of constraints defined on an academic application. Constraints state that only a researcher may coordinate a project (r_1) and that he must also be a member of this project (r_3). All researchers are required to be a member of at least one project (r_2). Researchers are professors (r_4), professors cannot be students (r_5) and are required to teach at least one course (r_6). Finally, people receiving research grants should be researchers (r_7). Constraints are defined as rules, where the left-hand side is called the *body* of the rule, and the right-hand side is its *head*. In this example we consider this set of constraints and analyze how a database instance is updated according to successive update requirements which may generate side-effects *w.r.t.* $\mathcal{C}$. Consider the database instance in Fig. 1, which is consistent *w.r.t.* $\mathcal{C}$.

The insertion of the fact $CI(Ann, Professor)$ cannot be executed by simply adding this new fact in D because it provokes the violation of r_5 and r_6. The following side effects should be considered: (i) rule r_5 generates $\neg CI(Ann, Student)$, which corresponds to the deletion of $CI(Ann, Student)$, and (ii) rule r_6 produces $PI(Ann, N_1, teaches)$ where N_1 a new fresh null, indicating that Ann teaches a course, although it is not yet known which one. The new updated database is: $D_1 = (D \cup \{CI(Ann, Professor), PI(Ann, N_1, teaches)\}) \setminus \{CI(Ann, Student)\}$. Notice that D_1 contains a null value produced during the side effect computation.

Now, consider the deletion of $f = PI(Bob, Jupiter, isMember)$ from D_1. To avoid the violation of r_2 we cannot just eliminate f from D_1. The usual solution (also proposed by [8]) is to delete all facts generating f, in order to obtain $D_{trad} = D_1 \setminus \{PI(Bob, Jupiter, isMember), CI(Bob, Researcher), PI(Bob, CNPq, grantFrom)\}$. Such a solution seems too radical. Rule r_2 says that if someone is a researcher, there should *exist* a project having this person as a member. Deleting f only indicates that Bob is not a member of project $Jupiter$ any more (he is perhaps a member of an another project which we do not know yet). Therefore, in this situation, our proposal is to replace $PI(Bob, Jupiter, isMember)$ by $PI(Bob, N_3, isMember)$ where N_3 is a fresh null, a placeholder indicating that, for the moment, we do not know on which project Bob is working. The new updated database is $D_2 = (D_1 \setminus \{PI(Bob, Jupiter, isMember)\}) \cup PI(Bob, N_3, isMember)$. Notice that in this way we limit cascading deletions.

The latter reasoning is not appropriate for every situation. As a last example, consider the deletion of $CI(Bob, Professor)$. Rule r_4 states that *all* researchers must be professors. Clearly, if Bob is not a professor, he cannot be accepted as a researcher. Also if he is not a researcher, he cannot receive a research grant (r_7). In this case, replacing Bob by a null in $CI(Bob, Professor)$ is meaningless. To perform this update, cascading deletes are necessary. The new database is: $D_3 = D_2 \setminus \{CI(Bob, Professor), CI(Bob, Researcher), PI(Bob, CNPq, grantFrom)\}$. Notice that $PI(Bob, N_3, isMember)$ is still in D_3. Indeed, a rule such as r_2 does not impose members (of a project) to be researchers.   □

An important aspect of our update proposal is to use nulls when a deletion concerns instantiations of existential variables in the head of a constraint. To the best of our knowledge this is the first work that proposes an automated mechanism to introduce null nodes in RDF datasets to limit cascade updates.

We deal with two kinds of constraints separately: application constraints ($\mathcal{C}$), imposed by a user to personalize his context or to establish the particularity of his application and the RDF intrinsic semantic constraints ($\mathcal{A}$). Algorithms to compute side-effects from $\mathcal{A}$ and $\mathcal{C}$ are developed as distinct steps. Our goal is to compute $\mathcal{C}$'s side effects without consulting the database. The computation of $\mathcal{A}$'s side-effects, however, requires knowledge of the underlying data.

The rest of the paper is organized as follows. Section 2 defines our data model, constraints and update operations. Section 3 considers the computation of side-effects. Section 4 reports our experimental results. Related work and final remarks (Sects. 5 and 6) conclude the paper.

2 RDF Constraints and Updates

Let $\mathbf{A}_C = \{a, b, \ldots, a_1, a_2, \ldots\}$, be a countably infinite set of constants, $\mathbf{A}_N = \{N_1, N_2, \ldots\}$, a countably infinite set of nulls and $var = \{X_1, X_2, \ldots, Y_1, \ldots\}$ be an infinite set of variables ranging over elements in $\mathbf{A}_C \cup \mathbf{A}_N$. We use $\mathbf{X}$ as an abbreviation for $X_1 \ldots X_k$ where $k \geq 0$. A *term* is a constant, a null or a variable. Extending the logical formalism of [8] to deal with nulls, we classify predicates into two sets: (i) SchPred $= \{Cl, Pr, CSub, Psub, Dom, Rng\}$, used to define the database schema, standing respectively for classes, properties, sub-classes, sub-properties, property domain and range, and (ii) InstPred $= \{CI, PI, Ind, BN\}$, used to define the database instance, standing respectively for class and property instances, individuals and blank nodes (or nulls). An *atom* has the form $P(u)$, where P is a predicate, and u is a list of terms. When all the terms of an atom are in $\mathbf{A}_C \cup \mathbf{A}_N$, we have an instantiated atom.

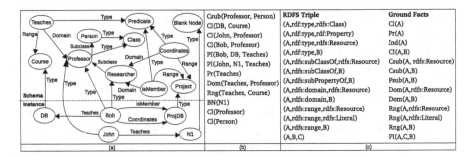

Fig. 2. (a) RDF instance and schema. (b) Dataset defined as a set of facts. (c) Correspondence of RDF/S triples and facts [8]

When they are all in $\mathbf{A}_C$, we have a fact. Figure 2 illustrates our model. Figure 2(a) shows an RDF instance and schema as a graph, and Fig. 2(b) a subset of its representation as positive atoms. Figure 2(c), reproduced from [8], shows the correspondence between triples in RDF/S and facts.

Definition 1 (Database). An RDF database is a triple $\Delta = (D, D_{Sch}, \Sigma)$ where D is the database instance (a set of instantiated atoms with predicates in INSTPRED), D_{Sch} is the database schema (a set of facts with predicates in SCHPRED) and $\Sigma = (\mathcal{A}, \mathcal{C})$ is a set of constraints, where $\mathcal{A}$ is a set of RDF semantic constraints and $\mathcal{C}$ is a set of application constraints. □

2.1 Constraints

A constraint is a logical rule r whose left-hand side is denoted as $body(r)$, while the right-hand side is denoted as $head(r)$. Application constraints personalize the context on which an RDF database is treated while RDF constraints ensures the intrinsic RDF/S semantics.

Definition 2 (Application Constraints). Let c_1, c_2 be class labels, and p_1, p_2 be property labels in $\mathbf{A}_C$. Application rules in $\mathcal{C}$ have the forms presented in Table 1. Moreover, the following restrictions are imposed on $\mathcal{C}$: (1) for constraints r_1 of Type 2 there exists no constraint $r_2 \in \mathcal{C}$ such that $head(r_1)$ and $body(r_2)$ are unifiable; (2) for constraints r_1 of Type 3 there exists no constraint $r_2 \in \mathcal{C}$ such that $body(r_1)$ and $head(r_2)$ are unifiable. □

Our constraints are special cases of tuple generating dependencies (TGDs). We refer to [14] to recall that TGDs are database dependencies represented by the logical formula $\forall \mathbf{X} \ \phi(\mathbf{X}) \rightarrow \exists \mathbf{Y} \ \psi(\mathbf{X}, \mathbf{Y})$; where ϕ and ψ are conjunctions of atoms, all with variables among those in $\mathbf{X}$– every variable in $\mathbf{X}$ appears in $\phi(\mathbf{X})$ but not necessarily in $\psi(\mathbf{X}, \mathbf{Y})$. Although we restrict ourselves to the so-called linear LAV (local-as-a-view) TGDs [1] *i.e.*, to rule's body and head with a single atom, our constraints allow a negative atom in their heads. Restrictions imposed to our constraint rules aim at avoiding null

propagation and at guarantee-
ing deterministic updates. In
this context, side-effects generation does not deal with
the well-known chase problems (considered, for instance,
in [6]). For instance, consider r_2 of Example 1 and
r_2' : $PI(X_1, X_2, isMember)$ →
$PI(X_2, X_3, postulate4Grants)$.
In such a context, the insertion of $CI(Bob, Researcher)$
would generate $PI(Bob, N_1, is$
$Member)$ and $PI(N_1, N_2,$
$postulate4Grants)$. Our con-

Table 1. Types of application constraints.

Type 1:	$CI(X_1, c_1) \rightarrow CI(X_1, c_2)$ or
	$CI(X_1, c_1) \rightarrow \neg CI(X_1, c_2)$ or
	$PI(X_1, X_2, p_1) \rightarrow PI(X_1, X_2, p_2)$ or
	$PI(X_1, X_2, p_1) \rightarrow \neg PI(X_1, X_2, p_2)$
Type 2:	$CI(X_1, c_1) \rightarrow PI(X_1, X_2, p_1)$ or
	$CI(X_1, c_1) \rightarrow \neg PI(X_1, X_2, p_1)$ or
	$CI(X_2, c_1) \rightarrow PI(X_1, X_2, p_1)$ or
	$CI(X_2, c_1) \rightarrow \neg PI(X_1, X_2, p_1)$
Type 3:	$PI(X_1, X_2, p_1) \rightarrow CI(X_1, c_1)$ or
	$PI(X_1, X_2, p_1) \rightarrow \neg CI(X_1, c_1)$
	$PI(X_1, X_2, p_1) \rightarrow CI(X_2, c_1)$ or
	$PI(X_1, X_2, p_1) \rightarrow \neg CI(X_2, c_1)$

straints bypass controversial aspects related to the generation of nulls from nulls
in an update context by avoiding their propagation (*i.e.*, refusing a set $\mathcal{C}$ where
both rules, such as r_2 and r_2', exist).

Our choice in separating application constraints from RDF/S semantic constraints allows us to impose the above restrictions to $\mathcal{C}$ without interfering with
the well-known RDF/S constraints. With such restrictions we are able to build
a simple and efficient algorithm to compute $\mathcal{C}$'s side-effects without dealing with
some tricky aspects of the chase and without consulting the database instance.

2.2   Updates

An *update set* is a set of operations where each operation is a positive or negative
instantiated atom corresponding, respectively, to insertions and deletions which
are performed in just *one* transaction. Only instance-level updates (*i.e.*, involving
predicates in INSTPRED) are treated by our approach.

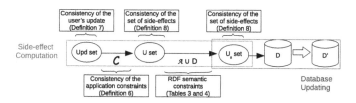

Fig. 3. Computing update side effects: consistency requirements at each step.

We distinguish three update steps, summarized in Fig. 3. Each step refers to
a distinct update set respecting specific consistency requirements. To update a
database instance D, a user gives as input his update requests in the set *upd*
where only *facts* are allowed. Facts in *upd* may trigger constraints in $\mathcal{C}$ allowing

the computation of the set U. Finally, by using the RDF constraints in $\mathcal{A}$ and information in D, the set U_S is generated. The updated database D' is obtained by applying updates in U_S on D.

Before considering the computation of U and U_S, we establish our update semantics by showing how the updates in U_S are applied to D.

Table 2. Semantics of update operations on a database instance D.

$[\![CI(a,c)]\!]_D = \{CI(a,c)\}$	$[\![\neg CI(a,c)]\!]_D = \{\neg CI(a,c)\}$
$[\![\neg Ind(a)]\!]_D = \{\neg Ind(a)\}\cup$ $\{\neg PI(X,a,p) \mid PI(X,a,p) \in D\}\cup$ $\{\neg PI(a,X,p) \mid PI(a,X,p) \in D\}$	$[\![Ind(a)]\!]_D = \{Ind(a)\}$
$[\![BN(N_1)]\!]_D = \{BN(N_1)\}$	$[\![\neg BN(N_1)]\!]_D = \{\neg BN(N_1)\}$
$[\![PI(a,b,p)]\!]_D = \{PI(a,b,p)\}\cup$ $\{\neg PI(N_1,b,p),\neg BN(N_1) \mid PI(N_1,b,p) \in D\}\cup$ $\{\neg PI(a,N_1,p),\neg BN(N_1) \mid PI(a,N_1,p) \in D\}$	$[\![\neg PI(a,b,p)]\!]_D = \{\neg PI(a,b,p)\}$
$[\![PI(N_1,b,p)]\!]_D =$ if there exists X such that $PI(X,b,p) \in D$ then $\{\}$ else $\{PI(N_1,b,p),BN(N_1)\}$	$[\![\neg PI(N_1,b,p)]\!]_D = \{\neg PI(X,b,p) \mid PI(X,b,p) \in D\}$
$[\![PI(a,N_1,p)]\!]_D =$ if there exists X such that $PI(a,X,p) \in D$ then $\{\}$ else $\{PI(a,N_1,p),BN(N_1)\}$	$[\![\neg PI(a,N_1,p)]\!]_D = \{\neg PI(a,X,p) \mid PI(a,X,p) \in D\}$

Given an update set, a positive atom denotes an insertion while a negative one denotes a deletion. When dealing with a non-null database instance, the semantics of these basic operations is straightforward. Given update sets $\{CI(a,c)\}$ and $\{\neg CI(a,c)\}$ on D, the resulting databases are, respectively, $(D \cup \{CI(a,c)\})$ and $(D \setminus \{CI(a)\})$. The possibility of having nulls in the database, introduced during side-effect computation, imposes a new update semantics, defined in Table 2. Operations not listed in the table have no direct effect on the database. For each operation, we show what should be added (positive atoms) and removed (negative atoms) from a database D. Intuitively, nulls are inserted in D only if there exists no other resource in D that plays the same role. Insertion and removal of *property instances* are particularly impacted. An insertion of a property p without nulls ($[\![PI(a,b,p)]\!]_D$) *replaces* one that involves nulls that may already exist in the database. This is in accordance to our semantics for nulls as placeholders of unknown required data. Thus, an insertion of a property involving a resource a and a null N_1 ($[\![PI(a,N_1,p)]\!]_D$ or $[\![PI(N_1,a,p)]\!]_D$) only affects the database if a is not involved with any other resource or null by property p. Since inserted nulls are interpreted as existential variables, nulls in removal operations (negative atoms) are interpreted as universally quantified variables. Thus, $[\![\neg PI(a,N_1,p)]\!]_D$ results in deleting from the database all properties p of resource a. We denote by $D \uplus U_S$ the result of applying an update set

U_S on database D. Based on the semantics of these operations, we can define the notion of subsumption between update operations.

Definition 3 (Update Subsumption). Operation op_1 is subsumed by op_2, denoted as $op_1 \preceq op_2$ if for any database D, $(\llbracket op_1 \rrbracket_D \cup \llbracket op_2 \rrbracket_D) = \llbracket op_2 \rrbracket_D$. That is, the effect on the database of executing op_1 and op_2 is the same as executing only op_2. Given a set of update operations upd, we denote by $clean(upd)$ a subset of upd with no subsumed operations.   □

As examples, $PI(Bob, N_1, coordinates) \preceq PI(Bob, Proj\ DB, coordinates)$ and $\neg PI(Bob, ProjDB, coordinates) \preceq \neg PI(Bob, N_1, coordinates)$. Thenceforth, we only consider update sets containing non-subsumed operations.

3   Computing Side Effects

We develop algorithms for computing side-effects which only affect the database instance. They have the following guidelines: (i) from an input set of updates upd we use constraints to generate a new list of updates that should be performed to guarantee constraint validation; (ii) null nodes are generated during the inference process that computes side effects $i.e.$, the inference can be stopped by the use of nulls; (iii) side effects from application constraints are computed without consulting the database; and (iv) side effects from RDF semantic constraints are computed in a subsequent step and may require inspection of the database.

3.1   Side Effects Based on Application Constraints

Firstly, let us introduce some notations. Let I be an update set, then: (1) $I = I^+ \cup I^-$ where I^+ is the set of positive atoms and I^- is the set of negative atoms and (2) $I = \bullet(I) \cup \circ(I)$ where $\bullet(I)$ is a set of ground atoms (without nulls) and $\circ(I)$ is a set of atoms having nulls. Clearly we can write $I^+ = \bullet(I^+) \cup \circ(I^+)$ and $I^- = \bullet(I^-) \cup \circ(I^-)$. We recall that a *homomorphism* from the set of atoms A_1 to the set of atoms A_2, both over the same predicate P, is a a function (or a *substitution*) h from the terms of A_1 to the terms of A_2 such that: (i) if $t \in \mathbf{A}_C$, then $h(t) = t$, and (ii) if $P(t_1, ..., t_n) \in A_1$, then $P(h(t_1), ..., h(t_n)) \in A_2$. If h is a homomorphism, $P(h(t_1), ..., h(t_n))$ is simply denoted by $h(P(t_1, ..., t_n))$. The notion of homomorphism naturally extends to conjunctions of atoms.

Now, given I and $\mathcal{C}$ we introduce three operators used for computing side-effects. The first operator (T) computes the side-effects traversing the rule forward, from the body to the head while the second (ϑ) moves backwards, from the head to the body.

Definition 4 (Operators T and ϑ). Let $T_\mathcal{C}$ and $\vartheta_\mathcal{C}$ be operators over a set of application constraints $\mathcal{C}$ and an update set I. When the set $\mathcal{C}$ is understood, we write T (respectively, ϑ) instead of $T_\mathcal{C}$ (respectively, $\vartheta_\mathcal{C}$).

$T(I) = I \cup \{l \mid \exists c \in \mathcal{C} \land$ there is a homomorphism h such that $h(body(c)) \in I$
$\land\, l = \widehat{h}(head(c))$ where $\widehat{h} \supseteq h$ is an extension of h such that when c is of Type 2
and Y is the set of terms in $head(c)$ and not in $body(c)$, then for every $y_i \in Y$,
if $y_i \in var$, then $\widehat{h}(y_i) = N_i$ where $N_i \in \mathbf{A}_N$ is a fresh null.$\}$

$\vartheta(I) = I \cup \{l \mid \exists c \in \mathcal{C} \land$ there is a homomorphism h such that $h(head(c)) \in I$
$\land\, l = \widehat{h}(body(c)$ where $\widehat{h} \supseteq h$ is an extension of h such that when c is of Type 3
and Y is the set of terms in $body(c)$ and not in $head(c)$, then for every $y_i \in Y$,
if $y_i \in var$, then $\widehat{h}(y_i) = N_i$ where $N_i \in \mathbf{A}_N$ is a fresh null node.$\}$ □

Operators T and ϑ are monotonic and have a *least fixed point*. We denote by
$T^*(I)$ and by $\vartheta^*(I)$, the least fixed point of T and ϑ, respectively, with respect
to a set of ground atoms I. Since constraints in $\mathcal{C}$ have only positive bodies, we
have $T^*(I^-) = I^-$ and $T^*(I^-)^- = \vartheta^*(I^-)^- = I^-$.

Example 2. Consider $\mathcal{C}$ of Example 1. Let $I = \{PI(John, projDB, coordinates)\}$.
Applying Definition 4, we obtain $T^*(I) = \{\ PI(John, projDB, coordinates),$
$CI(John, Researcher), PI(John, N_1, isMember), PI(John, projDB, isMember),$
$CI\ (John, Professor), PI(John, N_2, teaches), \neg CI(John, Student)\}$. Now let $I = \{$
$PI(John, projDB, isMember)\}$. Applying Definition 4, we obtain $\vartheta^*(I) = \{$
$PI(John, projDB, isMember), CI(John, Researcher), PI(John, projDB, coordi$-
$nates), PI(John, N_1, coordinates), PI(John, N_2, grantFor)\}$.   □

We denote by $\vartheta_{|ty_{1,3}}$ the application of the operator ϑ restricted to con-
straints of Types 1 and 3 and we introduce a third operator, η, that only
applies to rules of Type 2. As seen in Example 1 (instance D_2), a special treatment is proposed when dealing with instantiations of existential variables on
a constraint's head. As defined below, $\eta(PI(Bob, Jupiter, isMember))$ contains
$PI(Bob, N_1, isMember)$.

Definition 5 (Operator η). Operator η (or $\eta_{\mathcal{C}}$) is applied only on Type 2
rules. For each rule c of Type 2, let us denote by (i) X the set of terms appearing
in both $body(c)$ and $head(c)$ and (ii) Y the set of terms appearing in $head(c)$
but not in $body(c)$. Let I be a set of facts.
$\eta(I) = \{l \mid \exists c \in \mathcal{C}$ such that c is of Type 2 $\land$ there is an homomorphism h such
that $h(head(c)) \in I \land$ there is an homomorphism h_2 such that $l = h_2(head(c))$
with $h_2(X) = h(X) \land$ for every $y_i \in Y$ if $y_i \in var$ then $h_2(y_i) = N_i$, where N_i
is a fresh null node.$\}$ □

Consistency. We now turn to the problem of defining different notions of consistency employed on each step of our method (as illustrated in Fig. 3). Firstly,
since rules in $\mathcal{C}$ may have positive or negative literals in their heads, it is important to determine when a set $\mathcal{C}$ is a consistent set of update rules (algorithms of
this kind can be found in [11]).

Definition 6 (Consistent set of application constraints). *A set $\mathcal{C}$ of appli-*
cation constraints is consistent if for every fact f, $(T^*(f))^+ \cap \neg.(T^*(f))^- = \emptyset$. □

Now let us establish the consistency definition for the user's update requests.

Definition 7 (Consistency of the user's update requests). *Given an update set I, we say that I is a consistent set of user's update requests if there are no two atoms $l_1 \in I$ and $l_2 \in I$ for which there exists a homomorphism from $A_N \rightarrow (A_N \cup A_C)$ such that $h(l_1) = \neg h(l_2)$.* □

The consistency introduced in Definition 7 is imposed to the set *upd* (Fig. 3) and also to its immediate consequences obtained by traversing rules forward (operator T). However, the notion of a *consistent set of side-effects*, such as U, which considers traversing backwards (operator ϑ and η) is more relaxed than the notion stated in Definition 7. Indeed, the consistency of sets U (obtained by Algorithm 1, which computes the side-effects imposed by application constraints) and U_S (obtained by Algorithm 2, which computes the side-effects imposed by RDF semantic constraints) follows the definition below.

Definition 8 (Consistency of the set of side-effects). *Given an update set I, we say that I is a consistent set of side-effects if for each positive atom $l_0 \in I$ there is no negative atom $\neg.l_1 \in I$ such that $l_0 = h(l_1)$ where h is a homomorphism from $A_N \rightarrow (A_N \cup A_C)$.* □

As an example, let us consider the following sets: $I_1 = \{\neg.PI(Bob, DB, teaches), PI(Bob, N_1, teaches)\}$; $I_2 = \{PI(Bob, DB, teaches), \neg.PI(Bob, N_1, teaches)\}$; $I_3 = \{PI(Bob, N_1, teaches), \neg.PI(Bob, N_1, teaches)\}$ and $I_4 = \{PI(Bob, DB, teaches), \neg.PI(Bob, DB, teaches)\}$. According to Definition 7 all of them are examples of inconsistent update sets. However, according to Definition 8 set I_1 is consistent, while sets I_2, I_3 and I_4 are inconsistent. Indeed, I_2, I_3, and I_4 are direct consequences from the semantics of the operations. The discussion of I_1 consistency is more subtle. First, consider Definition 7, which must be satisfied after traversing rules forward. Observe that if $PI(Bob, N_1, teaches)$ is in I_1 it must be the case that there exists a rule imposing the insertion, such as rule r_6 in Example 1. Thus, the update set must contain an operation that triggered the insertion of $PI(Bob, N_1, teaches)$, such as $CI(Bob, Professor)$. On the other hand, $\neg.PI(Bob, DB, teaches)$ is also in I_1. In the traditional (cascading) semantics, this update would required the deletion of $CI(Bob, Professor)$, according to r_6. But this is inconsistent with the fact that triggered the insertion of $PI(Bob, N_1, teaches)$. Intuitively, it is not clear what the user's intentions are. Since Definition 7 concerns the consistency of the *user* update requests, we consider this demand inconsistent. Inconsistent sets of user's update requests are rejected. The process of computing side-effects continues for consistent update sets, traversing rules backwards (operators ϑ and η). Definition 8 applies to the result of this process. Consider again rule r_6 of Example 1 and a single user update request $\neg.PI(Bob, DB, teaches)$. The application of operator η generates as side-effect $PI(Bob, N_1, teaches)$ as a *technique* to stop cascading updates. Thus, according to Definition 8, set I_1 is consistent because it started with a consistent set of user's update requests (and thus without the aforementioned ambiguity) and the existence of both $\neg.PI(Bob, DB, teaches)$ and $PI(Bob, N_1, teaches)$ must be the result of applying the η operator.

Algorithm. We now introduce Algorithm 1 to compute side effects imposed by *application* constraints. In this algorithm, Line 1 applies operator T over an update set *upd* and obtains a set of positive and negative atoms which are the side effects imposed by C moving forward. Once update requests are cleaned, consistency is verified (Line 2). Notice that at this step we still require consistency *w.r.t.* Definition 7. However, for the result set U consistency requirements is relaxed (Definition 8). We handle positive and negative atoms separately. When rules of Type 2 (Definition 2) are used by T, fresh null nodes appear and thus the result in $T^*(upd)$ is a set of atoms, not necessarily grounded.

Algorithm 1. Side effects due to C

Input: A set of (positive or negative) facts *upd* and a consistent set of constraints C.
Output: The set U of side effects of *upd* w.r.t. C.
 1: Compute $T^*(upd) := clean(T^*(upd))$, where $T^*(upd) = T^*(upd)^+ \cup T^*(upd)^-$
 2: **if** $T^*(upd)$ is consistent *w.r.t.* Definition 7 **then**
 3:    $Inv := \emptyset; U := \emptyset;$
 4:    **for all** $f \in \circ(T^*(upd)^-)$ or $f \in T^*(upd)^+$ **do**
 5:       $Inv := Inv \cup \vartheta^*(\neg f)$
 6:    **for all** $f \in \bullet(T^*(upd)^-)$ **do**
 7:       $Inv := Inv \cup \vartheta^*_{|ty1,3}(\neg f);$ $\circ(U^+) := \eta(\neg f);$
 8:    $U^+ = U^+ \cup T^*(upd)^+;$ $U^- = T^*(upd)^- \cup \neg.Inv^+;$
 9:    **return** U
10: **else**
11:    Error Exception: update requests are not consistent

Example 3. Analyzing $T^*(upd)$ of Example 2 we notice that: (i) cleaning $T^*(upd)$ implies annulling the update request $PI(John, N_1, isMember)$, since it is subsumed by $PI(John, projDB, isMember)$ and (ii) if $CI(John, Student)$ is in the database, it must be removed. Now, each positive atom obtained in Line 1 of Algorithm 1 represents a required insertion and thus, cannot be false in the database. Operator ϑ is used to find atoms which generate $\neg f$ (the inverse of f). Given $f_1 : \neg PI(John, projDB, isMember)$, the result of $\vartheta^*(\neg f_1)$ is depicted in Example 2 and contains $f_2 : PI(John, projDB, coordinates)$, obtained when moving backwards on r_3. Observe that f_1 is in $T^*(upd)^-$ and f_2 is inserted in Inv (Line 7). Subsequently, in Line 8, $\neg f_2$ is inserted in the resulting update set U^-. The same treatment is given for positive atoms in $T^*(upd)^+$, and negative atoms with null nodes ($\circ(T^*(upd)^-)$) (Lines 4 and 5). Notice that $f_3 : CI(John, Researcher)$ is also in $\vartheta(\neg f_1)$ by rule r_2. However, r_2 is of Type 2. In this case, the algorithm does not include f_3 in Inv (since the ϑ operator is restricted to rules of Types 1 and 3), but applies the η operator (Line 7) on f_1 and includes $f_4 : PI(John, N_1, isMember)$ in Inv. As consistency requirements at this step are different from those in Definition 7, both f_1 and f_4 are accepted in the resulting set U and we stop the backward chaining.   □

Proposition 1. *Let upd be a consistent set of user's update requests (Definition 7) and $\mathcal{C}$ a consistent set of constraints (Definition 6). The update set U obtained by Algorithm 1 satisfies the following properties: (1) $upd \subseteq U$; (2) U is consistent w.r.t. Definition 8; (3) U satisfies $\mathcal{C}$ and (4) for each atom $l \in (U \backslash upd)$ the set $U \backslash \{l\}$ does not satisfy $\mathcal{C}$.*   □

3.2   Side Effects Based on RDF Semantic Constraints

An RDF database should respect the intrinsic RDF semantic constraints in $\mathcal{A}$. Thus, following the same reasoning used for constraints in $\mathcal{C}$, our approach proposes to generate additional updates in order to maintain consistency w.r.t. $\mathcal{A}$. Constraints in $\mathcal{A}$ are those presented in Tables 3 and 4. Table 3 borrows from [8] a subset of the RDF semantic constraints, and Table 4 presents the rules that have been modified and added in order to consider the existence of nulls.

In Table 4, we consider that null nodes are distinct from predicates, entities and classes (rules b_1-b_3) and that they can be the subject or object of a property (rules b_4-b_5). They are also used to stop the propagation of properties required by particular classes. This is done by allowing properties to connect to null nodes even when the type of the subject and the object are defined (rules b_6 and b_7). The following example illustrates this point.

Example 4. In Example 1 we assume the existence of a class *Person* which is the domain of property *isMember*. The insertion of $PI(N_1, projDB, isMember)$ satisfies the RDF semantic constraints without any additional side-effects. This is because rule b_6 allows a null node to be the subject of *isMember* even when the schema defines that $Dom(isMember, Person)$. However, in [8], b_6 is defined as $PI(x,y,z) \wedge Dom(z,w) \rightarrow CI(x,w)$. In order to satisfy this constraint, $CI(N_1, Person)$ should be inserted as side-effect. As a consequence, the null node becomes "typed" and all properties for *Person* would be required for N_1, possibly generating several additional null nodes.   □

The above example shows that, similarly to the restrictions introduced to constraints in $\mathcal{C}$, the goal of the adapted rules b_1-b_7 is to avoid null propagation.

Algorithm 2 computes U_S, an extension of U which includes: (1) the interpretation of each update on D, according to Table 2 and (2) side-effects obtained by applying $\mathcal{A}$ on U, on the basis of D_{Sch}. The complete algorithm is in [18].

Line 3 rejects update sets that contain schema changes or that are inconsistent w.r.t. Definition 8. As an example, the insertion of $CI(CNPq, RInst)$, where $RInst$ is a non existent class, triggers m_7 and produces the insertion of this class in the schema $(Cl(RInst))$. Since we do not support schema changes, the entire update set is rejected by Algorithm 2. Consider now a consistent update set $U = \{CI(Bob, Student), \neg PI(Bob, N_1, hasSalary), \neg CI(Bob, Person)\}$. Assuming that *Student* is a subclass of *Person*, rule m_9 imposes the insertion of $CI(Bob, Person)$ in U, which becomes inconsistent, and thus rejected by the algorithm.

Table 3. Subset of rules from [8]

m_1: Cl(x) $\wedge$ Pr(y) $\rightarrow$ (x $\neq$ y)
m_2: Cl(x) $\rightarrow$ Csub(x, rdfs:Resource)
m_3: Ind(x) $\rightarrow$ CI(x, rdfs:Resource)
m_4: Csub(x,y) $\wedge$ Csub(y,z) $\rightarrow$ Csub(x,z)
m_5: Pr(x) $\rightarrow$ Dom(x,y) $\wedge$ Rng(x,z)
m_6: CI(x,y) $\rightarrow$ Ind(x)
m_7: CI(x,y) $\rightarrow$ Cl(y) $\vee$ (y=rdfs:Resource)
m_8: PI(x,y,z) $\rightarrow$ Pr(z)
m_9: CI(x,y) $\wedge$ Csub(y,z) $\rightarrow$ CI(x,z)

Table 4. Rules that involve blank nodes

b_1: Pr(x) $\wedge$ BN(y) $\rightarrow$ (x $\neq$ y)
b_2: Ind(x) $\wedge$ BN(y) $\rightarrow$ (x $\neq$ y)
b_3: Cl(x) $\wedge$ BN(y) $\rightarrow$ (x $\neq$ y)
b_4: PI(x,y,z) $\rightarrow$ Ind(x) $\vee$ BN(x)
b_5: PI(x,y,z) $\rightarrow$ Ind(y) $\vee$ Lit(y) $\vee$ BN(y)
b_6: PI (x,y,z) $\wedge$ Dom(z,w) $\rightarrow$ CI(x,w) $\vee$ BN(x)
b_7: PI(x,y,z) $\wedge$ Rng(z,w) $\rightarrow$ CI(y,w) $\vee$ (Lit(y) $\wedge$ (w=rdfs:Literal)) $\vee$ BN(y)

Algorithm 2. Side effects due to $\mathcal{A}$

Input: A set of updates U, a database instance D, application constraints $\mathcal{C}$
Output: The set U_S of side effects of U *w.r.t.* $(\mathcal{A} \cup \mathcal{C})$.
1: $U_S := \emptyset$;
2: **repeat**
3:     **if** $ChangeSchema(U, D)$ or U is not consistent *w.r.t.* Definition 8 **then**
4:         Error Exception;
5:     $U_0 := U$;
6:     $U_S := \{ [\![op]\!]_{D \cup U_S} \mid op \in U \wedge op$ has a predicate in INSTPRED$\}$;
7:     $U_{BN} := \{ op \mid op \in U_S \wedge op = \neg PI(u) \wedge$ u contains null $N_1 \wedge \neg BN(N_1) \in U_S \}$;
8:     $U_S := U_S - \{ [\![op]\!]_{D \cup U_S} \mid op \in U_{BN} \}$;
9:     $U := U \cup \{ l \mid l = ResultTrigRule(r, op, D \uplus U_S)$ such that $r \in \mathcal{A} \wedge op \in U_0$
        is an insertion or a deletion without nulls $\}$;
10:    **for all** deletions of the form $\neg CI(a, c)$ in U_0 **do**
11:       $U := U \cup \{ l \mid l$ is of the form $PI(N_1, b, p)$ (or $PI(b, N_1, p)$), s.t. there exists
          $PI(a, b, p)$ ($PI(b, a, p)$, respect.) in $(D \uplus U_S)$ whose deletion violates $r \in \mathcal{C} \}$;
12: **until** $U = U_0$;
13: **return** U_S;

As in Algorithm 1, insertions (positive atoms) activate rules forward and their instantiated heads are added to U while deletions (negative atoms) trigger them backwards, inserting the inverse of their instantiated bodies to U. On line 9, the set U is completed in this way by function *ResultTrigRule*. Consider $\mathcal{C}$ of Example 1 and database instance $\{CI(Bob, Professor), CI(DB, Course), PI(Bob, DB, teaches)\}$. The update $upd = \{\neg CI(DB, course)\}$ does not produce any application-level side-effects. However, b_7 imposes the deletion of $PI(Bob, DB, teaches)$. Such a deletion violates r_6 (line 11). Thus, our algorithm adds $PI(Bob, N_1, teaches)$ to U. Note that this operation will *effectively* insert a null if DB was the only course taught by Bob, according the the semantics of $[\![PI(Bob, N_1, teaches)]\!]$ as defined in Table 2.

Proposition 2. *Let $\Delta = (D, D_{Sch}, \Sigma)$ be a consistent database w.r.t. $\Sigma = (\mathcal{C}, \mathcal{A})$. Given an update set upd, let U_S be the set of side-effects computed according to Algorithms 1 and 2. Let U_S^{Sch} be the subset of U_S with facts in SCHPRED. If U_S is consistent (Definition 8) and $U_S^{Sch} \subseteq D_{Sch}$ then the result of $D \uplus U_S$ is a new database instance D' which satisfies the following properties: (i) $upd^+ \subseteq D'$ and $upd^- \not\subseteq D'$ and (ii) D' is satisfies $\mathcal{C}$ and $\mathcal{A}$.* □

Complexity. In Algorithm 1 the computation of the fix-point of T corresponds to the immediate consequence operator used in Datalog, since only positive rules may iterate. It is known that the number of iterations is bounded by the number of rules plus one ($O|C|$). As in our approach instantiation of application constraints is bounded by $O(|C| \times |upd|)$, the size of U is $O(|C|^2 \times |upd|)$. Cleaning U is $O(|C|^2 \times |upd|)^2$. Algorithm 2 nowadays works on a simple file, a non-optimized version where the most expensive task is limited by $O(|U|^2 \times |D|^2)$ (where $|D|$ is the database instance size). Thus, its complexity is $O(|C|^4 \times |upd|^2 \times |D|^2)$.

4   Experimental Study

We have implemented the BNS system, based on Algorithms 1 and 2, using the Standard ML of New Jersey compiler. In this section, BNS is compared with the FKAC system. FKAC is an implementation of the approach proposed in [8]. Among the related work, it is the most similar to our proposal. To provide a fair comparison, the FKAC approach has been modified in the following aspects: (i) We do not allow it to compute schema changes. As the original system selects the smallest side-effect set among all possible ones, this modification reduces considerably the search space and thus the execution time for side-effects computation. (ii) We do not allow it to compute deletions as insertion side-effects. Deleting atoms with a null value is much more expensive than inserting them, since the removal requires a database traversal to determine all possible null instantiations while the insertion needs just one instantiation. Thus, these implementation choices are advantageous for the FKAC system.

An important difference between FKAC and BNS concerns the capability of storing null values. As the FKAC system *does not accept nulls*, when an insertion imposes the existence of an unknown required data, an arbitrary instantiation is performed. This instantiation is *not* a user's choice, and thus cannot be considered as semantically meaningful. For instance, consider the insertion of $CI(Bob, Researcher)$ in the context of Example 1. Rule $r2$ requires Bob to be a member of some project. If the range of $isMember$ is the class $Project$, one of the possible side-effects proposed by the FKAC system is to *choose* an arbitrary instance of $Project$, say $Jupiter$, and insert $PI(Bob, Jupiter, isMember)$.

Worst still, when in a later time, the user associates Bob to a *real* project, the previous arbitrary fact is not removed. Contrary to that, BNS stores null values, provided that it is not subsumed by an existing fact in the database. An atom with a null value can then be replaced by facts introduced by the user in later time. Since BNS never introduces arbitrary facts in the database we can say that its strategy is semantically more meaningful.

In our experiment, we first generate U using Algorithm 1 with $\mathcal{C}$. Then, before forwarding U to the FKAC system, we replace or expand atoms involving nulls.

We use Berlin [5](BSBM) and LUBM [10], two benchmarks frequently adopted in RDF experiments. A dataset with 500 products were generated using Berlin (220,000 triples, 900,000 facts in our model). Based on the benchmark specification, we have identified 19 application constraints, which were also translated to our model ([18]). For LUBM, a dataset with 198,668 triples (1,784,296 facts) were generated. We have identified 10 application constraints in the benchmark specification. As the motivation for introducing nulls in side-effect operations is to limit cascade updates, we compare BNS and FKAC systems *w.r.t.* the size of the update set. That is, given an input update set *upd*, randomly generated, we compare the final size of U_S. We consider sets *upd* of increasing sizes, and with three different compositions: only insertion operations, only deletion operations, and mixed sets with insertion/deletion percentage of 70/30, 50/50, and 30/70.

Figure 4 presents the results for the LUBM and Berlin benchmarks. The bars show the final update size (which in Algorithm 2 are denoted as U_S), while the lines present the percentage of operations in this set that *effectively* modify the database, by inserting or removing facts (following Definition 3). Figure 4(a) show that when considering only insertion operations, the number of side-effects in the BNS system is larger than in the FKAC system. This is caused by the way we handle insertions with nulls in both systems. For instance, in FKAC, the insertion $PI(a, N_1, p)$ is a single operation which replaces N_1 by an existing instance in the database. However, in BNS, besides the insertion with N_1, we also create as side-effects: $CI(N_1, rdfs : resource)$ and $BN(N_1)$. In the graph, for 40 insertions in LUBM, 16 operations with nulls have been created as side-effects, resulting in a set U_S with 32 more updates for BNS than for FKAC. This number corresponds to the two additional atoms for each null created by BNS.

With respect to the percentage of side-effects that affect the database, the lines for both systems follow the same pattern. Indeed, the number of SCH-PRED predicates (which have no effect on the database) in the resulting U_S is similar for both systems. Additionally, as Berlin datasets are more densely populate the percentage of effective operations in the BNS system is smaller.

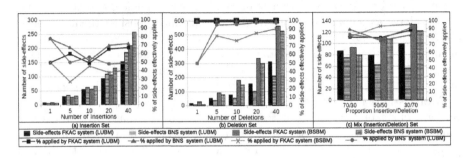

Fig. 4. Results for BSBM and LUBM benchmarks

To see why, consider again the operation $PI(a, N_1, p)$. According to Table 2, this operation has no effect if a is already linked to some other instance (or null node) through p. Thus, more densely populated databases tend to create fewer nulls.

BNS system avoids deletion propagation, as shown in Fig. 4(b). As the FKAC system generates only facts to be deleted by finding the nulls' instantiations, its percentage of *effectively* executed updates is always close to 100%. For BNS, on the other hand, a deletion may require the insertion of a null value and this null value may require a not allowed schema change. Thus updates may be rejected more often. Consider for example the LUBM benchmark with 20 deletions. The number of updates with side-effects is 155 for FKAC system and 103 for the BNS system. While FKAC performs 100% of the operations, BNS performs only 83%. The results with sets of mixed operations show that even when 70% of the operations are insertions, the reduction on the number of side-effects resulting from the deletions, by limiting cascading updates, overcomes the overhead of creating null nodes. Thus, our approach is effective on reducing the size of the update set, while generating semantically meaningful operations as side-effects. Making an analogy with the semantics of deletion operations in the relational model, the FKAC approach is similar to the 'on delete cascade' while we adopt the 'on delete set null' semantics.

5   Related Work

The co-existence of constraints, as in a CWA, with inferences, as in a OWA, has recently inspired some works on RDF data management. In [7] and in Stardog [17], a new knowledge graph platform, coincide in their capability of considering both types of rules, reliving the proposal in [11]. Technologies such as ShEx [16] and SHACL [13], deal with the validation of the shape of an RDF graph. Although their focus in on schema and ours is on integrity constraints, a study on their interaction with our work deserves further investigation. However, none of them deals with consistency maintenance due to updates. Mechanisms to control frequent updates on RDF are desirable [15], but the update and consistency maintenance approaches in [8,15] do not consider nulls. As stated in [4], the standard semantics for blank nodes comes from first order logic and interprets them as existential variables. However, blank nodes are now treated in different ways, implying different semantics (such as [9]). This paper focus on its standard semantics and to avoid confusion, we denote them as *nulls*. Our update approach falls into the category Sem_2^{mat} of [2]. Adapting the exception viewpoint in [11] to our current work is a future perspective which approaches our work to [3]. Even if we consider updates as changes in the world; and not as a revision in our world's knowledge ([12]), it is possible to relate results of our method to the core principles of belief revision.

6   Conclusion

We present algorithms to determine the set of side-effects, consisting of additional updates required to keep the database consistent in the presence of application and RDF semantic constraints. Our approach differs from previous works because side-effects may introduce nulls in order to reduce cascade updates. Our experimental study shows that, although insertions tend to generate a larger set of updates (justified by null nodes definitions), deletions tend to generate a smaller set of updates, since null nodes interrupt cascade deletions. Future work perspectives include a more expressive class of application constraints, while keeping the ability to deterministically compute the set of side-effects and experiments with a larger number of updates and datasets.

Acknowledgements. This work is partially funded by APR-IA Girafon and PEPS-INS2I Multipoint.

References

1. Afrati, F.N., Kolaitis, P.G.: Repair checking in inconsistent databases: algorithms and complexity. In: Proceedings of the International Conference on Database Theory, pp. 31–41 (2009)
2. Ahmeti, A., Calvanese, D., Polleres, A., Savenkov, V.: Dealing with inconsistencies due to class disjointness in SPARQL update. In: Proceedings of the 28th International Workshop on Description Logics (2015)
3. Ahmeti, A., Calvanese, D., Polleres, A., Savenkov, V.: Handling inconsistencies due to class disjointness in SPARQL updates. In: Sack, H., Blomqvist, E., d'Aquin, M., Ghidini, C., Ponzetto, S.P., Lange, C. (eds.) ESWC 2016. LNCS, vol. 9678, pp. 387–404. Springer, Cham (2016). doi:10.1007/978-3-319-34129-3_24
4. Arenas, M., Barceló, P., Libkin, L., Murlak, F.: Foundations of Data Exchange. Cambridge University Press, Cambridge (2014)
5. Bizer, C., Schultz, A.: The Berlin SPARQL benchmark. Int. J. Semant. Web Inf. Syst. **5**(2), 1–24 (2009)
6. Calì, A., Gottlob, G., Kifer, M.: Taming the infinite chase: query answering under expressive relational constraints. In: Proceedings of the 11th International Conference on Principles of Knowledge Representation and Reasoning, pp. 70–80 (2008)
7. Chabin, J., Halfeld-Ferrari, M., Nguyen, T.B.: Querying semantic graph databases in view of constraints and provenance. Technical report, LIFO- Université d'Orléans, RR-2016-02 (2016)
8. Flouris, G., Konstantinidis, G., Antoniou, G., Christophides, V.: Formal foundations for RDF/S KB evolution. Knowl. Inf. Syst. **35**(1), 153–191 (2013)
9. Frommhold, M., Piris, R.N., Arndt, N., Tramp, S., Petersen, N., Martin, M.: Towards versioning of arbitrary RDF data. In: Proceedings of the 12th International Conference on Semantic Systems (2016)
10. Guo, Y., Pan, Z., Heflin, J.: LUBM: a benchmark for OWL knowledge base systems. Web Semant. **3**(2–3), 158–182 (2005)
11. Halfeld-Ferrari, M., Laurent, D., Spyratos, N.: Update rules in datalog programs. J. Log. Comput. **8**(6), 745–775 (1998)
12. Hansson, S.O.: Logic of belief revision. In: Zalta, E.N. (ed.) The Stanford Encyclopedia of Philosophy, 2016 edn. Metaphysics Research Lab. Stanford University (2016). Winter
13. Knublauch, H., Ryman, A.: Shapes constraint language (SHACL). W3C first public working draft, w3c (2017). http://www.w3.org/TR/2015/WD-shacl-20151008/

14. Liu, L., Özsu, M.T. (eds.): Encyclopedia of Database Systems. Springer, US (2009)
15. Magiridou, M., Sahtouris, S., Christophides, V., Koubarakis, M.: RUL: a declarative update language for RDF. In: Gil, Y., Motta, E., Benjamins, V.R., Musen, M.A. (eds.) ISWC 2005. LNCS, vol. 3729, pp. 506–521. Springer, Heidelberg (2005). doi:10.1007/11574620_37
16. Solbrig, H., Hommeaux, E.P.: Shape expressions 1.0 definition. W3C member submission (2014). http://www.w3.org/Submission/2014/SUBM-shex-defn-20140602
17. Stardog5: Enterprise knowledge graph (2017). http://www.stardog.com/docs/
18. Uber, F.: RDF constraint satisfaction with blank nodes. Ph.D. dissertation Proposal, UFPR, Brazil (2016). http://www.inf.ufpr.br/fruber/BNS

Ephedra: Efficiently Combining RDF Data and Services Using SPARQL Federation

Andriy Nikolov[✉], Peter Haase, Johannes Trame, and Artem Kozlov

Metaphacts GmbH, Walldorf, Germany
{an,ph,jt,ak}@metaphacts.com

Abstract. Knowledge graph management use cases often require addressing hybrid information needs that involve multitude of data sources, multitude of data modalities (e.g., structured, keyword, geospatial search), and availability of computation services (e.g., machine learning and graph analytics algorithms). Although SPARQL queries provide a convenient way of expressing data requests over RDF knowledge graphs, the level of support for hybrid information needs is limited: existing query engines usually focus on retrieving RDF data and only support a set of hard-coded built-in services. In this paper we describe representative use cases of *metaphacts* in the cultural heritage and pharmacy domains and the hybrid information needs arising in them. To address these needs, we present Ephedra: a SPARQL federation engine aimed at processing hybrid queries. Ephedra provides a flexible declarative mechanism for including hybrid services into a SPARQL federation and implements a number of static and runtime query optimization techniques for improving the hybrid SPARQL queries performance. We validate Ephedra in the use case scenarios and discuss practical implications of hybrid query processing.

1 Introduction

SPARQL has emerged as a standard formalism for expressing information requests in Semantic Web applications where the goal is to retrieve the data stored as RDF. However, in many practical knowledge graph management use cases there is a need to address *hybrid information needs*. Such needs can be characterized by the following dimensions:

- *Variety of data sources.* There is often a need to integrate data stored in several physical repositories. These repositories can include both native RDF triple stores as well as datasets in other formats presented as RDF (e.g., a relational database exposed using R2RML mappings).
- *Variety of data modalities.* Graph data in RDF often needs to be combined with other data modalities, e.g., textual, temporal or geospatial data. A SPARQL query then needs to support corresponding extensions for full-text, spatial, and other types of search.

© Springer International Publishing AG 2017
P. Różewski and C. Lange (Eds.): KESW 2017, CCIS 786, pp. 246–262, 2017.
https://doi.org/10.1007/978-3-319-69548-8_17

- *Variety of data processing techniques.* Retrieved data often has to be further processed using dedicated domain-specific services: e.g., graph analytics (finding the shortest path or interconnected graph cliques), statistical analysis and machine learning (applying a machine learning classifier, finding similar entities using a vector space model), etc.

The main motivation for this work comes from our experience with the *metaphactory* knowledge graph management platform[1], which is used in a variety of application domains (e.g., cultural heritage, life sciences, pharmaceutics, and IoT infrastructure). Typical application scenarios often require dealing with a multitude of the above-listed aspects simultaneously: e.g., an example request like "give me the artists who collaborated with Rembrandt and others similar to them" involves (a) keyword search for an RDF resource based on the keyword "rembrandt", (b) structured search over the RDF graph for collaborators, and (c) applying an external model (vector space similarity) to find other similar entities.

To handle such use case scenarios involving hybrid information needs, we developed Ephedra: a federated SPARQL query processing engine targeted at processing hybrid queries. SPARQL 1.1 with its SERVICE clauses provides a convenient data retrieval formalism: a complex information request over several data sources can be expressed using a single query. However, the existing level of tools support for expressing and processing hybrid information needs using SPARQL is often limited. SPARQL federation implementations usually focus on the first dimension: they assume that federation members are data stores containing RDF triples. Some triple stores also contain built-in implementations of alternative search modalities: e.g., "magic" predicates or even custom language constructs to handle keyword search. With Ephedra we overcome these limitations: while adopting the SPARQL 1.1 federation mechanism, we broaden its usage to include custom services as data sources and optimize such hybrid queries to be executed efficiently.

Serving hybrid information requests using SPARQL raises challenges both at the level of configuring the federation and executing the query: (a) how can we use SPARQL queries to combine retrieval of RDF data with invoking additional hybrid services and (b) how can we execute hybrid SPARQL query services in an efficient way. Expressing a complex hybrid information request using a SPARQL query can be non-trivial due to the variety of potential types of services and the limitations of the SPARQL syntax: e.g., a service can take as input a single set of parameters or a list of arbitrary length; it can return as output one value, several values or a table of multiple records, etc.

Processing such service calls in the same way as genuine SPARQL endpoints would result in sub-optimal query runtimes or even failing queries. This is caused by the specific features of the service calls such as:

- *Input and output parameters.* Triple patterns that express a service call do not refer to the actual RDF data structures, but merely denote the values

---

[1] http://www.metaphactory.com/.

that should be passed as service inputs and the variables to which the service outputs should be bound. It has serious implications for the query evaluation, as they do not follow SPARQL semantics: e.g., join operands cannot be reordered, because a service call must only be scheduled after all its input variables are bound.

- *Data cardinalities.* Estimating the cardinality of a tuple expression is important for performing static query optimization and ordering the join operands. While for genuine SPARQL endpoints the selectivity of graph patterns depends on the distribution of data in the underlying repository, for service calls this often depends on the service itself: e.g., a keyword search always takes one input and returns a limited set of search results as output.

With Ephedra we take these factors into account and use them to support practical hybrid federation use cases of the *metaphactory* platform and address hybrid information needs in an efficient way. In this paper, we are making the following contributions:

- We describe two representative use case scenarios from different domains (cultural heritage and pharmaceutics) which present hybrid query processing challenges.
- We propose a reusable architecture in which hybrid services can be easily plugged in, described in a declarative way, and invoked using federated SPARQL queries.
- Based on the explicit descriptions of services, we propose a number of static SPARQL query optimization techniques to construct a valid and efficient query plan.
- We utilize dynamic query optimization to execute hybrid queries with SERVICE clauses in minimal time, improving the performance by up to an order of magnitude.
- We validate our approach in two use cases and discuss the practical implications.

The rest of this paper is structured as follows. Section 2 presents two representative use cases and the requirements for our work. Section 3 describes the general architecture of Ephedra and the meta-level descriptions of services available via SPARQL. Section 4 discusses the query algebra extensions for hybrid SPARQL queries and the relevant static query optimization techniques aimed at generating an optimal query execution plan. Section 5 outlines the techniques Ephedra applies at runtime to modify the execution plan on-the-fly based on the actual execution progress. Section 6 reports the experiments we performed in order to validate our approach. In Sect. 7, we provide an overview of existing solutions for processing federated hybrid SPARQL queries. Finally, Sect. 8 concludes the paper and discusses directions for future work.

2   Use Cases and Challenges

The *metaphactory* platform is used in production in a variety of scenarios involving knowledge graph management in different application domains. Retrieval of stored RDF data using structured SPARQL queries is often insufficient to satisfy the application requirements without invoking additional services. In the following we consider two practical use cases from the cultural heritage and pharmaceutics domains. For reproducability, they are based on publicly accessible data sets and resources, they are however representative examples akin to production use cases based on closed, proprietary data.

2.1 Use Case: Cultural Heritage

The CIDOC-CRM ontology[2] became a popular standard for exposing cultural heritage information as linked data. The *metaphactory* platform is utilized in the context of the ResearchSpace project[3] to manage the British Museum knowledge graph and help the researchers explore meta-data about museum artifacts: historical context, associations with geographical locations, creators, discoverers and past owners, etc. A crucial piece of functionality is *structured search*, where the user can construct a query request like "give me all bronze artifacts created in Egypt between 2500BC and 2000BC" that gets translated into SPARQL and answered using backend data. However, beyond querying the British Museum collections, the use case requirements also involve addressing hybrid information needs:

- *Variety of data sources:* Datasets of other museum collections structured using CIDOC-CRM as well as linked public RDF data sources (e.g., Wikidata[4]).
- *Variety of data modalities:* Some relevant data sources are also associated with keyword search services (e.g., Wikidata search API or an external Solr index) and geospatial search indices. The Wikidata search API, for example, provides a higher quality of retrieved results than a built-in triple store index, but is only available as a REST web service not directly accessible via SPARQL.
- *Variety of data processing techniques:* A custom semantic similarity search service based on a word2vec vector space model.

This scenario includes a specialized data processing service, which applies a trained machine learning model to find entities similar to a given set of other entities. This service utilizes the word2vec vector space model [1] trained on the English Wikipedia corpus. Each Wikidata entity is represented as an *embedding vector* of length 50. Similarity defined as a distance in the embeddings vector space serves as a means to indicate relatedness between entities and complements the explicit relations stored in the RDF triple store. An example request that can be run over such a federation can look like: "Give me other artists similar to the ones who collaborated with Ibaya Kyubei".

```
# Example query (Q1_CH) from the cultural heritage use case
SELECT ?collabWikidataIRI ?label ?artist WHERE {
  SERVICE metaphacts:wikidataSearch {
```

---

[2] http://www.cidoc-crm.org/.
[3] http://www.researchspace.org/.
[4] http://www.wikidata.org.

```
    ?wikidataIRI wikidata:search"ibaya".
  }
  SERVICE <http://public.researchspace.org/sparql> {
    ?bmIRI skos:exactMatch ?wikidataIRI.
    ?collabBMIRI rs:Actor_created_Thing ?artifact.
    ?bmIRI rs:Actor_created_Thing ?artifact.
    ?collabBMIRI skos:exactMatch ?collabWikidataIRI.
    FILTER (?bmIRI != ?collabBMIRI)
  }
  ?collabWikidataIRI rdfs:label ?label.
  SERVICE metaphacts:wikidataWord2Vec {
      ?collabWikidataIRI word2vec:similarTo ?artist.
  }
}
```

Such a query requires the query engine to federate over two different data repositories (ResearchSpace and Wikidata) as well as include information from two external services: Wikidata keyword search API and word2vec semantic similarity search.

2.2   Use Case: Pharmaceutics

Another scenario in which the *metaphactory* platform is employed is related to the pharmaceutics domain. The internal knowledge graph of the customer contains interlinked data about genes, proteins, and associated diseases. This information, however, has to be augmented with other sources which makes this use case another example of hybrid information needs:

- *Variety of data sources*: Additional relevant data sources include an Oracle relational database exposed as a SPARQL endpoint using R2RML/Ontop as well as relevant public RDF data sources (Wikidata and Nextprot[5]).
- *Variety of data modalities*: Full-text indices enable custom keyword search in addition to the SPARQL structured search.
- *Variety of data processing techniques.* Domain-specific services include the trained machine learning models realized in the KNIME[6] data science platform as well as the BLAST [2] web service to find similar entities based on the genome sequence data.

An example hybrid search request can look like "Give me the gene encoding the reelin protein and other genes having the most similar sequences". Such a simple request would involve all three hybrid search dimensions: keyword search service to retrieve the id of the "reelin" protein, query over the RDF graph to find an associated gene and the ID of the sequence, and a call to the BLAST service to find similar sequences.

```
# Example query (Q1_PH) from the pharmaceutics use case
SELECT * WHERE {
  SERVICE metaphacts:wikidataSearch {
      ?uri wikidata:search"reelin".
  }
```

---

[5] https://www.nextprot.org/.
[6] https://www.knime.org/.

```
  ?uri wdt:P702 ?gene.
  ?gene wdt:P639 ?refseqID.
  SERVICE ncbi:BLAST {
      ?refseqID blast:hasSimilarSequence ?y.
  }
}
```

These common challenges arising in diverse use cases necessitate the use of a generic approach for hybrid query processing. Extending the SPARQL 1.1 federation mechanism for that has important advantages:

- A hybrid information need can be expressed in a fully declarative way without the need for use case-specific custom implementations.
- The use of the standard SPARQL 1.1 syntax makes the approach compatible with third-party tools (e.g., client-side SPARQL processing libraries).

The solution we present focuses on keeping these advantages while maintaining the system reusable in different scenarios and processing the hybrid queries in the most efficient way.

3 Hybrid SPARQL Federation

To support the use cases described above, we developed the Ephedra query processing engine as a part of the *metaphactory* platform. A crucial requirement for enabling the hybrid querying functionality is the ability to plug in additional services with minimal effort and reference them from SPARQL.

3.1 Ephedra Architecture

Figure 1 shows the generic architecture of the *metaphactory* platform. Ephedra is used as a hybrid query federation layer to access the data repositories and services. In the course of the DIESEL project [3], we developed a set of structured search components which allow the user's hybrid information need to be captured interactively: the user can define search clauses, explore partial results, incrementally add new clauses, while the system provides relevant suggestions. These interactions generate information requests that are expressed as SPARQL 1.1 queries by the UI components and given to Ephedra to process them.

As the basis for Ephedra implementation, we used the RDF4J Federation SAIL API[7] reusing the common functions such as query parsing and accessing remote SPARQL endpoints. However, Ephedra extends the RDF4J object model and overrides the static optimization and query execution strategies to deal with hybrid queries. The Ephedra query evaluation strategy sends the sub-clauses of the query to the corresponding data sources and invokes the relevant processing services, then gathers the partial results, combines them using the union and join operations, and produces the final result set. In this way, processing becomes transparent: hybrid information needs are processed in the same way as ordinary SPARQL queries to an RDF triple store without the need to integrate related processing services at the UI level.

---

[7] http://docs.rdf4j.org/sail/.

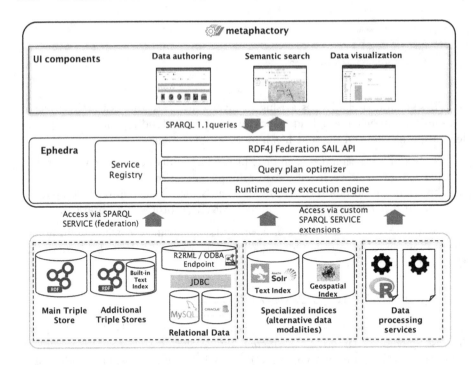

Fig. 1. Ephedra in the *metaphactory* platform architecture.

3.2 Describing and Configuring Hybrid Services

In order to configure the services as federation members, the system requires relevant information about the *service type* as well as *service instances*.

Ephedra includes two types of hybrid services: *extension services* and *aggregate services*. Extension services take as input a partial query solution (binding set) and extend it with additional variable bindings. Extension services are called in the query via a SPARQL SERVICE clause. On the contrary, aggregate services operate over a set of multiple query solutions as the SPARQL aggregate functions (e.g., AVG, MIN, MAX) do: they take as input a list of records and produce one or more resulting binding sets. As with the SPARQL aggregates, aggregate services are referenced as function calls in the SELECT clause.

Relevant meta-level information about the hybrid service types is summarized using the service descriptors structured according to the *service description ontology*. The ontology expands the well-known SPIN[8] ontology for SPARQL query engines to capture the relevant parameters of services.

A service descriptor contains the following information:

– Input parameters and their expected datatypes. An input parameter is described using the SPIN ontology vocabulary as a *spl:Argument* resource.

---

[8] http://spinrdf.org/.

- Output parameters and their expected datatypes. An output parameter is described as a *spin:Column* resource in the SPIN ontology.
- Expected graph pattern. The special triple patterns expected by the service are expressed using the SPIN SPARQL syntax[9]. The placeholders for input/output parameters are expressed as resources which are referenced from the input/output parameter descriptors.
- Input and output cardinalities of a service call (optional).

```
:WikidataTextSearch a eph:Service;
  rdfs:label "A wrapper for the Wikidata test search.";
  eph:hasSPARQLPattern (
      [ sp:subject :_uri;
          sp:predicate wikidata:search;
          sp:object :_token ] );
  spin:constraint
  [ a spl:Argument;
    rdfs:comment "Input token";
    spl:predicate :_token;
    spl:valueType xsd:string ];
  spin:column
  [ a spin:Column;
    rdfs:comment "URI of the Wikidata resource";
    spl:predicate :_uri;
    spl:valueType rdf:Resource ].
```

A descriptor for an aggregation service declares the input and output parameters in a similar way, but instead of the list of triple patterns it defines a custom aggregate function which will be referenced by its URI.

```
:word2vec a eph:AggregateService;
  rdfs:label "A wrapper for the word2vec similarity aggregate service.";
  eph:hasAggregateFunction
  [ a sp:Aggregate;
    sp:expression :_uri;
    sp:as :_similar ; ];
  spin:constraint
  [ a spl:Argument;
    rdfs:comment "Entity URI";
    spl:predicate :_uri;
    spl:valueType rdfs:Resource ];
  spin:column
  [ a spin:Column;
    rdfs:comment "URI of the similar entity";
    spl:predicate :_similar;
    spl:valueType rdf:Resource ].
```

3.3   Implementing Service Extensions

To simplify the integration of new hybrid services into the framework, the architecture provides a generic API to wrap arbitrary services and include them as SPARQL federation members. To this end, Ephedra reuses and extends the RDF4J SAIL API. A service is represented as a SAIL module which is responsible for extracting the values of input parameters from a given SPARQL tuple expression, executing the actual service call, and returning the results by binding resulting values to the output variables. Ephedra provides abstract implementations for a generic service SAIL as well as a specific wrapper for REST services.

---

[9] http://spinrdf.org/sp.html#sp-TriplePattern.

The common routines, such as extracting the input values and output variables and wrapping the results as binding sets do not depend on the actual service and are performed in a generic way using the declarative service descriptor.

A service instance is thus configured in the same way as a standard RDF4J repository: its descriptor contains a pointer to the service type as well as the specific parameters of the service installation (e.g., the URL by which the REST service can be accessed).

4   Adapting SPARQL Algebra for Hybrid Queries

On receiving a federated query, Ephedra has to create a suitable query plan for its execution. Although Ephedra adheres to the SPARQL 1.1 syntax, processing of hybrid queries requires introducing special algebra elements to handle the extension and aggregate services. Based on these, Ephedra is able to construct a suitable query plan that would avoid failing queries and reduce the subsequent execution time.

4.1   Basic Definitions

In a SPARQL query, the WHERE clause defines a *graph pattern* to be evaluated on an RDF graph G. An atomic graph pattern is a *triple pattern* defined as a tuple P from $(I \cup V) \times (I \cup V) \times (I \cup L \cup V)$, where I, L, and V correspond to the sets of IRIs, literals, and variables respectively. Triple patterns are combined by means of JOIN, UNION, FILTER, and OPTIONAL operators to construct arbitrary graph patterns. A *mapping* is defined as a partial function $\mu : V \rightarrow (I \cup L \cup B)$ (B is a set of blank nodes) [4], and the domain of the mapping $dom(\mu)$ expresses a subset of V on which the mapping is defined. Then, the semantics of SPARQL queries is expressed by means of a function $[\![P]\!]_G$, which takes as input a graph pattern P and produces a set of mappings from the set of variables $var(P)$ mentioned in P to elements of the graph G. The binding of the variable $?x$ according to the mapping μ is denoted as $\mu(?x)$. The basic query algebra then defines the standard operations (Selection σ, Join $\bowtie$, Union $\cup$, Difference $\setminus$, and Left Join $\bowtie\!\!\!\!\!\!\!\!\!\text{\tiny M}$) over the sets of mappings, and query evaluation involves translating the query into a *query tree* composed of these operations. For simplicity, in this paper we use the notation $P_1 \bowtie P_2$ to refer to the join operation over sets of mappings produced by the patterns P_1 and P_2. In order to allow hybrid queries to be optimized, we need to introduce hybrid service calls at the level of SPARQL algebra.

4.2   Service Clauses in SPARQL Algebra

To handle extension services, Ephedra introduces the notion of a *service call pattern* as a special graph pattern in the query tree.

Definition 1: *A service call pattern Σ^S is a tuple $(id, P^S, f_e^S, m^i, m^o)$ identified by an IRI id and characterized by the following parameters:*

- *Graph pattern P^S: a SERVICE clause by which the service call is expressed in the SPARQL query.*
- *Function $f_e^S : D^i \rightarrow D^o$: the function implemented by the service, which takes a list of input parameters $D^i = \{d^i\}$ and produces a list of output results $D^o = \{d^o\}$*
- *Input parameter mappings m^i: set of mappings from the elements of P^S to elements of D^i, which extract the values of service input parameters from the graph pattern P^S.*
- *Output variable mappings m^o: set of mappings from the elements of $D^o \cup d_{rank}$ to the elements of $var(P^S)$. One special case of a service output is d_{rank}: the rank of the returned result, which can be added implicitly to each result returned by f_e^S.*

Aggregate services represent a different case: since they can be applied to multisets of partial results, they are expressed at the syntax level in the same way as standard SPARQL aggregate operations such as COUNT, AVG, or SUM. To include the aggregate service expressions, we extend the aggregate algebra construct defined in [5]:

Definition 2: *A service aggregate A^S is a construct of the form*
   $Aggregate(\mathbf{F}, f_a^S(\mathbf{D^i}), \Gamma)$, *where*

- $\Gamma = Group(\mathbf{E}, P)$ *is a GROUP operator over the graph pattern P using a list of expressions $\mathbf{E}$.*
- $f_a^S(\mathbf{D^i})$ *is an aggregation function implemented by the service, which takes as input a multiset of parameter assignments $\mathbf{D^i} = \{D^i\}$, where $D^i = \{d_1^i \ldots d_n^i\}$, and produces as output a set of output values $\mathbf{d^o}$.*
- $\mathbf{F}$ *is a list of expressions which are applied to the results of Γ to produce the inputs of f_a^S.*

In Ephedra we extended the standard SPARQL 1.1 semantics to incorporate aggregate services which can produce as a result a set of values as well as a single value. One example of such services is the word2vec service we use in the cultural heritage use case: it can take a set of several entities as input and produce a list of additional entities that are similar to the whole set.

4.3   Building a Query Plan

After processing the parsed hybrid query and replacing the default SERVICE clauses with service call patterns Σ^S and service aggregates A^S, the Ephedra query engine tries to build an optimal execution plan for the query. Ephedra focuses on two types of improvements for the query:

- Join order optimization
- Assigning appropriate executors for JOIN and UNION operators

Determining the order in which join operators have to be processed as well as choosing appropriate execution algorithms (e.g., nested loop join vs hash join) may have very significant impact on the execution performance and so have been in the focus of research when designing the federation query engines. Traditionally, sorting the join order operands takes into account the estimated selectivity of the join parameters.

In case of hybrid queries, this step carries additional importance: an unoptimized query may not be executable at all. The query engine must ensure that all input parameters of a service call group Σ^S are bound before the service is called. Moreover, some join operators can be inappropriate for use with hybrid service clauses: e.g., a service clause which has unbound input parameters must be executed as a second argument in a nested loop join and cannot be processed by a hash join.

When processing an n-ary join operation, Ephedra groups together the join operands which can be executed at the same source. After that, it uses the following criteria (starting from the most important) to determine whether an operand tuple expression P can be added to the pipeline:

1. Join operands which contain service call patterns Σ^S are added immediately when all their input dependencies become satisfied.
2. A join operand P_x which binds a variable v_{xi} cannot be added if there exists a join operand $P_y = \Sigma_y^S(P^S, f_e^S, m^i, m^o)$ such that $v_{xi} \in m^o(P^S)$: if the same variable can be bound in a service call pattern and in an ordinary graph pattern, the service call is executed first.
3. A join operand P_x is preferred over P_y if $var(P_x)$ contains a variable bound earlier in the pipeline and $var(P_y)$ does not.
4. Finally, the estimated selectivities are taken into account. Selectivity is estimated based on the number of free variables (for the ordinary graph patterns) or based on service descriptors (for hybrid service calls).

The second technique used by Ephedra involves processing of the *top-k* queries and handling of the results' ranking. To this end, Ephedra uses the $\mathcal{S}$PARQL-$\mathcal{R}$ANK algebra [6].

5　Optimizing Hybrid Queries Execution

While static query optimization helps to reduce query execution time, the resulting query performance still strongly depends on the way the operators are processed. The power of static optimization in a hybrid federation is particularly limited, because precise selectivity estimation is impossible. Reversing the order of operands in a nested loop join or replacing it with a hash join can significantly improve the performance. To further minimize the query execution performance, Ephedra uses two techniques: *synchronizing loop join requests* and *adaptive processing of n-ary joins*.

5.1 Synchronizing Loop Join Requests

Let us consider our example query Q1_CH from Sect. 2.1. The top-level N-ary join of the query contains 4 operands: Σ_1^S (Wikidata text search), Σ_2 (British Museum), Σ_3 (Wikidata) and Σ_4^S (word2vec). When executing a nested loop join, the engine first retrieves the answers μ_i for Σ_1^S. Iterating over μ_i, it will probe Σ_2 to receive bindings μ_{ij} from $(\Sigma_1^S \bowtie \Sigma_2)$ and continue doing this until receiving the complete answers from $(\Sigma_1^S \bowtie \Sigma_2 \bowtie \Sigma_3 \bowtie \Sigma_4^S)$. When iterating over partial result sets μ_i, there are several possible strategies which can be chosen for sending the probing queries q_{ij} to join the next operand Σ_{i+1}.

1. *Synchronous vs asynchronous.* The engine can parallelize sending of the probing queries q_{ij} so that the answers are added into the resulting queue as soon as they appear. Alternatively, it can synchronize the requests to guarantee that the results produced by the query q_{ij} will appear in the final result set before the results of q_{ik} if $j < k$.
2. *Separate requests vs batch.* The engine can send a separate query request for each input binding set μ_{ij}. Another strategy called *Bind Nested Loop Join (BNLJ)* [7] involves grouping together several mappings $\mu_{ij}, \ldots, \mu_{ij+K}$ and sending a single query which would contain all bindings from the group expressed using a VALUES clause.

The choice of the appropriate strategy depends on the type of the query as well as the types of each join operand. The batch processing strategy using the BNLJ operator helps to reduce the number of potentially expensive remote requests and was shown to improve the performance significantly [7]. However, it can only be applied if the right join operand is an RDF repository: by default, a hybrid service can only process one set of input parameters at a time and can even break otherwise. Ephedra uses the service descriptors to choose an appropriate strategy: e.g., in our example query it will use the BNLJ strategy to join the data from the British Museum and Wikidata repositories, but send the requests to the word2vec service separately.

One additional technique used by Ephedra that helps to reduce the number of expensive requests to remote hybrid services involves caching the probing queries. In our example, the last operation in the pipeline involves joining the results of $\Sigma_1^S \bowtie \Sigma_2 \bowtie \Sigma_3$ with the word2vec service Σ_4^S using the join variable *?collabWikidataIRI*. The results of the previous operations in the pipeline contain multiple binding sets which have the same value bound to *?collabWikidataIRI*: the same pair of artists (Ibaya Kyubei and Utagawa Kuniyoshi) has collaborated on many woodblock prints. Processing each binding set separately would result in many requests to the *word2vec* service for the same input value (Utagawa Kuniyoshi). To avoid this, when performing a join between $(\Sigma_i^S \bowtie \Sigma_2)$ and Σ_3^S, we only send queries for unique key combinations (variables present in both operands). The remote query results are then joined to each relevant binding μ_{ij} that share the same values for key variables. This is equivalent to applying the REDUCED operation implicitly to the set of key combinations from μ_{ij}.

5.2 Adaptive Processing of N-ary Joins

Sometimes, an n-ary join contains multiple service call groups that do not have unbound inputs. It means that the evaluation of the join can start from both these groups independently. Let us suppose that our query contains 3 join operands: Σ_1^S, Σ_2, and Σ_3^S, where both Σ_1^S and Σ_3^S have all their input parameters bound. This n-ary join can be executed using different plans, for example $(\Sigma_1^S \bowtie_{BNLJ} \Sigma_2) \bowtie_{HJ} \Sigma_3^S$ or $\Sigma_1^S \bowtie_{HJ} (\Sigma_2 \bowtie_{BNLJ} \Sigma_3^S)$. It is not always known in advance, which of these plans is preferable: an incorrect choice can results in big differences in execution time. The *parallel competing join* strategy originally presented in [8] to handle keyword search queries tries to avoid this by executing the competing query plans in parallel and making the final choice between them at runtime.

Algorithm 1 Parallel competing n-ary join

1: $\mathbf{P^s}$: seed operands
2: $\mathbf{P^d} \leftarrow \mathbf{P} \backslash \mathbf{P^s}$: other operands
3: **for all** $P_i^s \in \mathcal{P}$ **do**
4:     $Q_i \leftarrow (\{P_i^s\} \cup \mathbf{P^d})$
5:     start(Q_i)
6: ...: wait until $\mathbf{P^d} = \emptyset$
7: **if** $\mathcal{P}^{\overline{s}} = \emptyset$ **then**
8:     **return** HashJoin($\{P_i\}$)
9: ...
10: **procedure** PUSHRESULTS(P_{curr}, P_{prev}^d, $[\![P_{curr}]\!]_G$)
11:     $\mathbf{P^d} \leftarrow \mathbf{P^d} \backslash P_{prev}^d$
12:     **for all** Q_i **do**
13:         $Q_i \leftarrow Q_i \backslash P_{prev}^d$
14:     $P_{next}^d \leftarrow Q_{curr}.next$
15:     **if** joinInThisPlan(P_{curr}, P_{next}^d) **then**
16:         $P_{curr} \leftarrow$ NestedLoopJoin(P_{curr}, P_{next}^d)

The algorithm starts with selecting the "seed" join operands, which serve as starting points for alternative query plans (Σ_1^S and Σ_3^S in our example). For each of these seeds, an alternative join sequence Q_i is produced and triggered. Whenever any of the competing query plans Q_i completes the join of some operand P_{prev} and produces a partial result set, a re-evaulation takes place. In the default case, it checks if the partial result set is too large and continuing the plan Q_i is likely to be more expensive than an alternative plan. If the check passes, the next operand from Q_i is joined, otherwise the plan stops.

Once all partial query plans are finished and no operand P_i remains unprocessed, Ephedra joins the partial results of all query plans via an n-ary hash join. Alternatively, if the ranking must be preserved, the n-ary Pull/Bound Rank Join algorithm [9] can perform the final join.

6 Evaluation and Discussion

In order to validate Ephedra, we used data and queries from our two representative use cases from Sect. 2. In the cultural heritage setup, we used two RDF data repositories (British Museum (BM) and Wikidata), the Wikidata entity lookup REST API (WD-text), and the *word2vec* vector space model similarity REST API. The latter was used both as an extension service (to retrieve instances most similar to a single input one) and as an aggregation service (to retrieve instances similar to a group of input entities). The model was trained on the English Wikipedia corpus using gensim[10]. In the pharmaceutics setup, we used two public RDF repositories (Wikidata and Nextprot), Wikidata entity lookup API (WD-text), and a wrapper around the public BLAST API[11].

The test queries for two domains were selected in such a way that (a) they were representative examples of queries arising in practical use and (b) each one covered at least two hybrid query dimensions. We used four queries from the cultural heritage domain and three queries from the pharmaceutics domain[12]. We compared the average query execution runtimes with disabled and enabled optimization techniques described in Sects. 4 and 5. Table 1 shows the runtime performance for each query, averaged over five runs.

Table 1. Average execution time (sec) for test queries taken over 5 query runs.

Query	Sources	Baseline		Ephedra	
		Time (sec)	σ	Time (sec)	σ
Q1_CH	Wikidata, WD-text, BM, word2vec	10.20	0.42	1.40	0.19
Q2_CH	Wikidata, WD-text	1.56	0.16	0.73	0.25
Q3_CH	Wikidata, BM	13.17	0.13	4.52	0.90
Q4_CH	Wikidata, WD-text, BM, word2vec (aggregate)	4.40	0.49	1.28	0.03
Q1_PH	Wikidata, WD-text, BLAST	12.38	0.76	2.24	0.28
Q2_PH	Wikidata, WD-text	1.50	0.25	0.70	0.02
Q3_PH	Wikidata, WD-text, Nextprot	3.54	0.06	1.31	0.36
Geom. Mean		4.82		1.43	

As we can see, query optimization techniques of Ephedra led to improvements in the query evaluation runtimes for all test queries, sometimes by an order of magnitude. The factors which contributed the most were:

- Processing of loop join requests, where combining asynchronous processing with the bound join operator resulted in the best performance of n-ary joins.
- Competing nested loop join, which was beneficial for n-ary joins with several potential starting points (Q2_CH and Q2_PH).
- Caching of probing requests, which helped to avoid expensive redundant remote service calls (Q1_CH and Q1_PH).

---

[10] https://radimrehurek.com/gensim/.

[11] https://ncbi.github.io/blast-cloud/dev/api.html. In the tests we only measured the time required to register a search request, since complete processing of the request takes ~1 min and varies greatly depending on the public server workload.

[12] The queries are available online on https://github.com/metaphacts/ephedra-eval.

Our validation experiments have shown that having special optimization techniques that treat hybrid service calls differently from "native" SPARQL endpoints can substantially benefit performance. This enabled Ephedra to fulfill the requirements of the use cases by maintaining acceptable response times of the *metaphactory* platform.

Beyond that, some of the lessons we learned concern more pragmatic aspects. One such conclusion is that conformance to the SPARQL standard helps both to improve reusability and reduce the maintenance effort. Sometimes, tools and triple stores introduce syntax-level modifications of SPARQL to realize the hybrid query functionalities: e.g., special syntax for keyword search clauses or graph analytics queries. Given that SPARQL processing must be performed at different layers (client-side, query federation, backend repository) using different libraries, special syntax changes become particularly difficult to handle and severely increase the solution building costs. Instead, introducing special interpretations of standard language concepts (e.g., "magic" predicates, custom functions) without changing the language syntax is preferable.

7 Related Work

There are several approaches that focused on specific dimensions of hybrid query processing. Main triple store implementations, such as Blazegraph[13], Virtuoso[14], GraphDB[15], and Stardog[16], take the SPARQL 1.1 standard as the basis for supporting federated queries. Usually, they share the assumption that federation members represent remote RDF repositories and they do not maintain meta-level information about federation members. Some triple stores (e.g., Blazegraph) also provide interfaces for adding custom service extensions. However, to our knowledge, they are not treated differently from remote SPARQL endpoints. Alternative data modalities (e.g., full-text and geospatial search) are supported using specialized built-in indices and expressed in SPARQL using "magic" predicates or SPARQL syntax modifications (e.g., full-text search in Virtuoso and Stardog). This makes it difficult to develop reusable database-independent solution applications.

Specialized SPARQL federation engines focus on optimal processing of distributed SPARQL queries. They usually maintain meta-level information about federation members which helps them to build an optimal query plan (e.g., SPLENDID [10], ANAPSID [11], or HiBISCuS [12]) as well as use special runtime execution techniques targeting remote service queries (e.g., FedX [7]), but still focus on interacting with RDF repositories rather than services.

In contrast, SCRY [13] and Quetzal-RDF [14] deal with calling data processing services using SPARQL queries. Quetzal-RDF defines custom functions and table functions (generalized aggregation operations) and invokes them from a

---

[13] http://www.blazegraph.com.
[14] http://virtuoso.openlinksw.com.
[15] http://ontotext.com/products/graphdb/.
[16] http://www.stardog.com/.

SPARQL query, but does not follow the SPARQL 1.1 syntax. SCRY conforms to SPARQL 1.1 using special GRAPH targets to wrap service invocations, although it cannot distinguish between multiple input/output parameters. None of these systems, to our knowledge, applies optimizations targeted at reducing the hybrid queries' execution time.

8 Conclusion

Our approach to the problems of handling hybrid queries was motivated by the requirements of commercial use case scenarios in two different domains. The design choices of Ephedra were influenced by the need to maintain the platform reusability and minimize the effort needed to develop and deploy a solution for a new use case. In this respect, expanding the intended usage area of SPARQL queries to express hybrid information needs while maintaining the conformance to SPARQL 1.1 helped us to achieve this goal. The main directions for the future work concern further minimizing the adaptation effort needed to deploy *metaphactory* in a new use case. This involves, for example, building a library of reusable data analytics services (e.g., for common machine learning algorithms).

Acknowledgements. This work has been supported by the Eurostars project DIESEL (E!9367) and by the German BMWI Project GEISER (project no. 01MD16014).

References

1. Mikolov, T., Sutskever, I., Chen, K., Corrado, G.S., Dean, J.: Distributed representations of words and phrases and their compositionality. In: Advances in Neural Information Processing Systems, pp. 3111–3119 (2013)
2. Altschul, S.F., Gish, W., Miller, W., Myers, E.W., Lipman, D.J.: Basic local alignment search tool. J. Mol. Biol. **215**(3), 403–410 (1990)
3. Usbeck, R., Röder, M., Haase, P., Kozlov, A., Saleem, M., Ngomo, A.-C.N.: Requirements to modern semantic search engine. In: Ngonga Ngomo, A.-C., Křemen, P. (eds.) KESW 2016. CCIS, vol. 649, pp. 328–343. Springer, Cham (2016). doi:10.1007/978-3-319-45880-9_25
4. Pérez, J., Arenas, M., Gutierrez, C.: Semantics and complexity of SPARQL. ACM TODS **34**(3), 16:1–16:45 (2009). http://dblp.uni-trier.de/rec/bibtex/journals/tods/PerezAG09
5. Kaminski, M., Kostylev, E.V., Grau, B.C.: Semantics and expressive power of subqueries and aggregates in SPARQL 1.1. In: WWW 2016, pp. 227–238 (2016)
6. Magliacane, S., Bozzon, A., Della Valle, E.: Efficient execution of Top-K SPARQL queries. In: Cudré-Mauroux, P., Heflin, J., et al. (eds.) ISWC 2012. LNCS, vol. 7649, pp. 344–360. Springer, Heidelberg (2012). doi:10.1007/978-3-642-35176-1_22
7. Schwarte, A., Haase, P., Hose, K., Schenkel, R., Schmidt, M.: FedX: optimization techniques for federated query processing on linked data. In: Aroyo, L., Welty, C., et al. (eds.) ISWC 2011. LNCS, vol. 7031, pp. 601–616. Springer, Heidelberg (2011). doi:10.1007/978-3-642-25073-6_38

8. Nikolov, A., Schwarte, A., Hütter, C.: FedSearch: efficiently combining structured queries and full-text search in a SPARQL federation. In: Alani, H., Kagal, L., et al. (eds.) ISWC 2013. LNCS, vol. 8218, pp. 427–443. Springer, Heidelberg (2013). doi:10.1007/978-3-642-41335-3_27

9. Schnaitter, K., Polyzotis, N.: Optimal algorithms for evaluating rank joins in database systems. ACM Trans. Database Syst. **35**(1), 1–47 (2008)

10. Görlitz, O., Staab, S.: SPLENDID: SPARQL endpoint federation exploiting void descriptions. In: COLD2011, at ISWC 2011, Bonn, Germany (2011)

11. Acosta, M., Vidal, M.-E., Lampo, T., Castillo, J., Ruckhaus, E.: ANAPSID: an adaptive query processing engine for SPARQL endpoints. In: Aroyo, L., Welty, C., et al. (eds.) ISWC 2011. LNCS, vol. 7031, pp. 18–34. Springer, Heidelberg (2011). doi:10.1007/978-3-642-25073-6_2

12. Saleem, M., Ngonga Ngomo, A.-C.: HiBISCuS: hypergraph-based source selection for SPARQL endpoint federation. In: Presutti, V., d'Amato, C., et al. (eds.) ESWC 2014. LNCS, vol. 8465, pp. 176–191. Springer, Cham (2014). doi:10.1007/978-3-319-07443-6_13

13. Stringer, B., Meroño-Peñuela, A., Abeln, S., van Harmelen, F., Heringa, J.: SCRY: extending SPARQL with custom data processing methods for the life sciences. In: SWAT4LS (2016)

14. Dolby, J., Fokoue, A., Muro, M.R., Srinivas, K., Sun, W.: Extending SPARQL for data analytic tasks. In: Groth, P., Simperl, E., et al. (eds.) ISWC 2016. LNCS, vol. 9982, pp. 437–452. Springer, Cham (2016). doi:10.1007/978-3-319-46547-0_36

Managing Lifecycle of Big Data Applications

Ivan Ermilov[1]([⊠]), Axel-Cyrille Ngonga Ngomo[2], Aad Versteden[3],
Hajira Jabeen[4], Gezim Sejdiu[4], Giorgos Argyriou[5], Luigi Selmi[6],
Jürgen Jakobitsch[7], and Jens Lehmann[4]

[1] Instituts für Angewandte Informatik, Hainstraße 11, 04107 Leipzig, Germany
iermilov@informatik.uni-leipzig.de
[2] Paderborn University, Warburger Street 100, 33098 Paderborn, Germany
axel.ngonga@upb.de
[3] Tenforce, Havenkant 38, 3390 Leuven, Belgium
aad.versteden@tenforce.com
[4] University of Bonn, Römerstraße 164, 53117 Bonn, Germany
{jabeen,sejdiu,jens.lehmann}@cs.uni-bonn.de
[5] University of Athens, Ilisia, Panepistimioupolis, 15703 Athens, Greece
gioargyr@di.uoa.gr
[6] Fraunhofer IAIS, Schloss Birlinghoven, 53757 Sankt Augustin, Germany
luigi.selmi@iais.fraunhofer.de
[7] Semantic Web Company GmbH, Mariahilfer Strasse 70, 1070 Vienna, Austria
juergen.jakobitsch@semantic-web.com

Abstract. The growing digitization and networking process within our society has a large influence on all aspects of everyday life. Large amounts of data are being produced continuously, and when these are analyzed and interlinked they have the potential to create new knowledge and intelligent solutions for economy and society. To process this data, we developed the Big Data Integrator (BDI) Platform with various Big Data components available out-of-the-box. The integration of the components inside the BDI Platform requires components homogenization, which leads to the standardization of the development process. To support these activities we created the BDI Stack Lifecycle (SL), which consists of development, packaging, composition, enhancement, deployment and monitoring steps. In this paper, we show how we support the BDI SL with the enhancement applications developed in the BDE project. As an evaluation, we demonstrate the applicability of the BDI SL on three pilots in the domains of transport, social sciences and security.

Keywords: Big Data · Software methodologies · Microservice architecture

1 Introduction

Digitization is taking an increasingly important role in our everyday life. The analysis and interlinking of the large amounts of data generated permanently by our society has the potential to build the foundation for novel insights and

© Springer International Publishing AG 2017
P. Różewski and C. Lange (Eds.): KESW 2017, CCIS 786, pp. 263–276, 2017.
https://doi.org/10.1007/978-3-319-69548-8_18

intelligent solutions for economy and society. However, processing this information at scale remains the privilege of a chosen few, who can integrate and deploy the plethora of Big Data processing tools currently available. What is needed are innovative technologies, strategies and competencies for the beneficial use of Big Data to address societal needs. In the Big Data Europe (BDE) project, we address this need by developing the Big Data Integrator (BDI) Platform. This platform is designed to accommodate various Big Data components such as Hadoop[1], Spark[2], Flink[3], Hive[4] and others[5] available out-of-the-box. The BDI Platform is based on the Docker technology stack and seamlessly integrates components into complex architectures necessary to solve societal data challenges. The platform is being used in 7 pilots in the areas of Health, Food, Energy, Transport, Climate, Social Sciences, and Security. Each of the pilots builds upon available Big Data components[6] and develops new ones specific to the challenge.

The integration of the components inside the BDI Platform requires an approach to the homogenization of components. To support the development of the BDI components and their integration in the BDI platform, we propose a novel methodology of developing Big Data applications dubbed BDI Stack Lifecycle (SL). Our methodology is based on the experience of developing Docker images inside the BDE project and their integration in the pilots. It consists of the following steps: (1) development, (2) packaging, (3) composition, (4) enhancement, (5) deployment and (6) monitoring. The prime goal of BDI SL is the creation of complex Big Data applications dubbed BDI Stacks. Each instantiation of BDI Stack methodology is a complex use-case-driven application designed to address a particular challenge (e.g., processing sensor data from taxi drives in Thessaloniki).

The BDI Platform itself was described in the previous work [1]. In this paper, we focus on the methodology for developing BDI Stack applications. In particular, the contributions of this paper are as follows:

– We propose a methodology for developing and maintaining complex Big Data applications using Docker containers dubbed BDI Stack Lifecycle.
– We describe the supporting tools for the six steps of the BDI SL.
– We apply our methodology on the real-world pilots and show its applicability.

2   Related Work

The BDI Platform is a distribution of Big Data components available out-of-the-box for easy installation and setup. The idea of creating library of components is

---

[1] http://hadoop.apache.org/.

[2] https://spark.apache.org/.

[3] https://flink.apache.org/.

[4] https://hive.apache.org/.

[5] https://www.big-data-europe.eu/bdi-components/.

[6] At the time of writing, more than 30 components are available in BDE Components Library.

not novel and a lot of other distributions are available at the time of writing [8]: Hortonworks[7], Cloudera[8], MapR[9], Bigtop[10]. Most of distributions are provided as open source software under Apache 2.0 license. The key difference between the BDI Platform and other distributions are (1) the flexibility of the Platform, which enables creating custom stacks for the pilots, and (2) the availability of documentation and use cases. To the best of our knowledge, we are the first project, which summarizes the development process and proposes a methodology for developing big-data stack applications.

Data management lifecycles relevant to this work have been developed in the domain of data warehousing. CRISP DM [17] is a generic data mining methodology, which can be applied in any kind of data mining tasks. CRISP DM breaks down the lifecycle of a data mining project into six phases: business understanding, data understanding, data preparation, modeling, evaluation, and deployment. It is common to extend or modify CRISP DM for particular needs of a project. For example, in [5] the authors extend CRISP DM to process big scientific data. CRISP DM is data centric and does not tackle service composition or deployment problems in detail. In the scope of our BDI SL, CRISP DM only covers the first development step, where a single Spark or Flink application is being designed to address a particular data mining challenge.

The other relevant research field, which studies the software lifecycles is Continuous Integration/Continuous Delivery (CI/CD). CI/CD assumes that the new software needs to be developed and has the goals of fast and frequent release, flexible product design and architecture, continuous and rapid experimentation etc. (see the recent state-of-the-art survey [14]). The main difference between the methodologies represented in CI/CD studies and our approach is that we reuse existing components for assembling complex architectures.

Mesosphere DCOS[11] and Canonical Juju [18] simplifies the deployment of complex architectures. Juju operates with charms, which are packaged components with preprogrammed dependencies and connection logic. The charms can be written in any programming language and are available in Juju Store [12] for reuse. Juju creates a thin layer over resource provisioning. To implement an architecture using Juju, it is necessary to program deployment logic, when the required components are not available in Juju Store or the available implementation does not fit the user requirements. For example, in [15] the authors implement flexible architecture for deploying applications on Apache Storm. In our approach the deployment logic is encapsulated into Docker images, which allows user to employ default Docker facilities such as *docker-compose*.

The architectures for big data applications include Model Deployment and Execution Framework (MDEF) [7], Extended IoT Framework [16], Big Data

---

[7] https://hortonworks.com/.
[8] https://www.cloudera.com/.
[9] https://mapr.com.
[10] http://bigtop.apache.org/.
[11] https://dcos.io/.
[12] https://jujucharms.com/store.

Processing Framework for Healthcare Applications [13], and On Demand Data Analytics in High Performance Cluster (HPC) Environments [6] among others. These architectures challenge a particular problem, which is often specific to the infrastructure. For example, in [6] the authors claim that facility resources (i.e. High Performance Cluster) cannot be simply partitioned or repurposed for specific architectures. Therefore, they propose an on demand solution for creating Spark instances on fly for HPC.

3   BDI Stack Lifecycle

In this section, we describe BDI SL methodology (see Fig. 1), which supports the creation, deployment and maintenance of the complex Big Data applications dubbed BDI Stacks. The BDI SL consists of the following steps:

1. **Development** includes the engineering of BDI Stack components such as, for example, Spark or Flink applications.
2. **Packaging** is the dockerization and publishing of the developed or existing third-party BDI Stack components.
3. **Composition** stands for the assembly of a BDI Stack, where we defined a BDI Stack as the integration of several BDI components (most commonly to address a particular data processing task).
4. **Enhancement** step is a process of extending BDI Stack with enhancement tools such as, for example, *Init Daemon* for the creation of dataflow, and *Logging* facilities for monitoring the health of a BDI Stack.

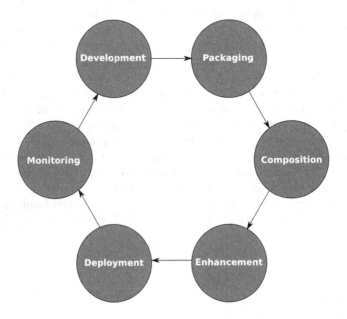

Fig. 1. BDI Stack Lifecycle.

5. **Deployment** is the instantiation of a BDI Stack on physical or virtual servers.
6. **Monitoring** consists of observing the status of a running BDI Stack and can lead to a reiteration of the BDI components and architecture of the BDI Stack.

To support the BDI SL methodology, we developed documentation and enhancement tools for each of the steps.

For the (1) **development** step we created application templates for Spark and Flink for the most common programming languages (Java, Scala, Python) [3]. We also provide Spark Notebook and Apache Zeppelin as a part of Spark/HDFS Workbench. Spark/HDFS Workbench is an integrated environment for developing, testing and running Spark applications. It can be deployed on any Docker host with only one CLI command [2]. Spark Notebook and Apache Zeppelin are the parts of the Spark/HDFS Workbench and can be used for interactive programming and execution of Spark jobs.

For the (2) **packaging** step we provide a library of ready-made components, which are created using best-practices on dockerization of Big Data technologies. More than 30 components are available in the BDE Github organization[13] at the time of writing. Also, we use the open source Docker technology, which has a community-driven components repository called Docker Hub. In case of a missing component or when BDI component does not meet project requirements, it is always possible to search the Docker Hub.

All the components from the BDE Github repository have example docker-compose snippets, which can be used for the (3) **composition** step. With this comprehensive library, the users are able to reuse existing BDI Stacks and create new ones by a simple extension. Additionally, we provide a graphical user interface for assembling BDI Stacks dubbed BDI Stack Builder.

For the (4) **enhancement** step we provide a set of enhancement applications aimed to improve usability of a BDI Stack. The enhancements include UI Integrator, Init Daemon, Healthchecks, Workflow Monitor, and Logging together with ELK Stack for logs visualization. We describe the enhancements in more detail in Sect. 4.

The (5) **deployment** step can be done by using both widely adopted *docker-compose* application or with our Swarm UI interface.

The (6) **monitoring** step is performed using Logging facility and visualized with ELK stack.

4 BDI Stack Assembly and Architecture

The central steps of BDI Stack assembly are (3) **composition** and (4) **enhancement** steps. In this section we describe them in more detail.

In Fig. 2 we show the process of BDI Stack assembly as well as resulting architecture of an example BDI Stack. As shown in Fig. 2, user employs *Stack*

---

[13] https://github.com/big-data-europe/.

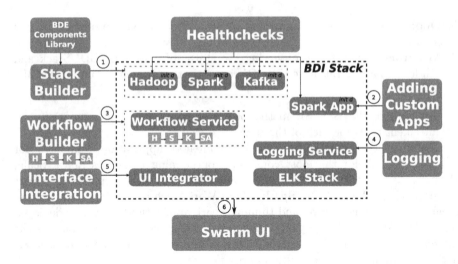

Fig. 2. BDI Stack Assembly and Architecture. (1) Picking up the ready-made BDI components from BDE Components Library. (2) Adding custom applications such as Spark applications. (3) Ensuring start-up of the BDI stack with BDI Workflows. (4) Adding logging facility. (5) Integrating interfaces with UI Integrator. (6) Deploying BDI Stack with Swarm UI.

Builder application[14] (1) to select required BDI components from *BDE Components Library* and have an initial version of BDI Stack with Hadoop, Spark and Kafka[15]. Then, the user adds custom applications (2) such as Spark App in the example, which performs the use case specific tasks. Often it is desired to ensure the execution order of a BDI Stack, for example, data processing should happen after data acquisition. The execution order can be controlled using docker native HEALTHCHECK Docker facility. However, due to the limitations of the native solution, we introduce an approach based on Init Daemon. With Init Daemon enabled components, it is possible to assemble a workflow (3) using Workflow Builder application[16]. The Workflow Service[17] executes a workflow during the BDI Stack deployment and ensures the correct execution order of the components of the stack. The logging facility[18] needs to be added as a Logging Service (4) together with ELK Stack for logging visualization. Interfaces from all the BDI Stack components (e.g. Hadoop, Spark, Kafka, Spark App) are aggregated using UI Integrator[19] (5) and can be accessed with a common GUI [4]. The resulting BDI Stack, which has docker-compose format, is deployed with Swarm UI[20] (6).

---

[14] https://github.com/big-data-europe/app-stack-builder.
[15] https://kafka.apache.org/.
[16] https://github.com/big-data-europe/app-pipeline-builder.
[17] https://github.com/big-data-europe/mu-pipeline-service.
[18] https://github.com/big-data-europe/mu-bde-logging.
[19] https://github.com/big-data-europe/app-integrator-ui.
[20] https://github.com/big-data-europe/app-swarm-ui.

The enhancement applications such as *mu-bde-logging* and *mu-pipeline-service* are developed using mu.semte.ch semantic technology stack [19].

5   Use Cases

In this section we present three selected pilots developed with the BDI SL methodology. Each of the pilots correspond to a societal challenge. As shown in the summary Table 1, the pilots process the data in various formats and have different number of components in a BDI Stack. In the following, we describe the pilots in more detail.

Table 1. Use cases summary.

Pilot name	SC4	SC6	SC7
Domain	Transport	Social sciences	Security
Data format	Sensor data	Tabular data (CSV)	Satellite images
Number of components in BDI Stack	9	8+	11

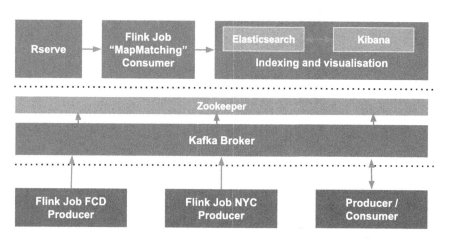

Fig. 3. Transport pilot architecture.

5.1   Transport

The H2020 Societal Challenge 4[21], Smart Green and Integrated Transport, covers a broad topic ranging from urban mobility, to safety, logistics, transport system integration, infrastructure monitoring and planning. Transport systems consume

---

[21] https://www.big-data-europe.eu/pilot-transport/.

huge flows of data to provide services, monitor infrastructures and discover the usage patterns in order to forecasting what will be the status in the near or distant future. All these systems consume streams of data from different sources and in different formats. In the SC4 pilot (Fig. 3) we have therefore decided to build a pilot that can ingest, transform, integrate and store streams of data that have spatial and temporal dimensions. One of the project partner, CERTH-HIT, is managing a system that monitors the traffic flow in Thessaloniki, Greece, using floating car data from a transport company. The legacy system is based on a relational database, stored procedures and R scripts to map-match the location of the vehicles to the road segments and compute the traffic flow and average speed among other statistical parameters. The result of the computation is used for monitoring and as input for forecasting the value of the parameters in the near future and is made available through a web service. The aim of the pilot is to address the scalability issues of the current system leveraging the availability of distributed frameworks and the containerization technology for the deployment of services in different environments (Fig. 4).

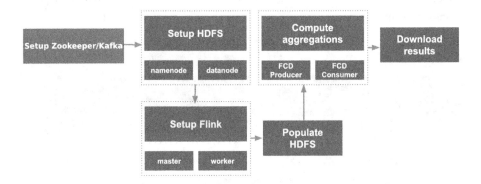

Fig. 4. Transport pilot initialization workflow.

The pilot is based on the microservices architecture where different software components, producers and consumers, communicate through a messaging system connecting data sources to data sinks. Producers and consumers are implemented as Flink jobs while Kafka has been chosen as the messaging system. The producer fetches the data every two minutes from the web service, stores the records sets into Hadoop HDFS, transforms the records into a binary format, using a schema shared with the consumer, and finally sends the records to a Kafka topic. The consumer reads the records from the Kafka topic and process them at event time applying the map matching function. The consumer must connect to an R server where an R script has been installed to perform the computation for the map matching using the road network data from Open Street Map stored in a PostGIS database. The consumer adds the identifier of the road segment as an additional field to the original record and finally aggregates the records per road segment and in time windows to compute the traffic flow and the average speed in each road segment. The result of the aggregation can be sent to Hadoop HDFS or to Elasticsearch[22]. From Elasticsearch different visualizations can be created easily with Kibana[23]. The records with the aggregated values stored in Elasticsearch will be used as input to a forecasting algorithm to predict the traffic flow. All the components are available as Docker images and a docker-compose file has been created adding the initialization service and the UI provided by the BDI Stack in order to start the services in the right sequence from the browser (e.g. Zookeeper before Kafka and PostGIS as well as Elasticsearch before the consumer).[24]

5.2   Social Sciences

The H2020 societal challenge 6[25], "Europe in a changing world - Inclusive, innovative and reflective societies" roughly covers topics improving the understanding of European societies in the context of the public sector. The Big Data Europe pilot implementation for this societal challenge exposes the foundation of making budget data comparable across European municipalities. Additionally the pilot architecture (see Fig. 5) can be used as a base for document processing at scale. The basic implementation is ingesting budget data from three Greek municipalities: Athens, Thessaloniki and Kalamaria. These datasets (in CSV format) are collected on a daily basis and transformed to RDF using ELOD's schema[26,27]. The transformed datasets are used to calculate financial ratios[28] and compare incomes/expenses of the municipalities.

The entry point for input data is Apache Flume agents[29], which can be configured according to the availability and format of budget datasets. All Flume agents are configured to store raw data into Hadoop HDFS and create an Apache Kafka message for each file, containing the name of the source document as key and the contents of the file as an array of bytes. Apache Spark acts as a consumer of Kafka messages. The Spark job consumes messages in parallel with the possibility of being scaled to any number of nodes. SC6 pilot makes use of SPI[30] to determine a suitable parser for a given file, which in turn will transform the source file into RDF and upload the resulting triples to a Virtuoso Triple Store[31]. After a successful upload financial ratios[32] are calculated for

---

[22] https://www.elastic.co/products/elasticsearch.

[23] https://www.elastic.co/products/kibana.

[24] https://github.com/big-data-europe/pilot-sc4-fcd-applications.

[25] https://www.big-data-europe.eu/pilot-social-sciences/.

[26] http://linkedeconomy.org/en.

[27] https://github.com/LinkedEcon/LinkedEconomyOntology-ELOD.

[28] http://www.accountingverse.com/managerial-accounting/fs-analysis/
financial-ratios.html.

[29] http://flume.apache.org.

[30] https://docs.oracle.com/javase/tutorial/sound/SPI-intro.html.

[31] https://virtuoso.openlinksw.com/.

[32] https://en.wikipedia.org/wiki/Financial_ratio.

the newly added data. Those aggregations are done using the SPARQL[33] query language. Both steps, the creation of the initial RDF dataset and the calculation of the financial ratios make use of PoolParty Semantic Suite's[34] capabilities as a clearing house for all literal terms that are involved. This means that financial terms, unknown to a wider public, can be translated, annotated and unified for municipalities of different languages. Finally a dashboard of financial ratios is created based on PoolParty's GraphSearch[35]. This way financial ratios are easily comparable. Financial ratios and periods of time as well as municipalities can be chose to be compared.

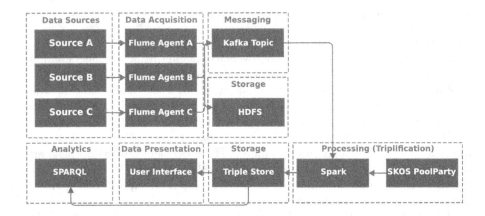

Fig. 5. Social sciences pilot architecture.

Note that the pilot setup can also be used as a template to get started with the relevant Big Data technologies. It comprises core Big Data tools such as Flume, Kafka, HDFS and Spark and shows how these can be setup to work together. The architecture is driven by the requirement of making the cooperation between a technical expert and a domain expert as easy as possible in that both have distinct points of control, for example, in the selection of source data and the transformation of the source files into RDF.

The pilot's stack is configured as a whole using a single docker-compose file, which can be run by Docker Swarm.[36] Although all involved docker images support BDE's Init-daemon, docker-compose "depends_on" feature is sufficient in the case of SC6. The pilot, however, makes use of Docker HEALTHCHECK[37] facility. The BDI Logging collects the health information about docker containers and store it in the ELK stack, thus the applications, that do not expose a

---

[33] https://www.w3.org/TR/sparql11-query/.

[34] https://www.poolparty.biz/.

[35] https://www.poolparty.biz/poolparty-semantic-graph-search-server/.

[36] https://github.com/big-data-europe/pilot-sc6-cycle2.

[37] https://docs.docker.com/engine/reference/builder/#healthcheck.

graphical user interface can be monitored (e.g. Apache Kafka). Apache Spark's BDE UI as well as unified logging[38] make sure the pilot's general working can be followed easily.

5.3 Security

The security pilot combines data relating to security domain[39], the H2020 Societal Challenge 7 (SC7). Sources of such data are Earth Observation products and the combination of news articles with user-generated messages from social media. In the first case, the pilot processes satellite images in order to detect changes in land cover or land use. In the second one, it processes news from the web sites and social media in order to detect events. Combining the outcomes, we achieve an integration of remote sensing sources with social sensing ones.

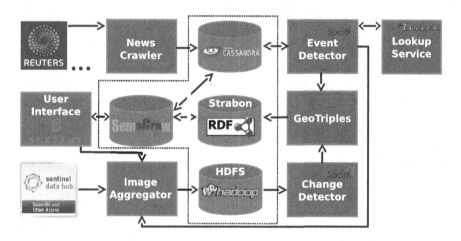

Fig. 6. Security pilot architecture.

The high-level architecture is depicted in Fig. 6 and represents three workflows. the top components (News Crawler, Cassandra, Event Detector and Lookup Service) implement event-detection workflow. The middle layer (Sextant, SemaGrow, Strabon and Geotriples) implements the activation-workflow. The bottom layer (Image Aggregator, HDFS and Change Detector) implements the change-detection workflow.

For the event-detection workflow, the News Crawler runs periodically and ingests data from various public news streams like Twitter and Reuters. All these news are stored in Cassandra[40] where they are processed by the Event Detector. Event Detector periodically executes a Spark job to cluster the news

[38] https://github.com/big-data-europe/mu-bde-logging.
[39] https://www.big-data-europe.eu/security/.
[40] http://cassandra.apache.org/.

items into events and to associate them with one or more geo-locations with the help of the Lookup Service. The Lookup Service is based on Lucene index. It accepts location names in plain text and returns their geo-coordinates by indexing 180,000 location names from the GADM dataset[41].

The change-detection workflow processes satellite images from ESA?s Copernicus Open Access Hub[42] (Open Hub). The Image Aggregator downloads large files that contain images with their metadata. These images cover a certain area of interest during a specific time period as requested by the user. The images are stored in Hadoop HDFS[43] being accessible to Spark nodes and thus available to the Change Detector. After each processing request[44], two satellite images are ingested in Change Detector and areas of highly possible changes are returned.

Sextant [12] is the basic component of the activation workflow and the entry point for the pilot. It has been widely extended for the pilot's needs and provides a graphic interface for the user to select either an event or a change detection triggering the corresponding workflow. SemaGrow, Strabon and GeoTriples provide support for the event- and change-detection workflows and complement the activation one. Geotriples [11] receives descriptions of areas (the output of Change Detector) or summaries of events (the output of Event Detector) and converts them into RDF. The output of the Geotriples is then stored in Strabon [10] which is a spatio-temporal triplestore that efficiently executes GeoSPARQL and stSPARQL queries. Sextant provides access and visualization the data that are stored either in Cassandra and Strabon through Semagrow [9].

All components of the pilot are provided as Docker images.[45] They run as Docker containers within Docker Swarm in order to be enhanced as a BDI Stack. The whole pipeline is deployed in the BDE Platform by running a docker-compose file that describes all the services of the pilot.

Executing a Docker Compose application will launch all Docker containers simultaneously. In order to avoid the immediate launch of all Docker containers and to control the execution order we enrich each service of the pilot with the native HEALTHCHECK Docker facility. Thus, we manage to keep a certain order of the initiation of the services and we make sure that every service will start running after it is confirmed that the services it is depended on, have completed a healthy start.

Ensuring the correct execution order of the components of the stack we have a ready pipeline for use. The user is able to select the provided functionality from Sextant choosing either of the two workflows described earlier. Aggregating the interfaces of the BDI Stack components (Sextant, Hadoop and Spark), which the SC7 pilot uses with the UI Integrator allows the user to monitor the progress of the selected workflow.

---

[41] http://www.gadm.org/.

[42] https://scihub.copernicus.eu/.

[43] The typical loading time for a set of images: 400 s.

[44] The typical Spark job execution time for Change Detector: 1000 s.

[45] https://github.com/big-data-europe/pilot-sc7-change-detector.

6   Conclusions and Future Work

In this paper, we presented a Big Data Integrator Stack Lifecycle methodology for creation, management and maintenance of Big Data applications. We showed how the six steps of the lifecycle are supported within the BDI platform. Three pilots are showcased as an evaluation of the presented lifecycle. The core of the BDI SL methodology is a composition step, which depends on the underlying technology (i.e. Docker). Thus, it will be hard to transfer the methodology to generic environments or adapt it for usage in High Performance Clusters (HPC) or vendor specific environments, which do not run on Docker or can not execute docker containers. The linear initialization workflows as presented in the pilots are tolerant to cascading failures. In the future, we will address this issue by further developing Docker images for all the components to be fault-tolerant and not dependent on the execution order. Additionally, we plan to test and evaluate not only three, but all 7 pilots and report on the evaluation results.

Acknowledgments. This work was supported by grant from the European Union's Horizon 2020 research Europe flag and innovation program for the project Big Data Europe (GA no. 644564).

References

1. Auer, S., et al.: The BigDataEurope platform – supporting the variety dimension of big data. In: Cabot, J., Virgilio, R., Torlone, R. (eds.) ICWE 2017. LNCS, vol. 10360, pp. 41–59. Springer, Cham (2017). doi:10.1007/978-3-319-60131-1_3. http://jens-lehmann.org/files/2017/icwe_bde.pdf
2. Ermilov, I.: Scalable spark/hdfs workbench using docker (2016), https://www.big-data-europe.eu/scalable-sparkhdfs-workbench-using-docker/. Retrieved 21 May 2017
3. Ermilov, I.: Developing spark applications with docker and BDE (2017), https://www.big-data-europe.eu/developing-spark-applications-with-docker-and-bde/. Retrieved 21 May 2017
4. Ermilov, I.: User interface integration in BDI platform (integrator UI application) (2017), https://www.big-data-europe.eu/user-interface-integration-in-bdi-platform-integrator-ui-application/. Retrieved 21 May 2017
5. Grady, N.W.: KDD meets big data. In: 2016 IEEE International Conference on Big Data (Big Data), pp. 1603–1608. IEEE (2016)
6. Harney, J., Lim, S.H., Sukumar, S., Stansberry, D., Xenopoulos, P.: On-demand data analytics in HPC environments at leadership computing facilities: Challenges and experiences. In: 2016 IEEE International Conference on Big Data (Big Data), pp. 2087–2096. IEEE (2016)
7. Heit, J., Liu, J., Shah, M.: An architecture for the deployment of statistical models for the big data era. In: 2016 IEEE International Conference on Big Data (Big Data), pp. 1377–1384. IEEE (2016)
8. Jabeen, H.: Bde vs. other hadoop distributions (2016), https://www.big-data-europe.eu/bde-vs-other-hadoop-distributions/. Retrieved 21 May 2017

9. Konstantopoulos, S., Charalambidis, A., Mouchakis, G., Troumpoukis, A., Jakobitch, J., Karkaletsis, V.: Semantic web technologies and big data infrastructures: SPARQL federated querying of heterogeneous big data stores. In: ISWC Demos and Posters Track (2016)

10. Kyzirakos, K., Karpathiotakis, M., Koubarakis, M.: Strabon: a semantic geospatial DBMS. In: Cudré-Mauroux, P., Heflin, J., Sirin, E., Tudorache, T., Euzenat, J., Hauswirth, M., Parreira, J.X., Hendler, J., Schreiber, G., Bernstein, A., Blomqvist, E. (eds.) ISWC 2012. LNCS, vol. 7649, pp. 295–311. Springer, Heidelberg (2012). doi:10.1007/978-3-642-35176-1_19

11. Kyzirakos, K., Vlachopoulos, I., Savva, D., Manegold, S., Koubarakis, M.: Geotriples: a tool for publishing geospatial data as RDF graphs using R2RML mappings. In: Proceedings of the 2014 International Conference on Posters & Demonstrations Track, vol. 1272, pp. 393–396. CEUR-WS. org (2014)

12. Nikolaou, C., Dogani, K., Bereta, K., Garbis, G., Karpathiotakis, M., Kyzirakos, K., Koubarakis, M.: Sextant: Visualizing time-evolving linked geospatial data. Web Semant. Sci. Serv. Agents World Wide Web **35**, 35–52 (2015)

13. Rahman, F., Slepian, M., Mitra, A.: A novel big-data processing framework for healthcare applications: big-data-healthcare-in-a-box. In: 2016 IEEE International Conference on Big Data (Big Data), pp. 3548–3555. IEEE (2016)

14. Rodriguez, P., Haghighatkhah, A., Lwakatare, L.E., Teppola, S., Suomalainen, T., Eskeli, J., Karvonen, T., Kuvaja, P., Verner, J.M., Oivo, M.: Continuous deployment of software intensive products and services: a systematic mapping study. J. Syst. Softw. **123**, 263–291 (2017)

15. Sebrechts, M., Borny, S., Vanhove, T., Van Seghbroeck, G., Wauters, T., Volckaert, B., De Turck, F.: Model-driven deployment and management of workflows on analytics frameworks. In: 2016 IEEE International Conference on Big Data (Big Data), pp. 2819–2826. IEEE (2016)

16. Sezer, O.B., Dogdu, E., Ozbayoglu, M., Onal, A.: An extended iot framework with semantics, big data, and analytics. In: 2016 IEEE International Conference on Big Data (Big Data), pp. 1849–1856. IEEE (2016)

17. Shearer, C.: The CRISP-DM model: the new blueprint for data mining. J. Data Warehouse. **5**(4), 13–22 (2000)

18. Tsakalozos, K., Johns, C., Monroe, K., VanderGiessen, P., Mcleod, A., Rosales, A.: Open big data infrastructures to everyone. In: 2016 IEEE International Conference on Big Data (Big Data), pp. 2127–2129. IEEE (2016)

19. Versteden, A., Pauwels, E.: State-of-the-dart web applications using microservices and linked data. In: Maleshkova, M., Verborgh, R., Keppmann, F.L. (eds.) 4th Workshop on Services and Applications over Linked APIs and Data (SALAD), vol. 1629, pp. 25–36. CEUR Workshop Proceedings, Aachen (2016). http://ceur-ws. org/Vol-1629/paper4.pdf

Semantic Web and Education

Ontology-Based Representation of Learner Profiles for Accessible OpenCourseWare Systems

Mirette Elias[1(✉)], Steffen Lohmann[2], and Sören Auer[3]

[1] University of Bonn, Bonn, Germany
melias@uni-bonn.de
[2] Fraunhofer IAIS, St. Augustin, Germany
steffen.lohmann@iais.fraunhofer.de
[3] Technische Informationsbibliothek (TIB), Hannover, Germany
soeren.auer@tib.eu

Abstract. The development of accessible web applications has gained significant attention over the past couple of years due to the widespread use of the Internet and the equality laws enforced by governments. Particularly in e-learning contexts, web accessibility plays an important role, as e-learning often requires to be inclusive, addressing all types of learners, including those with disabilities. However, there is still no comprehensive formal representation of learners with disabilities and their particular accessibility needs in e-learning contexts. We propose the use of ontologies to represent accessibility needs and preferences of learners in order to structure the knowledge and to access the information for recommendations and adaptations in e-learning contexts. In particular, we reused the concepts of the ACCESSIBLE ontology and extended them with concepts defined by the IMS Global Learning Consortium. We show how OpenCourseWare systems can be adapted based on this ontology to improve accessibility.

1 Introduction

Web accessibility has become a fundamental requirement in the development of web applications. It basically refers to the practice of making websites usable to people with disabilities. An example is the use of alternative text descriptions for images in order to enable screen readers to read aloud the text (using some text-to-speech technique) for people with vision impairments. A lot of knowledge, standards, guidelines, and techniques with regard to web accessibility are available in the literature that can be consulted when developing accessible web applications. In our previous work [10], we reviewed and classified the state of the art and identified the most relevant standards related to web accessibility in e-learning contexts.

In this paper, we present an ontology-based approach to reuse and integrate this accessibility knowledge in order to develop adaptable OpenCourseWare (OCW) considering the learners' accessibility needs and preferences. We are using ontologies to represent the accessibility concepts in a structured way in

© Springer International Publishing AG 2017
P. Różewski and C. Lange (Eds.): KESW 2017, CCIS 786, pp. 279–294, 2017.
https://doi.org/10.1007/978-3-319-69548-8_19

order to enable access to this knowledge for recommendations and adaptations in OpenCourseWare systems. In particular, we are reusing and extending concepts from the ACCESSIBLE ontology [11] to represent users with disabilities along with the accessibility specifications of e-learning systems as defined by the IMS Global Learning Consortium [2].

The remainder of this paper is organized as follows: In Sect. 2, we introduce the main components of OCW systems and outline related accessibility requirements. Section 3 presents our approach and methodology proposal for accessible OCW systems; it also defines the main components addressed in the paper. Section 4 describes the ontology, its concepts and structure. Section 5 presents a user dialogue we implemented to capture accessibility profiles of learners. In Sect. 6, we evaluate the approach and ontology with standard persona types defined by the W3C and examples of educational resources. Section 7 reviews the related work on learner profiles and ontologies, before the paper is concluded in Sect. 8.

2 Accessible OpenCourseWare

Open Educational Resources (OER) are openly licensed and freely accessible learning materials that can be used in e-learning contexts and beyond. Often, OER are published on the web in the form of OpenCourseWare (OCW) organized in courses and complemented by tools for collaboration and evaluation. OCW systems thus provide means for distributing free educational content to a wide range of learners over the web [4]. These learners include people with disabilities who have diverse needs, in terms of the type and severity of their disabilities, which must be addressed by OCW systems that aim to be *inclusive*. Designing one system that meets the needs of all learners is usually not possible, as learners have different needs and preferences, in particular disabled learners. For instance, one blind user might want to use a screen reader, while another blind user might prefer a braille display—or both might want to use the same device but with different configurations (e.g., different text reading speeds).

Basically, the components of most OCW systems can be divided into four categories, as illustrated in Fig. 1: (1) website, (2) digital resources, (3) assessments, and (4) communication and collaboration tools. When addressing accessibility of OCW systems, the individual peculiarities of each component must be considered.

Website. The website is the starting point which indexes, previews, and presents the OCW contents. The display and structure of the contents, metadata, and navigation are important elements of an accessible web interface. There are a number of accessibility standards and guidelines that can be followed (e.g., WCAG 2.0 [8] and WAI-ARIA [7]) in order to provide a website that is accessible by different types of users. For example, blinking icons and pop-up windows may be considered distracting elements by some people with cognitive disabilities and should therefore better be avoided in accessible OCW contexts.

Fig. 1. Typical components of OpenCourseWare systems

Digital Resources. Open Educational Resources (OER) are provided as digital resources in different types of media (e.g., slides, audio, video, etc.) used to represent the course material. The design, representation, and management of these files and formats are crucial for addressing the user needs and preferences. For example, if some educational resources are only available in auditory format and no alternative text transcripts are provided, they are not accessible to deaf learners.

Communication and Collaboration Tools. This category is concerned with providing accessible means for the communication and collaboration of learners in OCW systems to allow for an inclusive collaborative learning experience. Common features in OCW systems, such as commenting functions or discussion boards, should be accessible to all types of learners. For example, the navigation between the entries in a discussion board and the contribution of own entries should be possible with a keyboard only [9].

Assessments. Designing the assessment of the learning material in an accessible way is another challenge. Providing alternative forms of assessment and evaluation is helpful to address different learning styles and disabilities of learners. To give an example, time restrictions of assessments might cause problems to people with cognitive or motor disabilities, since they usually need more time than common learners. Thus, avoiding time restrictions or allowing time extensions should be considered for these types of learners [13].

3   Approach and Methodology

Our research is conducted in the context of the SlideWiki project[1], which is concerned with the development of an accessible OpenCourseWare system making use of semantic technologies. A central goal of the envisioned OCW system is that it can be adapted to the various learners' needs and preferences. Our first step to achieve this goal is to represent the learners' needs and references in an ontology. We are using profiles to describe the learners and recommend educational resources accordingly. The knowledge in the ontology will be used to infer

---

[1] https://slidewiki.eu.

and recommend the most appropriate resources and allow for adaptations of the website.

The overall system architecture of the envisioned adaptive accessibility component of the SlideWiki system is thus composed of an ontology (which we call the *AccessibleOCW ontology*), learner profiles, representations of the OCW components, and a personalization module, as it is illustrated in Fig. 2.

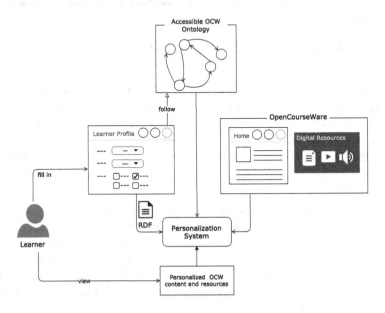

Fig. 2. Architecture of the OCW accessibility module

- **AccessibleOCW ontology:** our proposed ontology that contains the relevant accessibility knowledge required in OCW contexts (e.g., disability types, assistive technologies, accessibility guidelines, e-learning standards, etc.)
- **Learner profile:** a representation of learners, including information about their disabilities. The disabilities can be either automatically detected or manually entered by the learners (or their caregivers) using a form. Other needs and preferences of the learners can then be inferred from their disabilities using the ontology.
- **OCW components:** these are the website, educational resources, assessments, and collaboration tools, as illustrated in Fig. 1. The OCW components are adapted according to the learner profiles.
- **Personalization module:** processes the profile and ontology as an input to retrieve the preferences, assistive technology requirements, accessibility guidelines and standards related to the learners. Based on this information, it adapts the OCW content to the individual needs and preferences of the learners, and suggests the most appropriate educational resources to them.

Before we started development, we surveyed the available web accessibility and e-learning standards, guidelines, and ontologies [10]. We were particularly interested in the following aspects:

- Web accessibility guidelines and recommendations (e.g., WCAG 2.0, BBC);
- Disabilities, abilities, and limitations of users (e.g., ICF, FMA);
- E-learning and educational resources (e.g., IMS AfA, ISO24751);
- Assistive technologies and specifications (hardware and software).

We reviewed eleven web accessibility standards, guidelines, and techniques, and examined 20 ontologies that are available in the literature and on the web. As a result of the survey, we defined a set of OCW accessibility needs and requirements, and identified the most relevant standards and ontologies that meet the requirements. These standards and ontologies are the ACCESSIBLE ontology, IMS AfA, and WCAG2.

The overall methodology we follow can be divided into four phases:

1. *Defining and representing learner needs and preferences.* This is done by describing the different types of disabilities, their functional limitations, capabilities, and the related assistive technologies, also allowing the learners to enter any additional preferences. This is realized by representing the knowledge in the ontology and providing an input dialogue for learner profiles.
2. *Representing web accessibility needs.* The web accessibility needs (e.g., standards, techniques, approaches), which are related to the different disabilities, are also defined and represented in the ontology.
3. *Creating and managing educational resources.* This step is concerned with providing guidelines for creating accessible material with respect to the targeted learners, allowing for alternative representations of learning resources. Annotations are used for mapping suitable resources to the learners' needs.
4. *Adapting the website.* This concerns the adaptation of the content and structure of the website with respect to the learners' needs, as well as the validation of the communication and collaboration tools to be appropriate with the learners' capabilities.
5. *Assessment design.* Realizing alternative assessment types and styles to address different types of disabilities and assuring the support of assistive technologies.

This paper focuses on the learner profile and educational resources, and the AccessibleOCW Ontology we developed to represent and map this information.

4   AccessibleOCW Ontology

We are reusing the ACCESSIBLE ontology to represent domain knowledge of disability types, characteristics, and functional limitations. The ACCESSIBLE ontology contains knowledge about web accessibility standards and assistive technologies. For example, in the case of a color-blind user, success criterion 1.4.6 of WCAG 2.0 requires to check if the foreground and background color (or image) have a contrast ratio of at least 7:1. This could be automatically validated by the OCW system, and the colors could be adapted accordingly if they do not meet the requirements.

We extended the *User* concept of the ACCESSIBLE ontology with the *Learner* concept by adding properties that are needed to describe learners (e.g., education level, complexity level and preferred language of learning resources). We added properties and classes in accordance with the IMS AfA Personal Needs & Preferences (PNP) specification [2] to integrate accessibility needs in e-learning. The educational materials and resources are defined with respect to the IMS AfA Digital Resource Description (DRD) specification. The IMS AfA specifications implement the main guidelines of WCAG 2.0. For example, the access mode property represents the guidelines by describing the accessibility needs of a user with regard to digital resources (e.g., visual, audio). This property can be used for mappings between user needs and resources.

Our AccessibleOCW Ontology is currently composed of 16 classes.[2] We developed the ontology by parsing the IMS AfA specifications and schema documents to create the classes, properties, relationships, and individuals. The central **Learner** class describes the learners' properties together with their needs and preferences. It is a subclass of the *User* class of the ACCESSIBLE ontology. We extended this class to represent the disability information; thus, we can deduce the needs of users from their characteristics.

The learner properties are designed according to the IMS PNP specification. Some properties are directly added to the Learner class, whereas others are inferred from the ACCESSIBLE ontology, such as *isAtInteroperable* referring to the assistive technology used. This can be done by SWRL rules, such as the one shown below which describes an inference rule to conclude the *isAtInteroperable* property from the ACCESSIBLE ontology concepts. It states that if the user has a disability and this disability type is using an assistive device, then *isAtInteroperable* property is considered to be true.

```
User_has_Disability(?x,?y)^Disability_has_Device(?y,?z)
->isAtInteroperable(?x,true)
```

Yet other properties are a combination of both, i.e., the recommendation resulting from querying the ontology and the preference selected by the user, such as *RequiredAccessMode*. Consider the example of a visually impaired learner: by querying the ACCESSIBLE ontology, we can conclude that the learner might want to use a braille device, magnifier, or screen reader. Accordingly, we may recommend that the learner should use textual or audio resources, while the learner might decide to go for audio resources according to the preferences stated in her profile.

The **Learner** class is defined by a number of properties: *hasRequiredAccessMode* is the access mode required by a learner; here, the learners define their

---

[2] The ontology is available and deployed in a VoCol [12] environment at http://vocol.iais.fraunhofer.de/accessibilityOnto/. A visual representation created with Web-VOWL [16] is shown in Fig. 4 in the Appendix of this paper.

preference for the resource representation. For instance, a learner may define that if the resource is a visual one, she would prefer a textual representation.

The property *hasLanguageOfAdaptation* expresses the learner's language for the contents in the educational resources. The property *hasLanguageOfInterface* also represents the preferred language, but for the website itself and not for the educational resources, although the preferred language of both will mostly be the same.

The property *isAtInteroperable* has a boolean value indicating the usage of assistive technologies; when being true, all resources that are compatible with assistive technologies are returned.

The property *hasEducationalLevelOfAdaptation* indicates the educational level of the learners; at best, it should refer to a specific educational system definition (e.g., ASN Educational Level Vocabulary[3]). The property helps in guiding learners to the most appropriate educational resources with respect to their educational level.

The **Learner** class is linked with the **DigitalResource** class through an *hasAccess* relation, which is used to filter the accessible educational resources according to the user-defined disabilities and preferences. An example of using this relation will be given in Sect. 6.

The **DigitalResource** class defines the properties of an individual educational resource. The properties are designed based on the IMS DRD specification. They describe the resource access mode (e.g., visual, auditory) and define if a resource has adapted versions (e.g., if there is an alternative transcript file to an auditory file). It also defines properties like complexity of content, to provide authors the opportunity to offer a simplified version of the learning resource next to the normal one (*isAdaptationof, hasAdaptation, isAtInteroperable,* and *hasLanguageOfAdaptation*).

The property *hasAdaptedAccessMode* represents the adaptation format of a resource (e.g., visual); the type of adaptation is limited to the instances of AccessModeType.

The class *DisplayTransformabilityType* defines those components of a resource that can be easily adapted. The instances of this class include *backgroundColour, cursorPresentation, fontFace, fontSize, fontWeight, foregroundColour, highlightPresentation, layout, letterSpacing, lineHeight, structurePresentation,* and *wordSpacing.*

The following classes are related to both the **Learner** and **DigitalResource**:

The class *AccessModeType* defines the representation value, either for the resource or the requirements of the learner. The instances of this class include: *auditory, colour, itemSize, textual, visual, position, tactile,* and *textOnImage.*

The *RequiredAdaptationType* class describes the type of the adaptation required for special types of representations (e.g., if *auditory_caption* is set for an auditory format, an adapted caption is required).

---

[3] http://purl.org/ASN/scheme/ASNEducationLevel/.

The *AdaptationType* class represents the available types of adaptations. The instances of this class include: *alternativeText, audioDescription, captions, e-book, haptic, highContrast, longDescription, signLanguage*, and *transcript*.

The *ControlFlexibilityType* class describes the input requirements of the learners and resources, i.e., whether *fullKeyboardControl* or *fullMouseControl*.

The *EducationalComplexityType* class defines the level of complexity required by the learners and provided by the resources (e.g., *simplified, enriched*).

The *HazardType* class describes modes that should be avoided for some users, such as flashing, motionSimulation, olfactoryHazard, and sound.

5   Applying the Ontology

We designed an input dialogue to collect the disability information of learners. The learners are supposed to enter their profile information, which could also be automatically inferred from the context in parts. The form of the input dialogue is generated from our AccessibleOCW ontology using SPARQL queries. It asks the learners to input their disabilities, available devices, and personal preferences.

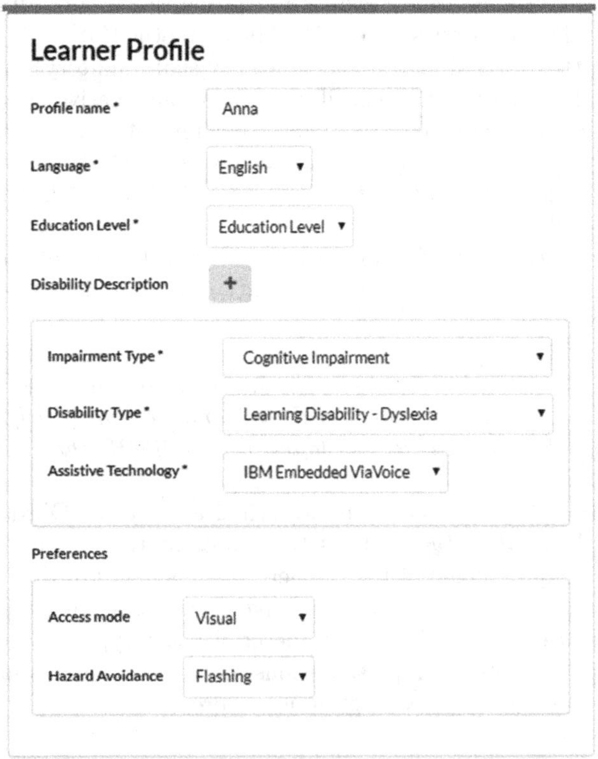

Fig. 3. Learner input profile

The learners create their personal profiles by selecting their preferences, as illustrated in Fig. 3. This example shows a learner profile for the user "Anna" who has dyslexia, which is a type of learning disability. The default language of Anna is English. She prefers visual representations over textual representations of the educational resources, while avoiding elements that are flashing. The user input is saved in an RDF file, which is compatible to the specifications of IMS PNP in order to use it for mapping to the properties of the educational resources represented in the IMS DRD format. We implemented a prototype of the dialogue using the JavaScript library React [5]. We decided for React as it already supports some of the accessibility concepts, such as ARIA attributes, and as it is widely used in web development nowadays. It furthermore supports the development of websites as separate components. This component structure will be useful when adapting the web interface.

We used SPARQL [6] to query the ontology, using Fuseki as the SPARQL server [1]. We created a SPARQL client in React that accesses our ontology on the Fuseki server. The SPARQL query in Listing 1 is an example of the queries that were used to feed the input fields of the profile dialogue. It retrieves the impairments, disabilities and assistive technologies that are related to each other with respect to the user's selection. As we mentioned before, several of the IMS properties can be automatically filled by SWRL rules based on the disability information entered by the users. In Listing 1, we added the IMS property *acc_ocw:isAtInteroperable*, making use of the SWRL rule defined in Sect. 4. According to the SWRL rule, the property is true whenever the selected type of disability has an assistive technology.

```
PREFIX acc: <http://www.Acces[...]ogy.com/GenericOntology.owl#>.
PREFIX acc_ocw: <http://purl.org/accessible_ocw#>.
SELECT ?impairment ?disability ?device ?ims_AT
WHERE {
?impairment a acc:Impairment.
?disability acc:Disability_belongsTo_Impairment ?impairment.
OPTIONAL  {?device acc:Device_belongsTo_Disability ?disability}.
OPTIONAL  {?disability acc:Disability_has_Device ?device}.
OPTIONAL {?ims_AT a acc_ocw:isAtInteroperable}.
}
```

Listing 1. SPARQL query retrieving impairments, their related disabilities and assistive technologies from our AccessibleOCW ontology

Finally, it must be considered that learners might have multiple disabilities. Thus, they must be enabled to enter all their impairments and the OCW system should adapt accordingly, which might require to perform some reasoning on the ontology to find the best combination and consistency of adaptations. In our learner profile dialogue shown in Fig. 3, we therefore allow for the input of multiple disabilities. In addition, a learner should also be able to define several profiles; for example, a person with visual impairments might prefer to use a braille device at work but a screen reader at home.

6   Evaluation

We designed a use case to evaluate our ontology and approach. It is composed of several examples from reliable sources. For the representation of learners, we used examples from W3C specifications, and for the educational resources, we used examples from the Accessibility Metadata Project [19]. This section will be divided into three parts: the first focuses on representing users profiles, the second focuses on educational resources, while the third addresses mappings between the both.

6.1   Representing Learners

We evaluated the developed ontology using the *personas* methodology. In particular, we used the personas that were created by W3C to test different types of user-centered systems.[4] We selected two personas for the evaluation: a hard hearing and a totally blind user. We used the descriptions to create corresponding instances in our ontology. The classes and properties used are a combination of the ACCESSIBLE ontology, representing the disability types and their characteristics, and the IMS concepts representing the accessibility needs in an e-learning system.

Persona 1 – Ms. Martinez is an old woman with hard hearing problems since her birth. She is trained in using sign language next to written language. She has problems with audio material; she requires audio contents to have a transcript and videos to have subtitles.

```
:Learner_1 a owl:NamedIndividual , :Learner;
   GenericOntology:hasName"Ms._Martinez"^^xsd:string;
   GenericOntology:hasAge"62"^^xsd:int;
   GenericOntology:User_has_Disability :Deafness;
   :hasLanguageOfAdaptation"English"^^xsd:string;
   :hasLanguageOfInterface"English"^^xsd:string;
   :hasReqAccessMode :auditory_textual;
   :hasReqAdaptationDetail :auditory_verbatim;
   :hasReqAdaptationType :auditory_caption.
```

Listing 2. Profile of a deaf user

Listing 2 depicts the representation of Ms. Martinez (Persona 1) in our ontology. The properties and classes of the ACCESSIBLE ontology start with the term *GenericOntology*. For example, we use the property "User_has_Disability" from the ACCESSIBLE ontology to define the user's disability. This property has a well-defined list of disabilities with respect to the ICF standard classification, as mentioned before. When this property is defined, other properties can be concluded from the ACCESSIBLE ontology, such as the devices that can be used by this type of disability, and the limitations resulting from this disability. The remaining properties are defined for IMS purposes; some of these properties require input by the user, while others can be concluded and recommended from

---

[4] https://www.w3.org/WAI/intro/people-use-web/stories.

the context or ACCESSIBLE ontology. For example, the *hasReqAccessMode* of this persona requires textual representations for resources that are of auditory type.

Persona 2 – Ms. Laitinen is a blind person. She uses a screen reader and only uses web browsers that can be fully controlled with a keyboard. She did not learn how to use a Braille device.

```
:Learner_2 a owl:NamedIndividual , :Learner ;
    GenericOntology:hasName"Ms._Laitinen"^^xsd:string;
    GenericOntology:hasAge"20"^^xsd:int;
    GenericOntology:hasJob"Chief_accountant"^^xsd:string;
    GenericOntology:User_has_Disability :Blindness;
    :hasReqAccessMode :visual_textual;
    :hasReqAdaptationType :visual_alternativeText,:visual_audioDiscription;
    :isAtInteroperable"true"^^xsd:boolean;
    :hasInputRequirements :fullKeyboardControl.
```

Listing 3. Profile of a blind user

Listing 3 shows the user profile of Persona 2. The attributes of the user are defined using the classes and properties of the ACCESSIBLE ontology. The needs of a blind person are defined by the IMS concepts, where *hasReqAccess-Mode* states that a textual representation is required for visual resources, and *hasReqAdaptationType* states to use this textual representation instead of the visual one.

In this subsection, we illustrated how to use the developed ontology to represent disabled learners on two example personas defined by W3C. We utilized the descriptions of disabilities from the ACCESSIBLE ontology to infer IMS properties using basic reasoning, whereas other properties are based on direct user input. Creating learner profiles with respect to disability description standards, web accessibility standards, and e-learning accessibility standards has not been done before and is still an open area of research to the best of our knowledge.

6.2 Representing Educational Resources

For evaluating our digital resources representation, we used examples from the Accessibility Metadata Project[5]. It provides several examples to represent various properties of digital resources based on the IMS AfA properties that are included in schema.org. In Listing 4, we give an example of a digital resource representation using our ontology. It shows one digital resource with three different representations: video, text, and audio. The source file *digitalResource1* is a video with a visual access mode property. This digital resource is available in two alternative forms for better accessibility: *digitalResource2* is a textual resource that can be tactually accessed, and *digitalResource3* is an auditory resource with full keyboard control.

The main idea is to store every educational resource with all the properties which it can support (e.g. keyboard access), together with all alternative

---

[5] https://wiki.benetech.org/display/a11ymetadata/Properties+Examples.

```
:digitalResource1 rdf:type owl:NamedIndividual , :DigitalResource ;
:hasAccessMode :visual ;
:hasControlFlexibility :fullKeyboardControl ;
:hasDisplayTransformability :backgroundColour ;
:hasEducationalComplexityOfAdaptation_dr :enriched ;
:hashazard :flashing ;
:isAtInteroperable_dr"true"^^xsd:boolean ;
:hasAdaptation"digitalResource2_URI"^^xsd:anyURI ,
"digitalResource3_URI"^^xsd:anyURI  .

:digitalResource2 rdf:type owl:NamedIndividual ,
:DigitalResource ;
:hasAccessMode :textual ;
:hasAdaptedAccessMode :tactile ;
:hasAdaptationMediaType :braille ;
:hasControlFlexibility :fullKeyboardControl ;
:hasEducationalComplexityOfAdaptation_dr :enriched ;
:isAtInteroperable_dr"true"^^xsd:boolean ;

:digitalResource3 rdf:type owl:NamedIndividual ,
:DigitalResource ;
:hasAccessMode :auditory ;
:hasControlFlexibility :fullKeyboardControl ;
:hasEducationalComplexityOfAdaptation_dr :enriched ;
:isAtInteroperable_dr"true"^^xsd:boolean ;
:isAdaptationOf"digitalResource1_URI"^^xsd:anyURI  .
```

Listing 4. Representation of a digital resource

resources and their properties with the structure described in our ontology. With this structure, we can use rules to filter the resources which can be used for a specific learner with respect to the described preferences.

6.3 Mapping Learners to Educational Resources

We can now use the above structure and data to map learners to the appropriate educational resources. In particular, we apply SWRL rules to represent these mappings. Listing 5 gives an example rule which defines the accessibility of resources in terms of validating the learners' access mode adaptations required with the available digital resources properties. The rule retrieves the type of resources which the learner can access together with the learners adaptation requests; it then maps them to the existing access mode of the educational resources or one of its appropriate adaptations.

For instance, Persona 2 requires a textual alternative for any visual resource; hence, only *digitalResource2* is an appropriate content format for this learner.

```
hasReqAccessMode(?x,?m)^accessMode_existingAccessMode(?m,?e)^
accessMode_adaptionRequest(?m,?a)^hasAccessMode(?y,?a)
->hasAccess(?x,?y)

hasReqAccessMode(?x,?m)^accessMode_existingAccessMode(?m,?e)^
accessMode_adaptionRequest(?m,?a)^hasAdaptedAccessMode(?y,?a)
->hasAccess(?x,?y)
```

Listing 5. SWRL rule mapping digital resources to disabled learners

In this section, we validated that the structure and annotations of the proposed ontology can be effectively used to represent common personas together with educational resource descriptions. More accessibility information for adaptations can be retrieved by querying the ontology to map learners to the most appropriate resources in an OCW system.

7 Related Work

This section reports on related work and is divided into two parts: The first discusses the state of the art in representing learner profiles in accessibility contexts, including the IMS AfA specification which our approach is based on. The second part reports on the most relevant ontologies addressing accessibility concepts in e-learning systems.

7.1 Standards for Learner Profiles

Since we are focusing on the accessibility needs and preferences of learners, two accessibility standards are most relevant: IEEE PAPI and IMS LIP. The "IEEE Standard for Learning Technology – Public and Private Information for Learners (PAPI Learner)" [17] was first published in 2001. It describes portable learner records that allow to exchange learner profiles among different systems. Accessibility is not explicitly addressed in the PAPI profiles, but related aspects can be implicitly represented in the preference category.

The IMS Learner Information Package (LIP) specification [3] is composed of a number of categories, including one for the accessibility aspects. This accessibility category is described in detail by the IMS Access For All (IMS AfA) specification [2]. IMS AfA is a guideline and metadata specification, based on the ISO/IEC 24751-1:2008 standard, for developing accessible e-learning applications and resources with respect to the learner needs and preferences. It links the accessibility preferences of a learner through the AfA Personal Needs & Preferences (PNP) model to the learning objects defined by the AfA Digital Resource Description (DRD).

We use similar properties and terms for representing learner preferences and features of digital resources in the ontology in order to ease their mapping. We decided to base our work on IMS LIP because it explicitly defines web accessibility concepts in accordance with the W3C WCAG standards and guidelines [8].

7.2 Accessibility Ontologies for e-Learning

A number of ontologies have been developed to represent accessibility knowledge and requirements. A comprehensive survey of such ontologies can be found in our previous work [10].

Among those ontologies, we identified the ACCESSIBLE ontology to be most suitable to represent learner profiles. It has been developed within the EU project

ACCESSIBLE[6] and comprises characteristics of disabled users according to the "International Classification of Functioning, Disability and Health (ICF)" of the WHO[7], descriptions of assistive devices and software applications, web accessibility standards and guidelines (WAI-ARIA and WCAG 2.0), as well as assessment rules for mapping user requirements and constraints.

Other ontologies have been developed specifically for e-learning contexts. They also take learner profiles into account: The Learning Object Context Ontologies (LOCO) are a group of ontologies developed for an e-learning framework to ease the exchange of data among multiple educational services [14]. Among the LOCO ontologies is also one for representing learning preferences in accordance with the aforementioned IMS LIP standard, but accessibility aspects were not explicitly addressed.

Another related ontology is ADOOLES (Ability and Disability Ontology for Online Learning and Services) that has been developed to annotate learning resources [18]. It is based on the ADOLENA ontology [15], which has been used to enhance search capabilities by Ontology-Based Data Access (OBDA). ADOOLES represents knowledge in the domain of e-learning and also includes a set of concepts describing disabilities. However, the number and types of disabilities covered by ADOOLES are very limited and given as a simple class hierarchy without any properties and further linking.

8   Conclusions and Future Work

In this paper, we presented an ontology that addresses accessibility in OCW systems. We reused and extended the ACCESSIBLE ontology to represent learner needs and preferences with respect to the accessibility requirements of the IMS AfA specifications. IMS AfA is concerned with annotating digital resources and learner preferences to achieve a better accessibility. Combining it with the ACCESSIBLE ontology makes it more extendable and does not limit it to special types of disabilities. The combination of IMS AfA and the ACCESSIBLE ontology provides more detailed descriptions of disabilities, assistive technologies, and user preferences. Furthermore, it allows to add concepts of other disabilities, such as cognitive impairments, which are relevant in learning contexts, and suggests mappings to educational resources.

In our future work, we will be focusing on two directions: First, integrating further accessibility guidelines and techniques into the ontology, also considering some concepts and best practices not (yet) included in the existing standards, such as recommendations for users with cognitive disabilities. Second, working on the personalization module to adapt the content presentation according to the learners' needs. Meanwhile, we are implementing the presented approach and ontology in the SlideWiki project with the goal to adapt the developed OCW system with respect to the learner preferences. The implementation of this approach will be evaluated in the trials of the SlideWiki project.

---

[6] http://www.accessible-eu.org.

[7] http://www.who.int/classifications/icf/en/.

Acknowledgments. This research has been supported by the EU project SlideWiki (grant no. 688095).

Appendix

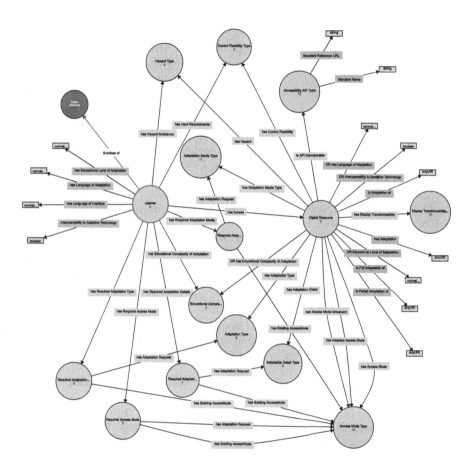

Fig. 4. AccessibleOCW ontology

References

1. Apache Jena Fuseki. https://jena.apache.org/documentation/fuseki2/
2. IMS Access for All (AfA). https://www.imsglobal.org/activity/accessibility
3. IMS Learner Information Package Specification. http://www.imsglobal.org/profiles/index.html

4. Open education consortium: What is open courseware? http://www.oeconsortium.org/faq/what-is-open-courseware/
5. React (JavaScript library). https://facebook.github.io/react/
6. SPARQL Query Language. https://www.w3.org/TR/rdf-sparql-query/
7. WAI-ARIA. https://www.w3.org/WAI/intro/aria.php
8. Web Content Accessibility Guidelines 2.0. https://www.w3.org/TR/WCAG20/
9. Calvo, R., Iglesias, B., Gil, A., Iglesias, A.: Accessibility evaluation of chats and forums in e-learning environments. In: Proceedings of the 2013 International Conference on Frontiers in Education: Computer Science & Computer Engineering (FECS 2013): WORLDCOMP 2013, pp. 133–139. CSREA Press (2013)
10. Elias, M., Lohmann, S., Auer, S.: fostering accessibility of OpenCourseWare with semantic technologies – a literature review. In: Ngonga Ngomo, A.-C., Křemen, P. (eds.) KESW 2016. CCIS, vol. 649, pp. 241–256. Springer, Cham (2016). doi:10.1007/978-3-319-45880-9_19
11. Grammati-Eirini, K., Lopes, R.: Deliverable 4.1 - a set of formalisms and taxonomies for accessibility assessment procedures and their inherent meta models. Technical report, ACCESSIBLE (Grant Agreement No. 224145) (2009)
12. Halilaj, L., Petersen, N., Grangel-González, I., Lange, C., Auer, S., Coskun, G., Lohmann, S.: VoCol: an integrated environment to support version-controlled vocabulary development. In: Blomqvist, E., Ciancarini, P., Poggi, F., Vitali, F. (eds.) EKAW 2016. LNCS, vol. 10024, pp. 303–319. Springer, Cham (2016). doi:10.1007/978-3-319-49004-5_20
13. Heiman, T., Precel, K.: Students with learning disabilities in higher education: academic strategies profile. J. Learn. Disabil. **36**(3), 248–258 (2003)
14. Jovanović, J., Gašević, D., Devedžić, V.: Dynamic assembly of personalized learning content on the semantic web. In: Sure, Y., Domingue, J. (eds.) ESWC 2006. LNCS, vol. 4011, pp. 545–559. Springer, Heidelberg (2006). doi:10.1007/11762256_40
15. Keet, C.M., Alberts, R., Gerber, A., Chimamiwa, G.: Enhancing web portals with ontology-based data access: the case study of South Africa's accessibility portal for people with disabilities. In: OWL: Experiences and Direction (OWLED 2008), vol. 432. CEUR-WS (2008)
16. Lohmann, S., Link, V., Marbach, E., Negru, S.: WebVOWL: web-based visualization of ontologies. In: Lambrix, P., Hyvönen, E., Blomqvist, E., Presutti, V., Qi, G., Sattler, U., Ding, Y., Ghidini, C. (eds.) EKAW 2014. LNCS, vol. 8982, pp. 154–158. Springer, Cham (2015). doi:10.1007/978-3-319-17966-7_21
17. LTSC: IEEE P1484.2.1/D8, Public and Private Information (PAPI) for Learners — Core Features. Technical report, ISO/IEC JTC1 SC36 Information Technology for Learning, Education, and Training (2002)
18. Nganji, J.T., Brayshaw, M., Tompsett, B.: Ontology-driven disability-aware e-learning personalisation with ONTODAPS. Campus-Wide Inf. Syst. **30**(1), 17–34 (2012)
19. Project, A.M: Making Accessible Content Discoverable: A Benetech Led Proposal for Accessibility Metadata in Schema.org (2014). http://www.a11ymetadata.org/

Towards the Semantic MOOC: Extracting, Enriching and Interlinking E-Learning Data in Open edX Platform

Dmitry Volchek[✉], Aleksei Romanov, and Dmitry Mouromtsev

Laboratory of Information Science and Semantic Technologies,
ITMO University, St. Petersburg, Russia
dvolchekspb@gmail.com, gloomspb@gmail.com, mouromtsev@mail.ifmo.ru
http://isst.ifmo.ru/en/

Abstract. In recent years, the educational technology market is growing rapidly. This phenomenon is explained by the increasing number of Massive Open Online Courses (MOOC) which provide learners an opportunity to study 24/7 at the top universities of the world. Information contained in such courses can be better structured, linked, and enriched by means of the semantic technologies and linked data principles. Given semantic annotations, discovery, and matching among learners, teachers, and learning resources can be made a lot more efficient. In this paper, we describe a method of metadata extraction from Open edX online courses for its subsequent processing. We solved the problem of a course representation at the formal and semantic levels, thus, both computers and humans could process and use the course following the ontology development. Also, we exploited NLP and RAKE technologies to integrate automatic concept extraction from course lectures. Triples are imported into RDF storage system allowing user the execution of SPARQL queries through the SPARQL endpoint. Moreover, plugin supports enriching and interlinking courses allowing users to learn the educational content of the courses on an individual trajectory. To summarize the above, it can be concluded that the considered data set is mapped at a satisfactory high level. The collected data can be useful for analyzing the relevance and quality of the course structure.

Keywords: Semantic web · Linked learning · EdX · Education · Metadata · Linked data in education · Educational ontology population · eLearning system · Semantic web technologies in education

1 Introduction

Massive Open Online Courses (MOOCs) form a part of a quickly growing sector of the educational technologies market. This concept was introduced in 2006 and turned into an influential and well-known trend in 2012. Nowadays, several large platforms offer courses from different universities. MOOCs contain a lot of useful information and open doors for the users willing to study different subjects.

© Springer International Publishing AG 2017
P. Różewski and C. Lange (Eds.): KESW 2017, CCIS 786, pp. 295–305, 2017.
https://doi.org/10.1007/978-3-319-69548-8_20

According to the official statistics, the audience of Coursera[1] is more than 25 million learners. The platform provides 2000+ courses of 149 universities. Open edX is an open-source platform that powers edX courses. With Open edX, educators, and technologists can create educational tools and contribute new features to the platform by means of innovative solutions that benefit learners around the world.[2] This platform is used by more than 150 projects in about 25 languages.[3]

The most definite advantages of e-learning are simple 24/7 access to the educational materials of leading universities, relevance, and completeness of information and a possibility to use external sources. All mentioned benefits can be extended even more if powered by semantic web technologies [1]. Additionally, these technologies allow to cross-link, process, refine, and reuse available information as well as apply the Linked Data principles [2]. Not only humans can use the rich data of MOOCs but also machines. Any of them can find suitable content, reuse parts of different MOOCs and generate a new data-based MOOC that answers both a learner's and a course designer's needs [3].

1.1 Motivation

The growth in the number of MOOCs certainly moves the online learning industry forward but such a rapid increase of the educational content entails higher difficulties. Having a large number of courses on various subjects, learners face troubles in terms of finding and highlighting the information that is needed specifically for them. Related problems lie in building cross-disciplinary links, reusing available educational content and paving the way for educational directions.

Semantic web and web intelligence provide means of representing, organizing, and interconnecting knowledge of human educators in a machine-understandable and machine-processable form as well as introducing intelligent web-based services for teachers and learners [4].

In our previous paper, we described the method of metadata extraction from Open edX online courses using dynamic mapping of NoSQL queries [5]. This method allows to extract information from MOOC and semantically represent it considering the designed ontology.

This article complements our previous paper and describes the method of automatic extraction of the domain concepts from the educational content using NLP algorithms and semantically searched open data sets, such as Wikidata[4], DBpedia[5], etc. The concept extraction solves a number of problems given below.

- **Construction of knowledge domain.** By analyzing the content and the order of appearance of the concepts used in the course, it is possible to build a semantic construction of subject areas.

---

[1] https://blog.coursera.org/about/.

[2] https://www.edx.org/about-us.

[3] https://openedx.atlassian.net/wiki/display/OPEN/Sites+powered+by+Open+edX.

[4] https://wikidata.org.

[5] http://wiki.dbpedia.org.

- **Navigation.** By analyzing the relationships between concepts, the intersection of different subject areas, it is possible to implement navigation taking into account other courses within the platform.
- **Binding.** If a learner encounters a concept she does not know, she can locate a concept domain, where else the concept is used or described, what other concepts are associated with it, what concepts are preceding or following it.
- **Individualization.** Having the concepts described in the semantic form as a base, a learner has not only one specific course but all courses within the platform. Learners can explore various sections of different courses in a convenient and simple way and gain knowledge in their own pace and to the required extent.
- **Recommendations.** Concepts can be extracted from the educational content and quizzes as well. Thus, it gives us a chance to generate recommendations for learners based on the insufficiently studied subject area and an individual need to catch up. The recommendations can be generated for the course creators as well, for example, if quizzes and educational materials are not correlated with the concepts.

1.2 Related Work

The use of semantic technologies in education is a fairly common practice. The affordances of semantic technologies are increasingly significant and potentially transformative for the higher education sector considering the amount of the online learning resources and the growing number of learners with access to collaboration tools and online repositories [6].

The ECOLE [7] system collects educational content from different sources and shares it with the university learning systems. The implemented ECOLE system allows to exchange the educational content between the universities and other institutions.

The research of Zablith [8] describes the semantic technology integration into the education by creating a linked data layer that serves as a conceptual connection between higher education courses. In the research, the author sets out the applications that reflect flows of the learning materials between different courses. These flows are based on the interlinked concepts in e-learning environments.

Aye Saliha Sunar's study showed the growing researching trend to personalize and adapt MOOCs in order to promote user engagement and subsequently reduce MOOCs dropout rate [9].

Concept extraction method is widely used in the medical field. For example, Manabu Torii in his research investigates the performance of machine learning taggers for clinical concept extraction, particularly the portability of taggers across documents from multiple data sources [10]. Siddhartha Jonnalagadda [11] uses NLP algorithms to extract medical concepts. mEducator that applies the principles of linked data [12] and standardizes medical information provides users an access to medical education resources.

2   Ontology Development

An important question related to the educational semantic web is how a course can be represented in a formal and semantic way for the successful processing and using by both computers and humans [13]. That's why an ontology should be well-structured and reflect all course components in detail. The main part of ontology was designed in our previous paper [5], so we enriched it to support links between course parts and the concepts they include.

Our ontology is based on the following top-level ontologies [14]. Firstly, AIISO[6] that provides classes and properties to describe the internal organizational structure of an academic institution. Secondly, BIBO[7] that provides main concepts and properties for describing citations and bibliographic references. Thirdly, FOAF[8] (an acronym of Friend of a Friend), the ontology describing people, their activities and their relations to other people and objects. Finally, TEACH (Teaching Core Vocabulary)[9], a lightweight vocabulary providing terms teachers use to relate objects in their courses.

We used Protégé[10] for ontology development and Ontodia[11] for visualization. A part of the ontology that describes the relations between concepts, knowledge domains, and course structure is shown in the Fig. 1.

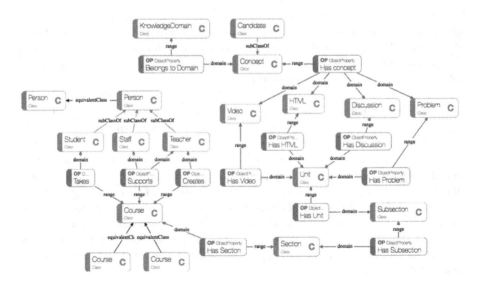

Fig. 1. Course and concept ontology

---

[6] http://purl.org/vocab/aiiso/schema.

[7] http://purl.org/ontology/bibo/.

[8] http://xmlns.com/foaf/spec/.

[9] http://linkedscience.org/teach/ns/teach.rdf.

[10] http://protege.stanford.edu.

[11] http://ontodia.org.

Ontology consists of classes that describe course structure and classes for concept and domain description:

- **Course** (equivalent to AIISO: Course and TEACH: Course) is the main class of the entire ontology. The data properties of this class include course image, start, and end date, number of hours per week (estimated time for successful completion of the course), title and overview.
- **Sections** is a class describing main sections of a course. It has the following data properties: title, start date and visibility. This derives from the fact that course sections appear step by step (usually every week).
- **Subsections** is a class containing a description of the main elements of a section. It has such data properties as title, start date, visibility, and deadline (it is necessary for a learner to finish the assignments by the due date in order to receive points).
- **Units** is a class describing Subsection elements. The class data properties include title and visibility.
- **HTML** is a class containing a description of content of the study course in HTML format;
- **Video** is a class used to describe educational materials in the video format. It has the following data properties: URL, hosting resource, subtitles;
- **Problem** is a class for different tasks and quizzes. Number of points, number of tries, and type data properties are described;
- **Person** (equivalent to FOAF: Person) is a class for the users engaged at the MOOC educational process. This class include subclasses:
  - *Teacher* is a class describing course creators;
  - *Staff* is a class for supporting the educational process persons description. They usually answer the questions at the discussion blocks, receive complains, and suggestions;
  - *Student* is a class used to characterize learners;
- **Concept**
  - *Concept Candidate* is a class to describe extracted concepts using NLP algorithm. It includes weight, the number of encounters in the block, and score data properties;
  - *Concept* is a class containing a description of the concept candidates corresponding to their entry in the external data sources (DBpedia, wikidata, etc.). The Concept candidates may be approved by the course staff, and in this case they become concepts;
- **Domain** is a class to describe domains of knowledge.

3   Method

The primary use case of developing and maintaining plugin is to interlink relevant materials by reducing the number of the material repetition and, consequently, to increase the course quality. The developed plugin should provide tools for tutors to check the quality of a course by means of assessing the relations between elements of the course and between different courses. The learners can use their personal statistics to fill their knowledge gaps.

To extract concepts from the educational content of a course, it is necessary to access the database that is used on the Open EdX platform. The system itself is comprised of a wide stack of technologies, but the required information is located in MongoDB. The Open EdX system contains details about the API that is used to interface the courses. Using such an approach, only the data about the structure of a course can be obtained, but not the information about its content.

Comparing with the last version of our Mongo-parse plugin, we changed the technology used in a newer one taking into account further development of the project. Moreover, the parsing script now works for Python instead of PHP as the last one does not require additional edX server settings. This script loads course data from the Open edX MongoDB and converts the data to N-Triples[12] and then it performs ontology mapping. For processing and handling RDF data as well as for SPARQL Endpoint deployment, RDF4J was chosen[13].

3.1 Overall Architecture

As shown in the Fig. 2 of the overall architecture, Python plugin connects to the Open edX storage using MongoDB queries returning courses with their structure, lectures data (HTML text), concepts that are derived by the way of NLP keyword extraction and concepts that are manually assigned to the lectures through the developed XBlock component. SPARQL Endpoint makes all this data accessible.

3.2 XBlock Component

Figure 3 relating to XBlock is one of the platform's extension points with its API. To store the entire data, it uses the XBlock Fields which are automatically stored at the SQL database. The database has been expanded manually due to the need for dynamic search through known concepts associated with other lectures. Additionally, the dynamic search interfaces Wikipedia using the MediaWiki action API. This solves the problem of linking lectures with concepts for further ontology mapping.

3.3 Mongo-Parse Plugin

Figure 4 shows a detailed diagram of the plugin data flow process. Three different data types are described below.

- **Structure.** Analysis of the storage structure and all the relations have been determined in the method described earlier. The courses have a specific storage structure but the required information (such as labels) is human readable. So, the data is easily mapped with the developed ontology.

---

[12] http://www.w3.org/TR/n-triples/.
[13] http://rdf4j.org/.

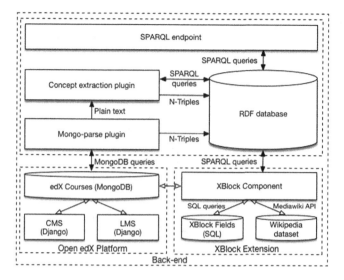

Fig. 2. The overall architecture

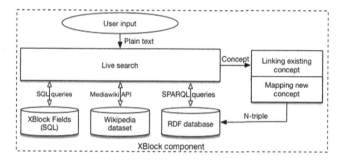

Fig. 3. XBlock component architecture

- **HTML text.** All lectures are stored in a text string with HTML tags. Therefore, plugin strips out all HTML tags and removes invalid characters for the further correct mapping and string import into RDF storage.
- **Concepts.** The concepts are divided into two types that are the concepts manually added from XBlock component and already linked with lectures and the concepts automatically extracted from lectures with concept extraction plugin (Fig. 5).

3.4   Concept Extraction Plugin

Linking lectures and concepts by hand rather than automatically is a time-consuming process if the course is completely developed. Therefore, we used NLP keyword extraction to make it fully automatic. The developed module combines

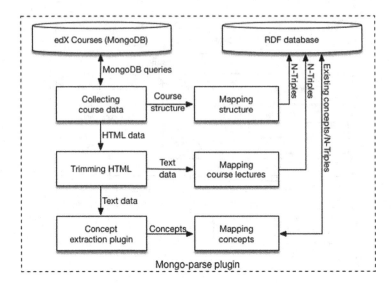

Fig. 4. Mongo-parse plugin architecture

existing technologies such as RAKE[14] (a python implementation of the Rapid Automatic Keyword Extraction) and NLTK[15] (Natural Language Toolkit). The combination of these technologies as well as changes therein have led to better HTML text trimming and keywords inflection (plural to singular, word endings). At this stage, keywords are defined as the concept candidates. After that, all the candidates are evaluated considering their entries in the external data sources (DBpedia, wikidata, etc.) and available concepts in RDF database.

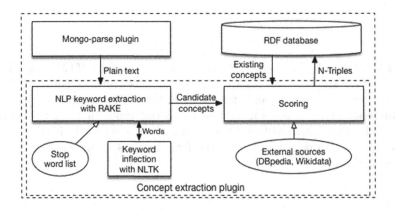

Fig. 5. Concept extraction plugin architecture

---

[14] https://github.com/csurfer/rake-nltk.
[15] http://www.nltk.org.

4   Evaluation

The proof of the concept was used to assess the attainment of the set goals. They are plugin correctness assurance aiming at finding critical errors, the assessment of conformity and completeness of the results and identification of discrepancies and inaccuracies. To confirm the functionality, the plugin was tested on 6 courses having the related subjects. The plugin runs without critical errors. The structure as well as content and ontology mapping was successful.

The quantitative metrics given below are proposed for the assessment: total courses: 6, sections: 64, subsections: 315, units: 941, static pages (lectures): 1005; script time (MongoDB queries and creating local triple store for course structure): average of 0.72 s after 1,000 executions; average script time (MongoDB queries, DBpedia SPARQL queries and NLP analysis) is 2 hours; number of triples after the ontology mapping is 32,590.

As an example, the query to count the number of courses, total, and unique concepts was considered. The obtained data was reflected in the Table 1, where the courses are marked with Roman numbering: Applications of Linear Algebra Part 1(I), Applications of Linear Algebra Part 2(II), Introduction to Differential Equations(III), Introduction to Differential Equations(IV), LAFF On Programming for Correctness(V), Pre-University Calculus(VI).

The Table 1 shows a significant number of repetitions of concepts within the course. Having made a more detailed analysis of the concepts, we can note their frequent occurrence in various courses.

Table 1. Course materials mapping

Course	Sections	Subsections	Units	Static pages	Concepts	Unique concepts
I	8	49	50	50	1072	559
II	9	58	86	86	1610	811
III	18	45	106	106	1890	639
IV	5	28	181	181	3346	1078
V	8	39	127	181	1871	654
VI	9	71	163	238	1950	770

Below is a query example that shows all the occurrences of the specified concept (Vector) in the courses and its units.

```
PREFIX  rdf:  <http://www.w3.org/1999/02/22-rdf-syntax-ns#>
PREFIX  Main:  <http://www.semanticweb.org/EdxOntology/Main#>
PREFIX  rdfs:  <http://www.w3.org/2000/01/rdf-schema#>
SELECT  ?course ?course_label ?unit
WHERE {  ?concept rdf:type Main:Concept.
          ?concept rdfs:label ?concept_label.
          ?unit Main:hasConcept ?concept.
```

```
?vertical Main:consistsOf ?unit.
?seq Main:consistsOf ?vertical.
?chapters Main:consistsOf ?seq.
?course Main:consistsOf ?chapters.
?course rdfs:label ?course_label.
FILTER (?concept_label="Vector"@en)
}
```
ORDER BY (?course)

Such queries allow us to analyze the significance of the concepts and implement direct navigation in the sections of other courses. This is possible because all the units have unique identifiers, and we use them while the ontology is mapped. As a result, navigation links are generated by applying the embedded Open edX platform methods in the format shown below. Where the IDs of a course and a unit are obtained from the triple store after the analysis of the current concepts of this lecture material.

```
/courses/{course_id})/jump_to/block@4{unit_id}
```

5 Conclusions

The set tasks related to the ontology development, data extraction from Open edX, implementation of Xblock component for users and automatic concept extraction with NLP were completed, and the main goal was attained.

Course structures and concepts are presented in a semantic form in accordance with the developed ontology that allows to use the obtained information for:

– establishing interdisciplinary connections;
– generating recommendations on the course completion order;
– simple and convenient navigation on the educational materials not only one course but all courses hosted on a specific Open edX platform and on any future platform using the developed plugin.

By analyzing the correlation of concepts in lecture materials and quizzes, it is possible to generate recommendations for creators of courses, for improving the quality of education, and for learners in order to provide the widest possible selection of relevant information.

Hence, this paper is expected to encourage the further development of data visualization for navigation in educational materials, machine learning to improve the quality of automatic concept extraction, recommendation service that can analyze terms and quiz results to create a flexible learning style of a particular person.

Recent links of the LMS of edX system, SPARQL Endpoint, ontology documentation, developed plugin and source code can be found at the Laboratory of Information Science and Semantic technologies GitHub: https://github.com/ailabitmo/edx-ontology

References

1. d'Aquin, M.: Linked Data for Open and Distance Learning (2012)
2. Bizer, C., Heath, T., Berners-Lee, T.: Linked data: the story so far. In: Semantic Services, Interoperability and Web Applications: Emerging Concepts, pp. 205–227 (2009)
3. Höver, K.M., Mühlhäuser, M.: LOOCs-linked open online courses: a vision. In: 2014 IEEE 14th International Conference on Advanced Learning Technologies (ICALT), pp. 546–547. IEEE (2014)
4. Deved, V., et al.: Semantic Web and Education, vol. 12. Springer Science & Business Media, Heidelberg (2006)
5. Mouromtsev, D., Romanov, A., Volchek, D., Kozlov, F.: Metadata extraction from open edX online courses using dynamic mapping of NoSQL queries. In: Proceedings of the 25th International Conference Companion on World Wide Web, pp. 501–506. International World Wide Web Conferences Steering Committee (2016)
6. Tiropanis, T., Millard, D., Davis, H.C.: Guest editorial: special section on semantic technologies for learning and teaching support in higher education. IEEE Trans. Learn. Technol. **2**, 102–103 (2012)
7. Mouromtsev, D., Kozlov, F., Kovriguina, L., Parkhimovich, O.: ECOLE: student knowledge assessment in the education process. In: Proceedings of the 24th International Conference on World Wide Web Companion, pp. 695–700. International World Wide Web Conferences Steering Committee (2015)
8. Zablith, F.: Interconnecting and enriching higher education programs using linked data. In: Proceedings of the 24th International Conference on World Wide Web Companion, pp. 711–716. International World Wide Web Conferences Steering Committee (2015)
9. Sunar, A.S., Abdullah, N.A., White, S., Davis, H.C.: Personalisation of MOOCs: the state of the art. In: 7th International Conference on Computer Supported Education (CSEDU 2015) (2015)
10. Torii, M., Wagholikar, K., Liu, H.: Using machine learning for concept extraction on clinical documents from multiple data sources. J. Am. Med. Inform. Assoc. **18**(5), 580–587 (2011)
11. Jonnalagadda, S., Cohen, T., Wu, S., Gonzalez, G.: Enhancing clinical concept extraction with distributional semantics. J. Biomed. Inform. **45**(1), 129–140 (2012)
12. Bamidis, P.D., Kaldoudi, E., Pattichis, C.: mEducator: a best practice network for repurposing and sharing medical educational multi-type content. In: Camarinha-Matos, L.M., Paraskakis, I., Afsarmanesh, H. (eds.) PRO-VE 2009. IAICT, vol. 307, pp. 769–776. Springer, Heidelberg (2009). doi:10.1007/978-3-642-04568-4_78
13. Koper, R.: Use of the semantic web to solve some basic problems in education: increase flexible, distributed lifelong learning; decrease teacher's workload. J. Interact. Media Educ. **2004**(1) Art-5 (2010)
14. Keßler, C., d'Aquin, M., Dietze, S.: Linked data for science and education. Semant. Web **4**(1), 1–2 (2013)

Linked Data

DBpedia Entity Type Detection Using Entity Embeddings and N-Gram Models

Hanqing Zhou[✉], Amal Zouaq, and Diana Inkpen

School of Electrical Engineering and Computer Science, University of Ottawa,
Ottawa, ON K1N6N5, Canada
HZHOU020@uottawa.ca

Abstract. This paper presents and evaluates a method for the detection of
DBpedia entity types (classes) that can be used to assess DBpedia's quality and
to complete missing types for un-typed resources. This method compares entity
embeddings with traditional N-gram models coupled with clustering and classification. We evaluate the results for 358 typical DBpedia classes. Our results
show that entity embeddings outperform n-gram models for type detection and
can contribute to the improvement of DBpedia's quality, maintenance, and
evolution. This is a step toward improving the quality of Linked Open Data in
general.

Keywords: Semantic web · DBpedia · Entity embedding · N-Grams · Type
identification

1 Introduction

Wikipedia [1] is one of the most widely used encyclopedias. The DBpedia knowledge
base represents structured information from Wikipedia and contains millions of entities
in more than a hundred languages [2]. DBpedia uses a unique identifier to name each
particular entity (i.e., Wikipedia page) and associates it with an RDF description
accessible on the Web [3, 4]. Additionally, DBpedia is based on an ontology that
defines the various classes (or types) used in the knowledge base. In the last few years,
DBpedia has been playing a central role in the development of the Linked Open Data
(LOD) cloud, and has been used for tasks as diverse as semantic annotation [5] and
knowledge extraction and information retrieval [6].

Given the automatic extraction framework of DBpedia, and its dynamic nature,
there may exist some missing, invalid or irrelevant types in the RDF description of an
entity. Identifying these invalid types manually is unfeasible and non-scalable. Similarly, given the size of DBpedia and its cross-domain nature, type information is
sometimes unavailable for some entities. Several resources are still un-typed, or not
typed with all the relevant classes from the DBpedia ontology. Finally, some ontology
classes still do not have any instances (or entities), as we show in this paper. The
enrichment of DBpedia with new type statements (through *rdf:type*) is thus an
important challenge [7].

The methods presented in this paper aim at assessing DBpedia quality by identifying erroneous statements about types, as well as completing missing types in

© Springer International Publishing AG 2017
P. Różewski and C. Lange (Eds.): KESW 2017, CCIS 786, pp. 309–322, 2017.
https://doi.org/10.1007/978-3-319-69548-8_21

DBpedia by using entity embeddings, which are similar to word embeddings [8]. Entity Embedding are based on vectors that describe entity URIs instead of words [9, 10]. To our knowledge, none of the available approaches that target DBpedia type detection rely on Word2Vec vectors representing DBpedia entities.

In particular, our research questions are:

RQ1: Can entity embeddings help detect the relevant types of an entity?
RQ2: How do entity embeddings compare with traditional n-gram models for type identification?

The first contributions of this paper is that we train classification models based on Entity Embedding and N-gram models. The second contribution is that we use these models to detect invalid or irrelevant DBpedia types and complete missing types of DBpedia resources. The third contribution is that we compare the results of different Entity Embedding models with N-gram models.

The paper is structured as follows: In Sect. 2, we give a brief overview of related work. In Sect. 3, we explain the methods used for clustering and classifying entities, including the steps for learning entity and word embeddings and n-grams to represent DBPedia entities. Section 4 evaluates both the clustering and the classification experiments. Finally, conclusions and future work are discussed in Sect. 5.

2 Related Work

Given the central role of DBpedia in the LOD, the quality and evolution of this knowledge base is currently a hot topic. Several approaches to enhance the quality of DBpedia and Linked Data were proposed, such as user-driven quality evaluation of DBpedia [11] which assesses the quality of DBpedia by both manual and semi-automatic processes. These processes include detecting quality problems, building individual evaluation resources through crowdsourcing, and generating and verifying schema axioms. Another approach is test-driven quality assessment [12] that is inspired by test-driven software development to discover data quality problems.

Similarly, there has been a growing interest for automatic approaches for type extraction [13]. Paulheim and Bizer [14] proposed two state-of-the-art algorithms based on a statistical approach: SDType, for adding missing type statements, and SDValidate, for identifying faulty statements. Both methods rely on the statistical distribution of types and properties in the DBpedia knowledge base. In this paper, we are interested in investigating how word and entity embeddings can contribute to such a task. Word embeddings are vector representations of word(s) [8] and have been successfully applied to several NLP tasks such as entity recognition [15], information retrieval [16], and question answering [17]. Here, our aim is to learn vector representations of DBpedia entities *(entity embeddings)* from Wikipedia, by using the information available in each Wikipedia page. The first successful word embeddings model was the Neural Probabilistic Language Model [18], which learned distributed representations of each word and probability functions for word sequences. One of the most popular word embedding tools is Word2Vec [8, 19] and it has been used in this work.

The literature behind word embeddings describes other ways to induce word representations, inferred from n-gram language modeling rather than continuous language models [20]. Generally, language models compute probabilities assigned to sequences of words based on their frequency. An n-gram model, where the n represents the number of words to consider in sequence, assigns probabilities to sequences of words [21, 22]. Given that in practice n-gram models have been shown to be effective in several natural language tasks, one objective of this work is to compare n-gram models to the entity embeddings models for the task of type identification.

3 Methodology

3.1 Word2Vec Model Learning

Word2Vec is one of the most popular tools for computing a vector representation of words [8]. Given that we are interested in entities, we need a way to obtain vector representations of these entities (resources). In fact, each Wikipedia page is represented by a DBpedia resource. For instance, the Wikipedia page https://en.wikipedia.org/wiki/Natural_language_processing is directly mapped to the DBpedia resource http://dbpedia.org/resource/Natural_language_processing. Wiki2Vec[1] is one of the implementations of Word2Vec, which enables us to build a DBpedia model from Wikipedia by using a DBPEDIA_ID to replace Wikipedia hyperlinks and running Word2Vec on this modified corpus to train the DBpedia Model.

Our first step is the processing of Wikipedia Dumps (in English)[2] to extract text. Hyperlinks, which designate Wikipedia pages, are identified by a specific "DBPEDIA_ID/". (e.g. DBPEDIA_ID/Barack_Obama replaces the hyperlink Barack_-Obama). After that, lemmatization is performed using NLTK[3] and this allows us to build one vector representation for the words that share the same lemma. Finally, the model is trained with Wiki2vec.

Once the training phase is finished, we obtain a continuous vector representation of single words and DBpedia entities. These vectors are then used to compute similarities between entities and types. For example, in the Word2Vec model, the similarity between "DBPEDIA_ID/Bill_Clinton" and "DBEPDIA_ID/President" can be computed. The obtained word vectors represent features for the classification and clustering tasks and are associated to the types that are already represented in DBpedia, when applicable. For example, the vector related to Bill_Clinton is associated to the type President.

To give a better idea about this notion of distance, a visualization plot based on t-SNE[4] is provided in Fig. 1.

The 2D plot is based on a 100-dimension vector representation and uses Principle Component Analysis [23] for dimensionality reduction. For example, "Bill_Clinton"

---

[1] https://github.com/idio/wiki2vec.

[2] https://dumps.wikimedia.org/enwiki/.

[3] http://www.nltk.org/.

[4] https://lvdmaaten.github.io/tsne/.

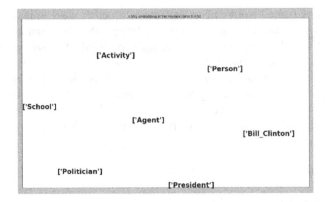

Fig. 1. Visualization of DBpedia types for the entity Bill_Clinton

has four related DBpedia types, "Agent", "President", "Politician" and "Person". Two unrelated types, "School" and "Activity" are added to the figure. Based on the plot shown above, the type "Agent", "President", "Politician" and "Person" are closer to the entity "Bill_Clinton" than the other two unrelated types.

Our trained model stores both entities and words; entities (i.e., DBpedia resources) are distinguished from words using their "DBPEDIA_ID/". DBpedia entities might be represented by compound words such as "Barack_Obama[5]". For single words, the dataset usually contains both DBpedia entities and words. For example, both "DBpedia_ID/Poetry" and "Poetry" will be represented in our model. This has some implications for the calculation of similarities, as the results will be different since DBpedia entities and words are represented by different vectors.

The process of building our DBpedia model is shown in Fig. 2.

Initially, we used prebuilt models from Wiki2Vec for the English Wikipedia (Feb 2015), without stemming, with a 10 Skip-Gram model. In fact, there are two model architectures in Word2Vec, a Skip-Gram model and a continuous Bag-of-Words model [8, 24]. These models output n-dimensional vectors to represent words; to increase the coverage, we learned our model with the following parameters: a minimum number of occurrences of 5, a vector dimension of 100 to limit the size of the model, and a window size set to 5, which means that the maximum distance between the current and predicted word within a sentence is 5.

3.2 N-Gram Model Learning

The N-gram models are learned using Weka n-gram tokenizer[6]. We first collect the DBpedia entities' abstracts through DBpedia public SPARQL endpoint[7] and save them into output files. The second step is to learn the n-gram models using Weka on the

---

[5] http://dbpedia.org/page/Barack_Obama.

[6] http://weka.sourceforge.net/doc.dev/weka/core/tokenizers/NGramTokenizer.html.

[7] http://dbpedia.org/sparql.

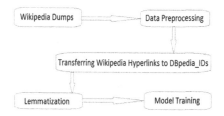

Fig. 2. The process of building the DBpedia model

output files by varying the number of word sequences (n) and using TF-IDF to weight the n-grams. We perform feature selection in order to reduce the number of features to be 1000. This number was determined after a set of empirical experiments.

3.3   Dataset Extraction and Preparation

There are 753 DBpedia types available in the current DBpedia ontology. Based on our experiments, only 457 classes currently have instances (through rdf:type), while the rest of classes, such as ArtisticGenre, TeamSport and SkiResort, still do not have any related instance (at the time of our experiments). In this work, we focus on the types that have instances. For any given type with instances, the minimum threshold for the number of examples is set to 20 to ensure a basic set of positive examples. Among the 457 classes, there are only 358 DBpedia types that contain at least 20 related entities. Other DBpedia types are ignored.

For each of these 358 DBpedia types, we retrieve entities with the specified type through the public DBpedia SPARQL endpoint. For example, the following query was run to find instances of the class (type) *Aircraft:*

```
PREFIX rdf: <http://www.w3.org/1999/02/22-rdf-syntax-ns#>
PREFIX ontology: <http://dbpedia.org/ontology/>
select distinct ?RelatedEntity
where { ?RelatedEntity  rdf:type ontology:Aircraft}
```

A sample of the retrieved entities is shown in Fig. 3.

To build the training dataset, for each DBpedia type (the test class), we first select at most 2000 entities after testing the availability of each of these entities in our trained Word2Vec model; that is, we tested the availability of a vector describing them. These entities are tagged as positive examples of that DBpedia type. Negative examples are chosen from a random selection of instances from all the remaining types except the test DBpedia types (see the test dataset description in the next section). Given the variations in the number of instances for some DBpedia classes (some classes have very few instances, while others may have more than a thousand entities), the number of negative entities depends on the corresponding number of positive entities in order to build a balanced dataset. Duplicates are removed from negative examples, to eliminate mis-clustering and mis-classification problems. For example, if "Bill_Clinton" is selected for the DBpedia

RelatedEntity
http://dbpedia.org/resource/Airbus_A300
http://dbpedia.org/resource/Boeing_B-17_Flying_Fortress
http://dbpedia.org/resource/Boeing_C-17_Globemaster_III
http://dbpedia.org/resource/Boeing_E-3_Sentry
http://dbpedia.org/resource/Boeing_KC-135_Stratotanker
http://dbpedia.org/resource/Boeing_OC-135B_Open_Skies
http://dbpedia.org/resource/Boeing_RC-135

Fig. 3. Sample Search Results

type President, this entity won't be selected as a negative example, which means that it won't appear in the negative class even if it has other types in DBpedia that could be used (e.g., Agent, Person, etc.). In short, once an entity is selected as a positive one, we eliminate any mention of this entity in the negative examples.

The process of building the test dataset is different from the training dataset as we randomly select around 5 to 6 entities for each type from all the 358 DBpedia types and remove their types (which represent our gold standard). In total, we obtained 2,112 entities in our test dataset, distributed among these various types. Each entity has around 5 types on average, and we store entities and all their related types. For example, we might have Bill_Clinton with types President and Person. Note that we make sure there are not any common entities in the training and test dataset.

While the training dataset already provides an idea on the accuracy of our approach to detect a DBpedia entity type, the test dataset is definitely a good use case for the identification of the type of new resources, for example, new Wikipedia/DBpedia pages, or pages that are still un-typed. We report results by cross-validation on the training data and we use them to find the most appropriate clustering and classification algorithms for our task. We also report our final results on a separate held-out test set. Note that the training and test datasets are available on our web site[8].

3.4 Clustering and Classification

This phase consists of two parts: performing clustering experiments and performing classification experiments using Skicit-learn[9] based on the prepared dataset for each DBpedia type. In both cases, we perform a binary clustering and a binary classification that represent two main categories for each type of interest: The Type category and the NOT Type category. For example, we want to classify examples as instances of President and instances of NOT President.

In terms of clustering, all the entities are clustered with the following standard algorithms: K-means, Mean Shift, and Birch. Clustering is performed by computing the

---

[8] http://www.site.uottawa.ca/~diana/resources/kesw17/.

[9] http://scikit-learn.org/stable/.

Euclidean distance between vectors. The number of clusters is set to two, one represents the positive cluster for the type of interest, and the other one the negative cluster.

For each DBpedia type, the related DBpedia entities are selected by `rdf:type` through DBpedia SPARQL endpoint for the training and test datasets and these entities are considered as examples of the positive class, the same number of negative entities is then selected from DBpedia except the entities from the positive class. Because the produced clusters are not labelled, the cluster with a greater number of positive entities is considered the positive class cluster and vice versa.

In the classification part, several classification algorithms are tested. We experimented with Random Forest (RF), Extra Tree, Naive Bayes (NB), SVM, Decision Tree, and KNN. We wanted to test tree-based algorithms because the trees could be helpful to track the decision process, Naive Bayes classifiers because they tend to work well on text data, and SVM classifiers because they are known to perform best for many tasks. We also wanted to test ensemble methods, RF in this case.

4   Evaluation

The Word2Vec model is trained with the following settings as previously explained: threshold of 5 for the minimum number of occurrences of a word in the corpus, vector dimension of 100, and window size of 5.

We used TF-IDF to assign a weight to the n-grams with a minimum frequency of occurrence of 5. Feature selection is performed using Random Forest in Scikit-learn, which can be used for ranking the features. A Random forest consists of a number of decision trees and every node in the decision trees is a condition on a single feature, designed to split the dataset into the two classes. A measure of impurity is computed for each feature, based on the number of training instances that end up in the wrong class. This is averaged over all the trees in which the feature appears [25]. We tried other feature selection methods, but the results of the classification were lower.

By varying n = 1, 2 and 3, the uni-gram, bi-gram and tri-gram models are generated, together with a mixed model including n = 1 ∼ 3.

The results are generated based on the confusion matrix with two classes/clusters. The positive class/cluster represents the target test type and the negative class/cluster represents the rest of types except the test type. For clustering, precision, recall, F-score and accuracy are calculated based on the confusion matrix [26], while classification is also measured with Area Under the Curve (AUC) based on the analysis of ROC curve [27].

4.1   Detecting DBpedia Entity Types on the Training Dataset

This subsection shows the results of detecting DBpedia entity types on the training dataset, by using a 10-folds cross-validation.

4.1.1   Detecting Types Based on Clustering Algorithms

With clustering algorithms, the results are generated based on two entity embedding models and four n-gram models.

Table 1 summarizes the average results for the three clustering algorithms based on 358 tested DBpedia types with Skip-Gram and CBOW entity embedding models. Overall, we can observe that both K-means and Birch obtain a good performance, while Mean Shift has some problems for discriminating negative types from positive types. Based on the results from Skip-Gram and CBOW model, we note that the Skip-Gram model outperforms the CBOW model.

Table 1. Summary of the results for clustering algorithms on 358 DBpedia types with Skip-Gram and CBOW entity embedding models

Model	Algorithm	Precision	Recall	F-Score	Accuracy
Skip-Gram	K-means	0.937	0.950	0.941	0.942
	Mean shift	0.560	0.991	0.713	0.713
	Birch	0.941	0.945	0.937	0.940
CBOW	K-means	0.826	0.653	0.669	0.667
	Mean shift	0.360	0.431	0.368	0.276
	Birch	0.815	0.753	0.733	0.708

Table 2 shows the average results for clustering algorithms on the 358 DBpedia types with n-gram models.

Table 2. Summary of the results for clustering algorithms on 358 DBpedia types with n-gram model

Model	Algorithm	Precision	Recall	F-Score	Accuracy
Uni-gram	K-means	0.819	0.657	0.669	0.663
	Mean shift	0.357	0.431	0.367	0.274
	Birch	0.811	0.745	0.726	0.702
Bi-gram	K-means	0.816	0.655	0.664	0.659
	Mean shift	0.361	0.429	0.365	0.275
	Birch	0.800	0.759	0.730	0.698
Tri-gram	K-means	0.818	0.661	0.670	0.664
	Mean shift	0.346	0.433	0.367	0.267
	Birch	0.810	0.743	0.721	0.698
Mix-gram	K-means	0.897	0.671	0.731	0.744
	Mean shift	0.407	0.391	0.345	0.288
	Birch	0.858	0.773	0.774	0.762

Among the n-gram models, the Mix-gram model gets the optimal results in terms of F-score and accuracy as shown in Table 2. However, when compared with the entity embedding model, the Skip-Gram entity embedding model gets much better results than the Mix-gram model with all clustering algorithms based on training dataset.

4.1.2   Detecting DBpedia Types Based on Classification Algorithms

We tested six classification algorithms to build binary classifiers for a specific type: Decision Tree (DT), Extra Tree, KNN, Random Forest, Naïve Bayes and SVM.

Table 3 shows the average results for the six classification algorithms based on the 358 tested DBpedia types with the Skip-Gram and CBOW embedding models. We note that all the algorithms exceed the DummyClassifier, the baseline classification in Sklearn.

Table 3. Average results for the classification algorithms on the training dataset with Skip-Gram and CBOW entity embedding models

Model	Algorithm	Precision	Recall	F-Score	Accuracy	AUC
Skip-Gram	Baseline	0.499	0.497	0.487	0.498	0.502
	DT	0.881	0.897	0.883	0.888	0.888
	Extra tree	0.962	0.950	0.954	0.958	0.988
	KNN	0.869	0.995	0.922	0.918	0.976
	RF	0.958	0.935	0.944	0.949	0.986
	NB	0.967	0.955	0.959	0.964	0.991
	SVM	0.958	0.986	0.970	0.973	0.995
CBOW	Baseline	0.496	0.500	0.486	0.497	0.503
	DT	0.908	0.896	0.881	0.890	0.896
	Extra tree	0.937	0.897	0.899	0.911	0.955
	KNN	0.941	0.641	0.739	0.802	0.898
	RF	0.936	0.857	0.887	0.901	0.955
	NB	0.837	0.925	0871	0.871	0.877
	SVM	0.959	0.836	0.883	0.903	0.955

Table 4 shows the average classification results on 358 DBpedia types with different n-gram models.

Similar to the clustering results, Skip-Gram obtained the best results among all models. Overall, K-means and Birch got much better performance than Mean Shift in clustering results for all entity embedding and n-gram models. In the classification part, with entity embedding models, Extra Tree, Random Forest, KNN, Naive Bayes and SVM get a similar performance, and their results are better than the other classifiers. Using the n-gram models, Extra Tree, Random Forest and SVM got better performance than others. One interesting point is Naïve Bayes, which performed well with entity embedding models, but whose performance dropped a lot when using n-gram models. In terms of comparison between entity embedding models and n-gram models, the Skip-Gram model obtains the best results, followed by mix-gram and uni-gram models. However, the difference between the mix-gram and Skip-gram models is quite small with the SVM classifier. Finally, the CBOW model gets results that are close to the bi-gram and tri-gram models, and performs worse than Skip-Gram, mix-gram and uni-gram models.

Table 4. Average results for the classification algorithms on the training dataset with n-gram models

Model	Algorithm	Precision	Recall	F-Score	Accuracy	AUC
Uni-gram	Baseline	0.498	0.496	0.485	0.497	0.500
	DT	0.907	0.874	0.884	0.894	0.898
	Extra tree	0.933	0.879	0.900	0.910	0.955
	KNN	0.933	0.640	0.737	0.801	0.896
	RF	0.932	0.858	0.886	0.901	0.951
	NB	0.835	0.927	0.871	0.869	0.876
	SVM	0.961	0.837	0.884	0.906	0.960
Bi-gram	Baseline	0.497	0.497	0.485	0.495	0.801
	DT	0.904	0.871	0.881	0.890	0.895
	Extra tree	0.935	0.876	0.898	0.910	0.957
	KNN	0.932	0.639	0.737	0.802	0.900
	RF	0.934	0.859	0.888	0.902	0.956
	NB	0.836	0.928	0.872	0.869	0.874
	SVM	0.956	0.833	0.879	0.902	0.955
Tri-gram	Baseline	0.501	0.501	0.489	0.499	0.501
	DT	0.906	0.872	0.883	0.893	0.897
	Extra tree	0.939	0.881	0.903	0.914	0.959
	KNN	0.934	0.640	0.737	0.802	0.897
	RF	0.938	0.860	0.891	0.905	0.957
	NB	0.840	0.930	0.875	0.872	0.880
	SVM	0.953	0.832	0.878	0.901	0.950
Mix-gram	Baseline	0.504	0.501	0.490	0.502	0.499
	DT	0.933	0.921	0.923	0.930	0.933
	Extra tree	0.955	0.926	0.936	0.941	0.976
	KNN	0.954	0.738	0.815	0.852	0.931
	RF	0.955	0.911	0.928	0.937	0.977
	NB	0.855	0.939	0.887	0.884	0.894
	SVM	0.964	0.895	0.921	0.936	0.976

The following subsections summarize the results of our trained models for clustering and classification when applied on the test dataset.

4.2   Detecting DBpedia Types on the Test Dataset

This subsection shows the results of detecting DBpedia entity types on the test dataset described above, by using the models trained on the training dataset. We show that the results on this unseen test set are comparable with the results presented above on the training data, in order to confirm that our models are general enough and not overfitted on the training data.

4.2.1   Clustering Experiments

Based on our results on the training datasets, K-means and Birch obtained the best performance with both entity and n-gram models; therefore, we report only these results on the test set. Table 5 shows the clustering results of K-means and Birch on the test dataset with entity embedding models. Table 6 shows the clustering results with different n-gram models.

Table 5. Average clustering results on the test dataset with CBOW model

Model	Algorithm	Precision	Recall	F-Score	Accuracy
Skip-Gram	K-means	0.820	0.935	0.856	0.830
	Birch	0.837	0.932	0.861	0.837
CBOW	K-means	0.543	0.924	0.667	0.538
	Birch	0.535	0.916	0.660	0.528

Table 6. N-gram models with clustering results

Model	Algorithm	Precision	Recall	F-Score	Accuracy
Uni-gram	K-means	0.505	0.940	0.657	0.509
	Birch	0.507	0.933	0.654	0.508
Bi-gram	K-means	0.503	0.937	0.654	0.505
	Birch	0.505	0.933	0.653	0.505
Tri-gram	K-means	0.513	0.930	0.655	0.511
	Birch	0.514	0.929	0.655	0.512
Mix-gram	K-means	0.507	0.935	0.656	0.509
	Birch	0.510	0.945	0.662	0.516

4.2.2   Classification Experiments

The following two tables show classification results based on the test dataset with Entity Embedding and n-gram models. Table 7 summarizes the average results for the three best classification algorithms, Extra Tree, Random Forest, and SVM, which have good performance in both entity embedding and n-gram models.

Table 7. Average classification results on the test dataset with Skip-Gram and CBOW entity embedding models

Model	Algorithm	Precision	Recall	F-Score	Accuracy	AUC
Skip-Gram	Extra tree	0.972	0.931	0.946	0.952	0.986
	RF	0.961	0.922	0.938	0.945	0.983
	SVM	0.966	0.967	0.962	0.965	0.991
CBOW	Extra tree	0.920	0.850	0.875	0.886	0.947
	RF	0.917	0.948	0.872	0.883	0.941
	SVM	0.926	0.890	0.897	0.904	0.962

Table 8 summarizes the average results for the best classification algorithms based on the 358 tested DBpedia types on n-gram models.

Table 8. Classification results on the test dataset with four n-gram models

Model	Algorithm	Precision	Recall	F-Score	Accuracy	AUC
Uni-gram	Extra tree	0.962	0.916	0.932	0.938	0.974
	RF	0.959	0.896	0.918	0.928	0.967
	SVM	0.982	0.874	0.912	0.927	0.970
Bi-gram	Extra tree	0.948	0.851	0.886	0.901	0.952
	RF	0.949	0.833	0.874	0.893	0.948
	SVM	0.968	0.778	0.842	0.879	0.953
Tri-gram	Extra tree	0.920	0.815	0.847	0.862	0.915
	RF	0.919	0.800	0.836	0.850	0.916
	SVM	0.941	0.697	0.766	0.817	0.913
Mix-gram	Extra tree	0.970	0.906	0.930	0.937	0.974
	RF	0.964	0.898	0.921	0.930	0.968
	SVM	0.987	0.962	0.907	0.924	0.965

Overall, the results of the Skip-Gram model are best among all models, and are much better than the results of the CBOW model in terms of f-score and accuracy. The classification results are usually better than the results of clustering algorithms no matter if entity embedding or N-gram models are used. Skip-Gram get the best results followed by mix-gram and uni-gram models. Overall, CBOW obtains a similar performance than Bi-gram and Tri-gram models in terms of f-score and accuracy.

A student t-test was calculated on the results (Accuracy) of the Skip-Gram model and Mix-gram model using the SVM classification algorithm based on the training and test dataset. The obtained t value is 7.722 with the p value is <0.00001 and the results are statistically significant at $p < 0.05$ for both one-tailed and two-tailed hypothesis on the training dataset. For the results based on the test dataset, the t value obtained is 0.109 with p value 0.457 for one-tailed hypothesis, and the t value is 0.109 with p value 0.913 for two-tailed hypothesis. Unfortunately, the result is not statistically significant at $p < 0.05$ for both one-tailed and two-tailed tests on this dataset.

5 Conclusion and Future Work

In this paper, we presented and evaluated clustering and classification algorithms that allow us to assess DBpedia types' quality and complete missing types with entity embeddings and n-grams. Overall, the results of the Skip-Gram entity embedding model are better than the results of n-gram models. In future work, we plan to train a larger Word2Vec model to contain more DBpedia entities and use the same approach to detect faulty statements.

Acknowledgements. We thank the Natural Sciences and Engineering Research Council of Canada (NSERC) for the financial support.

References

1. Krötzsch, M., Vrandečić, D., Völkel, M., Haller, H., Studer, R.: Semantic wikipedia. Web Semant. **5**(4), 251–261 (2007)
2. Morsey, M., Lehmann, J., Auer, S., Stadler, C., Hellmann, S.: DBpedia and the live extraction of structured data from Wikipedia. Program **46**(2), 157–181 (2012)
3. Lehmann, J., Isele, R., Jakob, M., Jentzsch, A., Kontokostas, D., Mendes, P.N., Bizer, C.: DBpedia - A large-scale, multilingual knowledge base extracted from wikipedia. Semant. Web **6**(2), 167–195 (2015)
4. Auer, S., Bizer, C., Kobilarov, G., Lehmann, J., Cyganiak, R., Ives, Z.: DBpedia: a nucleus for a web of open data. In: Aberer, K., Choi, K.-S., Noy, N., Allemang, D., Lee, K.-I., Nixon, L., Golbeck, J., Mika, P., Maynard, D., Mizoguchi, R., Schreiber, G., Cudré-Mauroux, P. (eds.) ASWC/ISWC -2007. LNCS, vol. 4825, pp. 722–735. Springer, Heidelberg (2007). doi:10.1007/978-3-540-76298-0_52
5. Zhang, Z., Chen, S., Feng, Z.: Semantic annotation for web services based on DBpedia. In: 2013 IEEE 7th International Symposium on Service Oriented System Engineering (SOSE), pp. 280–285 (2013)
6. Keong, B.V., Anthony, P.: Meta search engine powered by DBpedia. In: Proceedings of the 2011 International Conference on Semantic Technology and Information Retrieval, STAIR 2011, pp. 89–93 (2011)
7. Hulpus, I., Hayes, C., Karnstedt, M., Greene, D.: Unsupervised graph-based topic labelling using DBpedia. In: Proceedings of the Sixth ACM International Conference on Web Search and Data Mining (WSDM), pp. 465–474 (2013)
8. Mikolov, T., Corrado, G., Chen, K., Dean, J.: Efficient estimation of word representations in vector space. In: Proceedings of the International Conference on Learning Representations (ICLR 2013), pp. 1–12 (2013)
9. Hu, Z., Huang, P., Deng, Y., Gao, Y., Xing, E.: Entity hierarchy embedding. In: Proceedings of the Association for Computational Linguistics 2015 (ACL 2015), pp. 1292–1300 (2015)
10. Chen, T., Tang, L.A., Sun, Y., Chen, Z., Zhang, K.: Entity embedding-based anomaly detection for heterogeneous categorical events. In: Proceedings of the International Joint Conference on Artificial Intelligence (IJCAI 2016), vol. 2016, pp. 1396–1403, January 2016
11. Zaveri, A., Kontokostas, D., Sherif, M.A., Bühmann, L., Morsey, M., Auer, S., Lehmann, J.: User-driven quality evaluation of DBpedia. In: Proceedings of the 9th International Conference on Semantic Systems - I-SEMANTICS 2013, p. 97 (2013)
12. Kontokostas, D., Westphal, P., Auer, S., Hellmann, S., Lehmann, J., Cornelissen, R., Zaveri, A.: Test-driven evaluation of linked data quality. In: Proceedings of the 23rd International Conference on World Wide Web - WWW 2014, pp. 747–758 (2014)
13. Gerber, D., Hellmann, S., Bühmann, L., Soru, T., Usbeck, R., Ngonga Ngomo, A.-C.: Real-time RDF extraction from unstructured data streams. In: Alani, H., Kagal, L., Fokoue, A., Groth, P., Biemann, C., Parreira, J.X., Aroyo, L., Noy, N., Welty, C., Janowicz, K. (eds.) ISWC 2013. LNCS, vol. 8218, pp. 135–150. Springer, Heidelberg (2013). doi:10.1007/978-3-642-41335-3_9
14. Paulheim, H., Bizer, C.: Improving the quality of linked data using statistical distributions. Int. J. Semant. Web Inf. Syst. (IJSWIS) **10**, 63–86 (2014)

15. Seok, M., Song, H.-J., Park, C.-Y., Kim, J.-D., Kim, Y.-S.: Named entity recognition using word embedding as a feature 1. Int. J. Softw. Eng. Appl. **10**(2), 93–104 (2016)
16. Ganguly, D., Roy, D., Mitra, M., Jones, G.J.F.: Word embedding based generalized language model for information retrieval. In: Proceedings of the 38th International ACM SIGIR Conference on Research and Development in Information Retrieval, pp. 795–798 (2015)
17. Zhou, G., He, T., Zhao, J., Hu, P.: Learning continuous word embedding with metadata for question retrieval in community question answering. In: Proceedings of ACL (2015)
18. Bengio, Y., Ducharme, R., Vincent, P., Janvin, C.: A neural probabilistic language model. J. Mach. Learn. Res. **3**, 1137–1155 (2003)
19. Goldberg, Y., Levy, O.: Word2vec explained: deriving Mikolov et al. Negative-Sampling Word-Embedding Method. arXiv Preprint arXiv:1402.3722, **2**, 1–5 (2014)
20. Collobert, R., Weston, J., Bottou, L., Karlen, M., Kavukcuoglu, K., Kuksa, P.: Natural language processing (almost) from scratch. J. Mach. Learn. Res. **12**, 2493–2537 (2011)
21. Roark, B., Collins, M.: Discriminative n-gram language modeling. Comput. Speech Lang. **21**(2), 1–30 (2007)
22. Jurafsky, D., Martin, J.H.: N-Gram. Speech and Language Processing (2014). https://lagunita.stanford.edu/c4x/Engineering/CS-224N/asset/slp4.pdf
23. Abdi, H., Williams, L.J.: Principal component analysis. Wiley Interdisciplinary Reviews: Computational Statistics (2010)
24. Mikolov, T., Chen, K., Corrado, G., Dean, J.: Distributed representations of words and phrases and their compositionality. In NIPS, pp. 1–9 (2013)
25. Han, L, Embrechts, M., Szymanski, B., Sternickel, K., Ross, A.: Random forests feature selection with kernel partial least squares: detecting ischemia from MagnetoCardiograms. In: Proceedings of the European Symposium on Artificial Neural Networks, Burges, Belgium, pp. 221–226 (2006)
26. Han, J., Kamber, M., Pei, J.: Data Mining: Concepts and Techniques. 3rd edn. Morgan Kaufmann, San Francisco (2012)
27. Fawcett, T.: An introduction to ROC analysis. Pattern Recogn. Lett. **27**(8), 861–874 (2006)

Alignment: A Collaborative, System Aided, Interactive Ontology Matching Platform

Sotirios Karampatakis[1,2(✉)], Charalampos Bratsas[1,2], Ondřej Zamazal[3],
Panagiotis Marios Filippidis[1,2], and Ioannis Antoniou[1,2]

[1] Open Knowledge Greece, Thessaloniki, Greece
{karampatakis,cbratsas,filippidis}@okfn.gr
[2] School of Mathematics, Aristotle University of Thessaloniki, Thessaloniki, Greece
iantonio@math.auth.gr
[3] Department of Information and Knowledge Engineering, University of Economics,
Prague, Prague, Czech Republic
ondrej.zamazal@vse.cz

Abstract. Ontology matching is a crucial problem in the world of
Semantic Web and other distributed, open world applications. Diversity in tools, knowledge, habits, language, interests and usually level of
detail may drive in heterogeneity. Thus, many automated applications
have been developed, implementing a large variety of matching techniques and similarity measures, with impressive results. However, there
are situations where this is not enough and there must be human decision
in order to create a link. In this paper we present Alignment, a collaborative, system aided, interactive ontology matching platform. Alignment
offers a simple GUI environment for matching two ontologies with aid of
configurable similarity algorithms.

Keywords: Linked data · Ontology matching · SKOS · Thesauri

1 Introduction

The Web of Data is ever growing and very large amounts of data from a
wide range of domains such as publications, geographic, information, economic,
health, agriculture and many others become available. Diversity in data from different domains but also within the same or similar domains creates the need for
linking and integrating data from various sources. Due to their key features and
main advantages, Semantic Web technologies and RDF format tend to become
a lingua franca of data integration. Furthermore, one of the most important features Semantic Web has to offer is the ability to interconnect data in order to get
additional information. Thus, innovative and advanced approaches for link discovery in semantic resources can offer great possibilities and benefit significantly
the integration task of diverse data.

While there is a variety of tools for automated or semi-automated link discovery, there are cases where this approach is not enough or efficient. The potential

© Springer International Publishing AG 2017
P. Różewski and C. Lange (Eds.): KESW 2017, CCIS 786, pp. 323–333, 2017.
https://doi.org/10.1007/978-3-319-69548-8_22

complexity and versatility of vocabularies could easily lead to wrong links, while the appearance of many synonyms of a vocabulary term in another vocabulary is a common problem for creating accurate links between them [8]. An automatic approach of linking large vocabularies leads to only partial alignments between them and many of their concepts might have false or no links at all [7].

Consequently, linking different ontologies can be a very difficult and in some cases unfeasible task to get accomplished, with just a single approach. Moreover, some times the best algorithm for finding links between entities is human knowledge [12]. Especially when quality of mappings has higher value than quantity of links, which is the case for ontology matching. Thus, a mixed-methods approach is probably the most suitable for this complicated task.

The main text of the paper is organized as follows: in Sect. 2 we present the overall design philosophy of the Alignment platform, giving a short presentation of its main features. Next, in Sect. 3 we present the three experiments we conducted, in a form of workshops, in order to evaluate key features of the platform, collect feedback from the target audience and further improve the platform.

2   Alignment Platform Presentation

2.1   System Architecture

Alignment is a collaborative, system aided, user driven ontology matching platform. It offers a simple GUI environment for matching two ontologies based on a default or user defined configuration of similarity measures and algorithms. Users can select one of the suggested links for each entity, or they can choose any other link to the target ontology, based on their domain knowledge. Users can also customize the similarity variables that will be used for the comparison of the two ontologies and result in the suggested links, based on their preferences.

Multiple users can work on the same project and provide their own links simultaneously and interactively. The platform also offers evaluation and social features, as users can give a positive or negative vote or comment on a specific link between two entities, providing feedback on the produced linksets. The produced linksets are then available through both a SPARQL endpoint and an API. A typical workflow of a use case is shown in Fig. 1.

A user has to create a project within the platform. First it is needed to upload the ontologies he wants to produce a linkset[1]. The ontologies get validated and stored on the platform. Then the user has to define the source and target ontology. Also he needs to define which similarity algorithm configurations will be used for the system provided suggestions. The user can also choose if the project will be private or public. Then, upon creation of the project, the platform calculates similarities between the entities of the ontologies and renders the GUI. None of the suggestions provided by the system is realized as a valid link, unless some user decide to create the link. Finally, produced linksets can be exported, or send for crowd-sourced validation, through the Voting service. The platform

---

[1] A linkset is a set of links or mappings between entities of two ontologies/vocabularies.

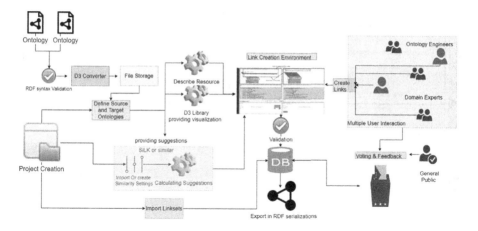

Fig. 1. Alignment workflow

is available for demonstration on http://alignment.okfn.gr and the source code is available under the MIT license[2].

2.2 GUI

As previous studies have shown [5], users should not be overwhelmed with too much information, but enough in order to decide if a mapping should be created or not. With this in mind, we designed our GUI to be as minimal as can be with enough utilities to aid users, either domain or ontology engineering experts on the linking workflow. The "Create Links" page presents both source and target graphs as expandable tree graphs. Detailed element description and system generated suggestions are presented as helpers for the linking process. Users can select a link type from predefined, grouped RDF links or custom. An overview of the created links is also presented, where user can edit or delete the produced links on the same page. A sample view of the GUI is shown in Fig. 2.

D3 visualization: The graph visualization is performed using D3 JavaScript library[3]. D3 reads data in JSON format using specific nested structure. We wrote a converter to transform RDF serializations or OWL ontology representations in that structure. The converter uses the SKOS semantic relationships properties (or OWL equivalent), such as `skos:broader` or `skos:narrower`, to realize the nested hierarchical structure.

The converter is executed whenever a new file is uploaded, and the resulted JSON formatted file is cached to further decrease loading times. Different coloring of elements denote different properties. For instance, blue elements contain child elements/nodes, while orange do not. Above each graph, there is a search bar, with autocomplete function enabled. On selection, the according element is highlighted on the graph and a detailed description is presented.

---

[2] http://github.com/okgreece/Alignment.
[3] https://github.com/d3/d3/wiki.

Source and Target Graph Presentation

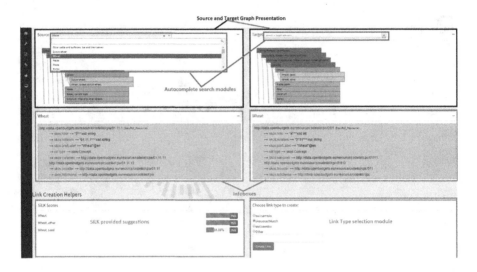

Fig. 2. Alignment GUI

On the left side of each element of the tree graph there is a circular indicator. This indicates the "linking" status of a particular element. Gray color indicate that there are no links for this element. Light blue color indicates elements that the system has found suggestions and finally green color indicates elements that have been linked with some element from the target graph. This is very helpful, especially when multiple users work on the same project in order to avoid already linked elements, focus on "easy" to link elements, or when a user wants to check what links have been created for a specific element. This functionality will be extended to indicate conflicts or validation errors.

Detailed Entity Info: Below the two graphs, user can find detailed information about the selected element of each graph. A SPARQL DESCRIBE query is run on the selected element and the result is presented as an infobox. Users can click on resources describing or linked with this element to get additional information. Thus, users can have an extensive description of the selected entity.

System Suggestions: Below descriptions section, there are the similarity scores which are calculated as described in Subsect. 2.3. On selection of a source element, this module gets the matched entities from the scores graph, if any, and presents them to the user, in score descending order. A button beside the score of each entity, selects the related target entity on the graph, and presents its description on target Infobox section. If there is not any suggested target entity found or the user decides that none of the proposed entities are related with the selected source element, the user can use the Target Graph module to select a target entity, as simple as click on it.

Link Type Option: On Semantic Web, different links express different semantics between entities. Sometimes two entities represent the same concept so could be linked via a skos:exactMatch, or owl:sameAs link, while sometimes concept A could be a broader match of concept B, so it could be connected via `skos:broadMatch`. The user should be able to choose the link type.

This is the role of the Link Creator module, where user can select the link type to establish between two entities from a short list of predefined link types, or define his own link type by simply insert it on the according text field. Link types are organized in discrete groups: SKOS, OWL and RDF related link types, in order to minimize confusion of the user. The Create Link button will create the new link, if the specified link passes the validation criteria.

On the bottom of the GUI, the user can overview the links created on the project in a simple searchable and orderable table. User can delete created links, export them in different RDF serializations or import linksets.

2.3   Calculating Similarities

Alignment integrates the Silk Linking Framework functionality [13] to calculate similarities between entities from different ontologies or vocabularies, to be used as aids for the manual linking part of the process. To do this, a Silk configuration file containing the comparison specifications in the Link Specification Language of Silk (Silk LSL) and XML form has to be created and executed by the Silk. The user can either select the default application settings as the comparison specifications, or can create user specified configuration file, customizing the similarity metrics to be used for the comparison or import a Silk LSL file.

In either case, the comparison process calculates similarities on predefined (defaulted to skos:prefLabel) or user provided properties of the entities of the source and target ontologies, using suitable similarity algorithms that are integrated in Silk, like Levenshtein distance, Jaro - Winkler distance, Dice coefficient or soft Jaccard similarity coefficient, as metric rules.

The comparison is exhaustive and for every entity of the source graph, all entities from the target graph are examined. This is the automatic part of the linking process and it has to be executed separately, before the user can manually select and link entities between the two graphs. The whole process is running on the background and user is notified upon completion. Silk is configured to store the calculated similarities using the Alignment format [6].

User Configuration: Besides the default comparison settings, a user can also guide the automatic part of the comparison process, by creating or import a configuration file for an Alignment project between two graphs. These specifications are related to the comparison metrics, the linkage rule and the output of the comparison. These user-customized settings are automatically saved to the user profile or shared in public, so they can be used in other projects as well.

2.4   Integration, Collaboration and Social Features

Users can log in the platform using a variety of OAUTH capable services such as Github, Google or Facebook accounts, or use the built-in registration system. Multiple users can work on the same project enabling faster completion of projects. Additionally, linksets can be exposed to open voting to get validation.

Integration with other Services: Created links can be exported as single file in different RDF serializations. The application offers two more options on how to provide linksets in different applications. Public linksets are available via a SPARQL Endpoint for Semantic Web applications and a generic API, as an access point for common applications.

SPARQL Endpoint: In order to be able to integrate with Semantic Web applications, a SPARQL Endpoint is established using D2RQ Platform [3]. By this approach we can reuse the existing MySQL database, without having to replicate the database in an RDF triple store. D2RQ platform offers access to SPARQL clients through its endpoint and Snorql SPARQL Explorer.

API: By design, the application should support integration with legacy applications using APIs, in a Semantic Web agnostic manner. In order to comply, a REST API interface was developed. The API accepts HTTP post requests on the URL http://{ip:port}/api/{function} where function can be string, IRI, ontology or ontologyIRI. The first two functions cover the situation where we need to lookup for a specific entity or a set of entities. We can provide for instance the label of an entity or it's IRI, and get all the matched links that the provided entity has on all or specific ontologies. The last two functions are used to get various information about a specific ontology that exists on the platform, or get all the matched links against a target ontology. User can select between various response formats, like CSV or JSON.

Working on the same project: Some projects may contain a huge number of entities to be linked, make them a bad candidate for manual linking. But effort needed may be reduced by employing multiple users [2]. One of the requirements of the platform was to have the ability to host multiple users per project, who could act simultaneously and interactively to quickly develop linksets between ontologies. Projects can be private or public. Public projects are visible to all users, so they can work simultaneously. Entities with existing links are marked to prevent double work. Links are connected with the creator to enable filtering and monitoring of work committed or to prevent vandalism [9].

Crowdsourcing Link Validation: For validation purposes, created links can be exposed on public voting, where users can upvote, downvote and comment on each link. A voter can request a chunk of the links (we call it a Poll) in order to start the voting process. Each chunk is different from any other as it is created upon request of the voter. The pool of links is firstly reduced by the links the voter may have already voted for and then picked randomly from the pool. The size of the chunk can be defined by the project owner, affecting the voting session duration and the overall user experience.

After creating the voter specific poll, the links composing the chunk are presented one by one to the voter. The interface is designed to be simple enough and hide the underlined RDF data modeling. Each entity and link type is presented by its label on the voters language if it is available, so actually the particular RDF link is represented as a sentence.

Each part is clickable pointing to the resource IRI, so the user can retrieve more information. This requires the IRIs to be dereferenceable. Information of the exact current score of each link is not shown by design, because this would bias the voter. The user can upvote, downvote or skip a link. In the case of downvoting a link, user feedback is requested, to categorize the cause of downvoting. There are three possible choices, (a) Unrelated, to mark false positive links, (b) Semantic error, to indicate a wrong link type (e.g. it should be `skos:narrowMatch` and not `skos:exactMatch`) and (c) Other, to cover the rest cases, where the user can provide feedback upon the cause of rejection. After the user has finished the poll, an overview is shown, in order to either change a vote, provide feedback, or revisit skipped links.

3 Evaluation

3.1 Link Creation Module Evaluation

As described in Subsect. 2.2 the Link Creation module was designed to cover the needs of both semantic web experts, or domain experts with no previous knowledge of ontology engineering. So we conducted two experiments, testing the user experience and the outcome of the linking project. The first was targeted to Junior Level Ontology Engineers (graduate students) and the second targeted to public officers, experts on the fiscal data domain. Both was supervised by experienced Senior Ontology Engineers.

Junior Level Ontology Engineers. The scope of the experiment was to provide mappings between two codelists, the Admin. Class. of the Municipality of Athens (hereafter **Codelist A** instances) and the Admin. Class. of Greek Municipalities (hereafter **Codelist B**) instances.[4] These codelists where used as dimension values in datasets described using the OpenBudgets.eu data model [11] an RDF DataCube [4] based fiscal data model. Thus the mapping would enable direct comparison of fiscal data across Greek Municipalities [7].

The group consisted of 21 graduate students. The test was separated in two parts with an overall duration of two hours. The scope was to compare the two by the time proposed methods to conduct links between SKOSified codelists. One was the use of advanced collaborative ontology editors, facilitating a platform like VoCol [10], and the other one was the use of a GUI with system provided suggestions, which led to the development of Alignment. Instead of actually using the VoCol platform, in order to overcome the additional time required to get used to the platform, we decided to simulate the VoCol user experience, leaving out the collaborative part. To simulate the VoCol user experience, participants had to open with a text editor the two codelists, and create a new file to commit the linksets. Both codelists were given in Turtle format for easier manipulation and commitments to the linkset file was asked also to be serialized in Turtle. So for each instance of the source Codelist A, the participants had to search for a

---

[4] Codelist A:https://goo.gl/UYrfPx, Codelist B: https://goo.gl/oxjjLn, Expert Mapping: https://goo.gl/yiCTUw.

compliant instance on the target Codelist B, and then commit the link on the linkset file on the following format:

$$< \texttt{clA\_instance\_IRI} > \texttt{skos} : \texttt{broadMatch} < \texttt{clB\_instance\_IRI} > .$$

Participants had a limited time of an hour to conclude their linking projects, validate, fix and finally send their linksets. Validation part was done by using the Rapper utility [1], as this is the case also for VoCol. Errors were fixed by the participants before sending the final linksets.

Afterwards, participants were asked to do the same task by using the Alignment platform. They created their own projects and then they had the same time limit to conclude their project. Each user was working separately, so we didn't tested the collaborative features of the platform. Bugs of the platform and notices where tracked by assistants to give feedback on the developers team.

Results. Out of the 21 submitted linksets produced with the text editor, 13 was unable to be parsed due to different kind of syntax errors. Some files contained invalid characters, other were missing the "<>" brackets, or the "." Turtle specific triple ending. Additionally, 5 of the linksets used concepts of the target codelist as subjects and vice versa, thus induced semantic errors. The Rapper utility provides syntax validation but the user has to understand and fix the error manually. On the other hand, by using Alignment platform, syntax errors are eliminated as serialization of the links is handled and validated by the platform. Additionally, semantic errors were not present, as the platform always sets as subject resource an instance of the source ontology and vice versa. The utilities that Alignment offers, helped the participants to create linksets in a more efficient way. The majority of the participants (76.19%) reported that they enjoyed the session with Alignment. A percentage of 90.47% answered that they would use again the platform for linking projects, but also there was some 61.9% of the participants stating that there is still room for improvements. We plan to conduct a more analytical evaluation of this trial, in order to evaluate overall improvement of user scores in terms of precision and recall, using the produced linksets by the students and the complete mapping provided by an Ontology Engineering Expert. The overall conclusion of the trial was that while VoCol is an excellent platform to maintain and collaboratively create an ontology, when it comes for Ontology Alignment and especially employing domain experts with no knowledge of RDF concept, it seems inadequate.

Domain Experts: The scope of the trial was to produce a linkset between EU categorization system for funded projects with those of individual EU countries, in order to enable straightforward fiscal analyses. This was realized by building a linkset between the Czech codelist (44 items) to the European one (142 items). In order to ensure the quality of the linkset we involved two domain experts, working separately, using the Alignment platform. They followed detailed guidelines. The guidelines also includes a brief manual how to use the Alignment GUI and instruction that experts should prefer certain types of links more, i.e. there was the following preference `skos:exactMatch`, then `skos:narrowMatch`, `skos:broadMatch` and then the rest.

Results. Both experts interlinked 32 same items where expert 1 linked 84% (37) items from source codelist and expert 2 linked 82% (36) items from source codelist. While the expert 1 employed all skos link types (out of all 53 links) more or less uniformly (21 narrowMatch, 11 closeMatch, 9 exactMatch, 8 relatedMatch, 4 broadMatch), the expert 2 created mainly narrowMatch links (116), additionally 8 exactMatch and 1 broadMatch, out of all 125 links. Both experts managed 32 times to linked the same two entities in one link and, more importantly, they managed to create the very same link 23 times where there are 7 exactMatch, 1 broadMatch and 15 narrowMatch. The resulted linkset of 23 links represents the nucleus of the reference linkset[5]. Since there are many links created by only one expert (57% in the case of expert 1 and 82% in the case of expert 2) we further plan to let experts discuss those not agreed links to extend the current reference linkset.

3.2   Link Validation Module Evaluation

To test the Voting module of the Alignment platform we conducted a workshop, engaging domain experts to validate a linkset between Wikidata and National Library of Greece Authority Records, produced by an automated procedure.

Trial Set up. Twenty two participants, varying from young undergraduate librarians to experienced public Library Cataloguers, worked on the same project to validate a linkset produced by an automated procedure. A sum of 11K links was created by using string similarity algorithms, provided by Silk. But in order to achieve high recall scores, the threshold was set to be low. So that configuration introduced also false positive links. Thus user interaction was needed. The linkset was imported on the platform within a public project created by the instructor. After a brief presentation of the concept and a demonstration of the basic Voting System utilities, the participants started to validate the links using the platform. The project was configured to offer chunks with a size of 25 links each time a user requested a new Poll. Each participant had to complete at least one Poll. Assistants were logging bugs of the system and feedback from the participants about possible design pitfalls or nice to have features.

Results: In almost 20 min a set of 395 unique links (3.5% of the total linkset) were validated using the Alignment Platform. The vast majority of the participants completed just one Poll while some more enthusiast participants started more Polls. The validation process helped us to improve our similarity algorithms configuration for the automated procedure. After the completion of the trial participants had to complete a questionnaire based on the usability of the platform and the overall experience. About 73% of the participants said that they would like include the platform on their workflow, while about 80% found that they felt it was easy to use the platform.

---

[5] Czech Codelist https://goo.gl/pKZpVR, European codelist https://goo.gl/9hCPZq, Guidelines: https://goo.gl/vRYc5r, Results: https://goo.gl/BEmzfb.

4 Summary

We presented Alignment, a new tool for manual link creation between entities, on the Semantic Web World. This tool was created based on the experience gained working in Horizon 2020 Openbudgets.eu project, in order to enhance domain experts of fiscal datasets, manually create and evaluate links between heterogeneous ontologies. It can also be used as a tool to crowdsource linkset creation and evaluation. We plan to extend the data model of linksets provided by the SPARQL endpoint. Moreover, a translation functionality is being developed to enable system aids for multilingual ontologies.

Acknowledgments. This work has been supported by the OpenBudgets.eu Horizon 2020 project (Grant Agreement 645833).

References

1. Beckett, D.: The design and implementation of the Redland RDF application framework. Comput. Netw. **39**(5), 577–588 (2002)
2. Cordasco, G., De Donato, R., Malandrino, D., Palmieri, G., Petta, A., Pirozzi, D., Santangelo, G., Scarano, V., Serra, L., Spagnuolo, C., Vicidomini, L.: Engaging citizens with a social platform for open data. In: Proceedings of the 18th Annual International Conference on Digital Government Research, DGO 2017, pp. 242–249. ACM (2017). doi:10.1145/3085228.3085302
3. Cyganiak, R., Bizer, C.: D2R Server: A Semantic Web Front-end to Existing Relational Databases. XML Tage, pp. 2–4 (2006)
4. Cyganiak, R., Reynolds, D., Tennison, J.: The RDF Data Cube Vocabulary. W3C Recommendation, January 2014 (2013)
5. Dragisic, Z., Ivanova, V., Lambrix, P., Faria, D., Jiménez-Ruiz, E., Pesquita, C.: User validation in ontology alignment. In: Groth, P., Simperl, E., Gray, A., Sabou, M., Krötzsch, M., Lecue, F., Flöck, F., Gil, Y. (eds.) ISWC 2016. LNCS, vol. 9981, pp. 200–217. Springer, Cham (2016). doi:10.1007/978-3-319-46523-4_13
6. Euzenat, J.: An API for ontology alignment. In: McIlraith, S.A., Plexousakis, D., van Harmelen, F. (eds.) ISWC 2004. LNCS, vol. 3298, pp. 698–712. Springer, Heidelberg (2004). doi:10.1007/978-3-540-30475-3_48
7. Filippidis, P.M., Karampatakis, S., Ioannidis, L., Mynarz, J., Svátek, V., Bratsas, C.: Towards budget comparative analysis: the need for fiscal codelists as linked data. In: SEMANTiCS (2016)
8. Filippidis, P.M., Karampatakis, S., Koupidis, K., Ioannidis, L., Bratsas, C.: The code lists case: identifying and linking the key parts of fiscal datasets. In: 11th International Workshop on Semantic and Social Media Adaptation and Personalization (SMAP), pp. 165–170. IEEE (2016). doi:10.1109/SMAP.2016.7753404
9. Geiger, R.S., Ribes, D.: The work of sustaining order in Wikipedia. In: Proceedings of the 2010 ACM Conference on Computer Supported Cooperative Work, CSCW 2010 pp. 117–126 (2010). doi:10.1145/1718918.1718941
10. Halilaj, L., Petersen, N., Grangel-González, I., Lange, C., Auer, S., Coskun, G., Lohmann, S.: VoCol: an integrated environment to support version-controlled vocabulary development. In: Blomqvist, E., Ciancarini, P., Poggi, F., Vitali, F. (eds.) EKAW 2016. LNCS, vol. 10024, pp. 303–319. Springer, Cham (2016). doi:10.1007/978-3-319-49004-5_20

11. Mynarz, J., Svátek, V., Karampatakis, S., Klímek, J., Bratsas, C.: Modeling fiscal data with the Data Cube Vocabulary, September 2016. doi:10.5281/zenodo.168588
12. Shvaiko, P., Euzenat, J.: Ontology matching: state of the art and future challenges. IEEE Trans. Knowl. Data Eng. **25**(1), 158–176 (2013). doi:10.1109/TKDE.2011. 253
13. Volz, J., Bizer, C., Gaedke, M., Kobilarov, G.: Silk a link discovery framework for the web of data. In: Proceedings of the 2nd Linked Data on the Web Workshop (2009). doi:10.1111/j.1467-9744.2007.00872.x

Semantic Technologies in
Manufacturing and Business

ODERU: Optimisation of Semantic Service-Based Processes in Manufacturing

Luca Mazzola$^{(\boxtimes)}$, Patrick Kapahnke$^{(\boxtimes)}$, and Matthias Klusch

DFKI - German Research Center for Artificial Intelligence,
Saarland Informatics Campus D3.2, 66123 Saarbrücken, Germany
mazzola.luca@gmail.com,
{Luca.Mazzola,Patrick.Kapahnke,Matthias.Klusch}@dfki.de

Abstract. A new requirement for the manufacturing companies in Industry 4.0 is to be flexible with respect to changes in demands, requiring to react rapidly and efficiently on the production capacities. Coupling it with the affirmed Service-Oriented Architectures (SOA) induces a need for agile collaboration among supply chain partners, but also between different divisions or branches of the same company. To this end, we propose a novel pragmatic approach for automatically implementing service-based manufacturing processes at design and run-time, called ODERU. It provides an optimal plan for a business process model, relying on a set of semantic annotations and a configurable QoS-based constraint optimisation problem (COP) solving. The additional information encoding the optimal process service plan produced by means of pattern-based semantic composition and optimisation of non-functional aspects, are mapped back to the BPMN 2.0 standard formalism, through the use of extension elements, generating an enactable optimal plan. This paper presents the approach, the technical architecture and sketches two initial real-world industrial application in the manufacturing domains of metal press maintenance and automotive exhaust production.

1 Introduction

As every other aspect of the everyday life, also the manufacturing domain is strongly influenced by innovations in the Information and Communication Technologies (ICT). Companies need to flexibly react to changing demands to remain competitive in a dynamic market. The impact of ICT in this domain is broadly known as Industry 4.0 and ranges from the application of artificial intelligence in robot-assisted production to the usage of Internet of Things (IoT) devices, always connected and controllable just-in-time. Along the same line, manufacturing business processes have to be designed and executed in a more dynamic production context, thus creating the need for adaptation and optimisation at design time as well as at run-time. As a consequence, the design of process models for business applications need to comprise representations for functional and non-functional requirements beyond what can be specified in traditional Business Process Modelling (BPM) systems, such as semantic representations of product

© Springer International Publishing AG 2017
P. Różewski and C. Lange (Eds.): KESW 2017, CCIS 786, pp. 337–346, 2017.
https://doi.org/10.1007/978-3-319-69548-8_23

models and manufacturing services as well as Key Performance Indicator (KPI) requirements and Quality of Service (QoS) aspects. Moreover, the tools need to be able to provide effective composition of services in the context of SOA and XaaS (Everything-as-a-Service) systems and reliable model optimisation to achieve the best executable service plans for business processes. Eventually, the provided process service plans (PSP) should be designed to support effectively a run-time incremental re-planning, in case an included service is temporarily failing or becomes unavailable.

Due to the unavailability of solutions to tackle these issues in an integrated way, we developed a novel pragmatic approach called ODERU (**O**ptimisation tool for **DE**sign and **RU**n-time). It is able to select the set of compliant services, available to implement the tasks, and subsequently to compose functionally correct plans based on semantic annotations, while optimising their non-functional aspects formalised in terms of a Constrained Optimisation Problem (COP). The resulting complete service plan (services used, their order, the variable bindings and the optimal environmental variables assignment) is encoded back into specifically developed BPMN 2.0 extensions, partially bridging the gap between models and executable plans, providing at the same time the best variable assignments to optimise the outcome of the plan execution.

The rest of paper is organised as follows: In Sect. 2, related work is briefly presented, then we describe the ODERU basics and algorithm in Sect. 3; while Sect. 4 introduces two use cases adopted as applications of ODERU. For each of them a short overview of the scenario is given, followed by a brief description of the design and runtime behaviours. The conclusions are given in Sect. 5.

2   Related Work

Process models are automatically implemented with semantic services by applying techniques of semantic service selection and composition planning, as for Semantic SOA (SemSOA). The key idea is to enable automated understanding of task requirements and services by providing semantic descriptions in a standardised machine-understandable way by using formal ontological definitions [1], for example in OWL2[1]. In [2], the authors propose SBPM, a framework to combine semantic web services and BPM to overcome the problem of automated understanding of processes by machines in a dynamic business environment. Similarly, the authors of [3] propose sBPMN, which integrates semantic technologies and BPMN to overcome the obvious gap between abstract representation of process models and actual executable descriptions in BPEL. [4] follows the same track with the proposal of BPMO, an ontology, which partly is based on sBPMN, while [5] takes sBPMN as basis for the Maestro tool, which implements the realisation of semantically annotated business tasks with concrete services by means of automatic discovery and composition. In [6], a reference architecture for semSOA in BPM is proposed, which aims to address the representation discrepancy business expertise and IT knowledge by making use of semantic web technologies.

---

[1] W3C standard; https://www.w3.org/TR/owl2-overview/.

All of these proposals rely on formalisation different from (although based on) BPMN or do not aim for a full integration from a formalism point of view. In the work [7] the authors propose an approach that uses BPMN extensions to add semantic annotations for automatic composition of process plan and to verify their soundness, but this approach does not consider QoS-aware or run-time optimisation. Adopting a similar approach, ODERU proposes a set of BPMN extensions that not only enable interoperability by offering process model composition, task service selection and process execution, but also provide a way to represent the best values to optimise the QoS and the quality values achieved.

ODERU applies state of the art semantic service selection technologies [8] for implementing annotated process tasks. Non-functional criteria, often referred to as QoS (e.g. costs, execution time, availability), can additionally be considered to find matching services in terms of functional *and* non-functional requirements [9,10]. In ODERU, optimality with respect to the non-functional QoS specifications is achieved on the process model level by solving (non-)linear multi-objective COP (muCOP) as an integrated follow-up to the pattern-based composition.

Most existing approaches to process service plan composition do not cover the combination of functional (semantic) aspects and non-functional (QoS-aware) optimisation. For example, [5,11,12] consider functional semantic annotations to implement business processes by means of a service composition plan. [13] provides a survey giving an overview of existing approaches and initiatives in this direction and highlights research questions. Integrated functional and non-functional optimisation has rarely been considered, with the notable exception of [14]. While composition typically includes the computation of possible data flows, ODERU additionally finds optimal service variable assignments that are also required for executing the resulting plans. This is a feature not yet considered by existing work. Moreover, ODERU performs re-optimisation of process service plans at run-time upon request, which is also a novel feature.

3   ODERU: Architecture and Overview

Given the problem at hand, we identified three main requirements for ODERU. It should support service selection and composition with non-functional optimisation based on flexible measures and objective functions, creating as output a complete plan, directly enactable by an execution environment. Eventually, the format used should simplify the re-use and adaptation of the created plans in a dynamic environment, at run-time. ODERU is a JAVA-based software implemented as a RESTful service. Figure 1 depicts its basic components and the interactions required for a fully functional Process Enterprise Execution Platform. Only the part enclosed inside dotted line composes the ODERU solution. To provide an optimal solution out of the set of possible functionally valid solutions, ODERU has to make particular choices driven by non-functional requirements, which are expressed as functions of the QoS measures provided by the services. Moreover, it computes concrete settings of service input parameter values, which yield optimal results in terms of the optimisation criteria.

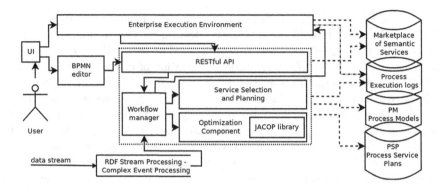

Fig. 1. The infrastructure and its interactions with a Process Enterprise Environment.

Analogous to the semantic service descriptions themselves, these process model annotations are structured in terms of IOPE and refer to domain knowledge in OWL2. Moreover, the BPMN should specify what QoS measures are to be optimised and how they are defined. This is done by specifying a COP at the process model level, whose solutions dictates what services to choose from and what parameter settings to use when calling services. The COP formulation includes information on how to map optimal parameter values to service inputs and service QoS to COP constants. The outputs produced by ODERU are process service plan encoded in the original BPMN itself by making use of extensions again. Besides the optimal services and input values for calling the services as described above, this also includes possible data flows with parameter bindings among services. Such a process service plan implementing the process model can then be instantiated at run-time by a process plan execution environment. To achieve this, ODERU works following two steps in a sequential manner: first it performs a (A) Pattern-based composition using semantic service selection for all semantically annotated process tasks and the computation of possible data flows. Then, ODERU executes a (B) QoS-aware non-functional optimisation by means of COP solving on the process model level. This second step selects particular services out of sets of functionally fitting services per tasks previously identified, and provides the optimal settings for service inputs.

3.1 Semantic Annotation of Tasks and Services

In order to be able to automatically compose functionally valid process service plans given a process model, we assume process tasks to be equipped with structured semantic descriptions. Following the SemSOA approach, IOPE of tasks are described in terms of formalised ontological domain knowledge. For the use cases described in this paper, we propose a reference domain ontology called CDM-Core [15], which provides OWL2 descriptions of concepts from the manufacturing domain, in particular for hydraulic metal press maintenance and car exhaust production. The semantic annotations are embedded in the BPMN model by

---

Algorithm 1. The pseudocode for the process service plan composition

---

Input: PM: semantically annotated BPMN model, **S**: set of available services
parameter : **Sim$_{min}$**: minimal similarity value accepted
Output: PSP: the computed process service plan
1 **forall** $s \in S$ **do**
2 $\quad |\quad IOPE_s \to IOPE_S$;
3 **end**
4 **forall** *task* $\in$ **PM** **do**
5 $\quad |\quad task \to T$;
6 **end**
7 **forall** $t \in T$ **do**
8 $\quad |\quad$ **forall** $s \in S$ **do**
9 $\quad |\quad |\quad$ **if** $SIM(IOPE_t, IOPE_s) >= $ **Sim$_{min}$** **then**
10 $\quad |\quad |\quad |\quad s \to CANDIDATES_t$;
11 $\quad |\quad |\quad$ **end**
12 $\quad |\quad$ **end**
13 **end**
14 **forall** $t \in T$ **do**
15 $\quad |\quad$ **forall** $s \in CANDIDATES_t$ **do**
16 $\quad |\quad |\quad$ **forall** *QoS* $\in$ **T** **do**
17 $\quad |\quad |\quad |\quad QoS \to Parameters_{s_t}$;
18 $\quad |\quad |\quad$ **end**
19 $\quad |\quad$ **end**
20 **end**
21 *Solutions* = COPSOLVER(Parameters);
22 **forall** *Solution* $\in$ *Solutions* **do**
23 $\quad |\quad$ COMPOSEVARIABLEBINDINGS(Solution) $\to$ *Plans*;
24 **end**
25 **PSP**=MERGEPMWITHSOLUTION(PM, *Plans*[0]);
26 **return PSP**;

---

making use of extension elements at the task level. Similarly, we assume that all services come with semantic annotations of IOPE. For this, the W3C recommendation OWL-S [16] is used, providing means for not only IOPE annotations, but also for the QoS aspect required for the non-functional optimisation.

3.2 Constraint Optimization Problem Definition

We defined an appropriate grammar to represent COPs, based on the requirements of the project use cases, but also taking into account its general reapplicability. We relied on a parser generator for this task, and the choice was antlr4 (http://www.antlr.org/). This decision allows the definition of complex aggregates of QoS and environment variables instead of mere lists of objectives for simple QoS, extending the expressive capability with respect to the nonfunctional optimisation problem definition.

3.3   Process Service Plan

The computation of the service plan is presented in Algorithm 1, which uses four helper functions. The first one is **SIM** ($IOPE_A, IOPE_B$) computing the similarity between two IOPE annotations based on a selected measure. A second helper function is the **COPsolve** (Parameters) for computing the set of Pareto-optimal solutions of the COP. This is a simple compiler that transform our COP definition into a running instance of a JaCoP solver (see http://jacop. osolpro.com/), using the set of parameters given. **ComposeVariableBindings** (Solution) takes care of computing a possible set of variable bindings for the data flow. It is based on the checking of the semantic compatibility of the variables, to ensure a meaningful assignment, going further the simple type compatibility checking. This ensures the direct executability of the computed service plan. Eventually, **MergePMwithSolution** (PM,Plan) takes care of adding the full metadata section into the original process model to create an executable PSP.

Functional Optimisation (Services selection). The first step for creating a Process Service Plan is to select all the possible candidates functionally valid for each task. We rely on functionally equivalent *exact* or on *plug-in* matches [11] limited to direct sub class relationships, in order to have a PSP whose logical properties (in term of IOPE) are conserved with respect to the given PM. Every task existing in the process model is considered, as the selection of a valid combination of the task to be actually implemented in the returned process service plan is left for the non-functional optimisation, based on the COP solution.

Non-Functional   Optimisation   (Optimal   Services   composition). Amongst all the possible combinations of services of the candidate pools of the tasks, the best (or Pareto-optimal in case of a multi-objective problem) option is chosen as part of the overall solution. This implies solving the COP problem associated to the process model. For an introduction to the BPMN extensions defined in CREMA and used by ODERU, we refer the reader to [17].

To achieve its objectives, during the CREMA project a set of functions was designed and implemented into ODERU. The provided calls allow to ask for an integrated composition and optimisation (meaning, considering both the functional and non-functional requirements specified into the input BPMN) or separately, in case when (a) the user is interested only in a functionally valid plan or (b) when exists already a composed plan that requires to be optimised based on the non-functional QoS measures and the user-defined objective function(s). This is valid both at design time (input is a process model) and at run-time (input is an instance of the process model, together with the execution log, if available). For accountability, then a functions to allow the user approval of the computed PSP is provided, together with a set of utility operations, such as for retrieving the ordered list of services found to implement a task and for fetching previously computed PSP, in case when other options would be useful or interesting to be explored.

4 Applications

In this section, two applications of the proposed approach are showcased, to demonstrate its applicability and the capability to cover different requirements.

– *Use Case A: Machine Maintenance.* This first use case refers to the maintenance of hydraulic metal presses, in particular the clutch-brake mechanism, that is its main active part in this scenario. To provide the necessary assistance to the press owner, the producer has some geographically distributed Technical Assistance Service (TAS) organised in teams, which can provide just-in-time on-site maintenance. The selected TAS Team usually requires also one or more replacement parts, in order to restore the full functionality of the press: these replacement parts are provided by some Spare Part (SP) Provider. This PM starts with a task for collecting (and remotely analysing) information about the signalled misbehaviour of the press, to decide if a maintenance is necessary and if the press has to be stopped instantly. Once the maintenance need is confirmed, customer requirements are collected: location, type and length of the warranty, maximum length of the press unavailability and maximum acceptable maintenance costs. Based on this information, the optimisation has to determine the best combination amongst all available TAS Teams and SP Providers, which respects the customer requirements and minimises the total costs. Based on the service selection for these tasks, the model continues computing the earliest possible date for the maintenance, and schedules the actual intervention by proposing and agreeing it with the customer. Eventually, after the maintenance has been executed, the model finishes by collecting customer feedback. At first the semantic service selection is applied, creating a ranked list of candidates for each task, as for Algorithm 1 and then optimisation happens by COP solving.

Design time optimisation is defined in this context as a simple reference, as the parameters used for the COP instantiation are some default values. Under this assumption, **run-time optimisation** means recomputing the actual costs and time based on the updates in the model, for example for considering the current scheduling of TAS Teams and the availability and offers for the required *spare part(s)* from the list of SP providers. As a result in both situations, a functionally equivalent process service plan was computed, with minimal value for the *objective function*, and the user partner confirmed that the proposed solution was effectively solving the given problem.

For the **Validation** phase, the user partner is expecting to obtain the following results, by the application of the optimisation to the process model: reduction of up to 60% of the unscheduled machine breakdown. At the same time, it is considering to achieve a reduction of up to 15% of the total machine breakdowns (increasing, consequently, up to about 18 % the machine availability for production operations). On the maintenance intervention, the expected benefit is a reduction of up to 50% in the intervention time and up to 25% in the costs.

– *Use Case B: OEE for Automotive parts production.* In the second use case, a process model for the production of car exhaust filtering systems is designed. The process starts with the responsible operator selecting a *production task*. This action triggers a sub-process devoted to check availability of and fix allocation of relevant resources and welding robots. Inside it, two tasks calculate the type and number of robots necessary and available in a loop, and computes for each of them the best parameter settings. At this point the production of the batch can start: the operator loads the required components, lets the welding operation execute, runs testing of the produced exhaust pieces and terminates, reporting any issue if present. Regarding the non-functional optimisation, after the selection step for sets of functionally equivalent services for each task and the following composition of a functionally optimal complete process service plan, in this scenario the COP is only relevant for a particular task, namely the "Allocate Robot". The general idea is to setup the welding robot such that it performs optimally with respect to the three main aspects of OEE (overall equipment effectiveness). OEE is composed of measurements for *availability*, *performance* and *quality*. Although these values are combined to give an overall indication for OEE, the three aspects are typically considered separately in order to provide more insight about the actual reasons for low effectiveness. To explore different (non-dominated) solutions and to better understand the actual behaviour with respect to availability, performance and quality, this optimisation problem is considered as multi-objective optimisation problem. So, the three aspects are separate objective functions with respect to the user-controlled parameters x, adjustable to explore the solution space. In this scenario, the Availability is defined as the Mean Time Before Failure (MTBF) of the robot cell: it depends only on the *welding current* $\mathbf{I}$, and can be further decomposed in four elementary components. The Performance aspect only directly depends on the *torch speed* $\mathbf{S}$, once divided by an ideal cycle time. Eventually, the Quality aspect is the most complex, as it relies on multiple independent dimensions that jointly determine the measure Defect Parts per Million (DPM). Besides the objective functions, a set of constraints is required to further characterise the problem.

Design time optimisation means here to compute the optimal parameter setting for the best robot cell existing in the pool of candidates. The settings for this optimisation are based on QoS measures computed from cumulated historical data of the robot cell. In this respect, this is the best possible configuration achievable, without considering constraints coming from robot unavailability or conflicting assignments of the same robot cell. **Run-time optimisation** here is considered for two different cases: (a) there are multiple batches of the same product to construct and after each one of them an analysis searching for better parameters setting can be performed, (b) there is a robot unavailability (e.g. hard failure) and consequently an alternative robot cell can (or should) be considered. As result, the computed process service plans maximize the three components of the *OEE* measure. As there is no natural automatic way of scalarizing the solutions on the Pareto-frontier, the user partner was in charge of selecting the preferred plan amongst the presented options. Its feedback indicated that the proposed solutions seems appropriate and can solve the given problem optimally.

For the **Validation** phase, the user partner is expecting to achieve the following results, by optimising the robot cell selection: in a first scenario to increase the speed to allocate production schedule to the manufacturing assets (from the current 6 hours to 1 hour), to reduce significantly the time for engaging additional manufacturing assets (from 6 months to 2 weeks) and, eventually, to increase the aggregated OEE measure from the current 60% to 70%. On another scenario, the plan is to increase OEE single components: "Quality" feature from the current 55% close to 75% and "Availability" ones from 60% to 70%.

5 Conclusions

In this work we presented our innovative flexible solution to optimal service composition of process models ODERU, which composes functionally correct plans and supports optimisation of non-functional aspects, in the form of a Constrained Optimisation Problem, using as measures generic QoS and supporting user-defined composed objective functions. To showcase the capabilities of the tool, we applied it for two scenarios in the manufacturing domain, with satisfactory results. ODERU will be publicly released at the end of the CREMA project under the Affero GPL v3.0 licence at https://oderu.sourceforge.io.

The main advantages of ODERU in respect of the existing approaches are manifold: the first improvement is the business process formulation: it allows a full integration of functional service selection and composition with non-functional optimisation based on user-defined QoS and objective functions arbitrarily complex in the COP. This is achieved through our BPMN extensions and thanks to the development of a grammar for the optimisation part. Secondly, the produced output is directly enactable by an execution environment, being a complete plan. This means that it is equipped with all the relevant information: service assignments, data flow (variable bindings) and optimal variable assignments for initialising the enactment environment. Eventually it, by encoding the computed PSP in an extended BPMN format, allows to maintain in a single place model and plan, together with the variables assignment and the optimality value achieved.

There are still open points we would like to tackle in the future. The most important ones affect (a) the internal ODERU work-flow and (b) the usage of data stream information for proactively directing and guiding the tool behaviour.

Acknowledgment. This work was partially financed by the European Commission within the CREMA project, agreement 637066; and by the German Federal Ministry of Education and Research (BMBF) within the project INVERSIV.

References

1. McIlraith, S.A., Son, T.C., Zeng, H.: Semantic web services. IEEE Intell. Syst. **16**(2), 46–53 (2001)

2. Hepp, M., Leymann, F., Domingue, J., Wahler, A., Fensel, D.: Semantic business process management: a vision towards using semantic web services for business process management. In: IEEE International Conference on e-Business Engineering (ICEBE) 2005, pp. 535–540. IEEE (2005)

3. Abramowicz, W., Filipowska, A., Kaczmarek, M., Kaczmarek, T.: Semantically enhanced business process modeling notation. In: Semantic Technologies for Business and Information Systems Engineering: Concepts and Applications, pp. 259–275. IGI Global (2012)

4. Dimitrov, M., Simov, A., Stein, S., Konstantinov, M.: A BPMN based semantic business process modelling environment. In: Proceedings of the Workshop on Semantic Business Process and Product Lifecycle Management (SBPM-2007), vol. 251, pp. 1613–0073 (2007)

5. Born, M., Hoffmann, J., Kaczmarek, T., Kowalkiewicz, M., Markovic, I., Scicluna, J., Weber, I., Zhou, X.: Semantic annotation and composition of business processes with maestro. In: Bechhofer, S., Hauswirth, M., Hoffmann, J., Koubarakis, M. (eds.) ESWC 2008. LNCS, vol. 5021, pp. 772–776. Springer, Heidelberg (2008). doi:10.1007/978-3-540-68234-9_56

6. Karastoyanova, D., van Lessen, T., Leymann, F., Ma, Z., Nitzche, J., Wetzstein, B.: Semantic business process management: applying ontologies in BPM. In: Handbook of Research on Business Process Modeling, pp. 299–317. IGI Global (2009)

7. Weber, I., Hoffmann, J., Mendling, J.: Beyond soundness: on the verification of semantic business process models. Distrib. Parallel Databases 27(3), 271–343 (2010)

8. Klusch, M., Kapahnke, P., Schulte, S., Lecue, F., Bernstein, A.: Semantic web service search: a brief survey. KI-Künstliche Intell. 30(2), 139–147 (2016)

9. Pilioura, T., Tsalgatidou, A.: Unified publication and discovery of semantic web services. ACM Trans. Web 3(3), 11 (2009)

10. Zhang, Y., Huang, H., Yang, D., Zhang, H., Chao, H.C., Huang, Y.M.: Bring QoS to P2P-based semantic service discovery for the universal network. Pers. Ubiquit. Comput. 13(7), 471–477 (2009)

11. Rodriguez-Mier, P., Pedrinaci, C., Lama, M., Mucientes, M.: An integrated semantic web service discovery and composition framework. IEEE Trans. Serv. Comput. 9(4), 537–550 (2016)

12. Klusch, M., Gerber, A.: Fast composition planning of OWL-S services and application. In: ECOWS 2006, 4th European Conference on Web Services, pp. 181–190. IEEE (2006)

13. Strunk, A.: QoS-aware service composition: a survey. In: 2010 IEEE 8th European Conference on Web Services (ECOWS), pp. 67–74. IEEE (2010)

14. Zou, G., Lu, Q., Chen, Y., Huang, R., Xu, Y., Xiang, Y.: QoS-aware dynamic composition of web services using numerical temporal planning. IEEE Trans. Serv. Comput. 7(1), 18–31 (2014)

15. Mazzola, L., Kapahnke, P., Vujic, M., Klusch, M.: CDM-Core: A manufacturing domain ontology in OWL2 for production and maintenance. In: Proceedings of the 8th International Joint Conference on Knowledge Discovery, Knowledge Engineering and Knowledge Management, vol. 2, KEOD, pp. 136–143 (2016)

16. Burstein, M., Hobbs, J., Lassila, O., Mcdermott, D., Mcilraith, S., Narayanan, S., Paolucci, M., Parsia, B., Payne, T., Sirin, E., et al.: OWL-S: Semantic markup for web services. In: W3C Member Submission (2004)

17. Mazzola, L., Kapahnke, P., Waibel, P., Hochreiner, C., Klusch, M.: FCE4BPMN: On-demand QoS-based optimised process model execution in the cloud. In: Proceedings of the 23rd ICE/IEEE ITMC Conference. IEEE (2017)

Why Enriching Business Transactions with Linked Open Data May Be Problematic in Classification Tasks

Eirik Folkestad[1]([email]), Erlend Vollset[1], Marius Rise Gallala[2], and Jon Atle Gulla[1]

[1] Department of Computer Science,
Norwegian University of Science and Technology, Trondheim, Norway
`eirik.ek.folkestad@gmail.com`
[2] Sparebank1 SMN, Trondheim, Norway

Abstract. Linked Open Data has proven useful in disambiguation and query extension tasks, but their incomplete and inconsistent nature may make them less useful in analyzing brief, low-level business transactions. In this paper, we investigate the effect of using Wikidata and DBpedia to aid in classification of real bank transactions. The experiments indicate that Linked Open Data may have the potential to supplement transaction classification systems effectively. However, given the nature of the transaction data used in this research and the current state of Wikidata and DBpedia, the extracted data has in fact a negative impact the accuracy on the classification model when compared to the Baseline approach. The Baseline approach produces an accuracy score of 88,60% where the Wikidata, DBpedia and their combined approaches yield accuracy scores of 84,99%, 86,65% and 83,48%.

Keywords: Classification · Bank transactions · Logistic Regression · Linked Open Data · Wikidata · DBpedia

1 Introduction

This project is carried out in cooperation with Sparebank1, which is Norway's largest regional bank, to gain insight into classification of real bank transactions. Progress in the domain at the intersection of finance and machine learning is important as it has a lot of potential applications like accurate consumption statistics, financial trend predictions, or financial crime detection to name a few. In the research we previously conducted on automatic classification of bank transactions [12], an approach that utilizes transaction description texts to classify transactions was developed. This approach has proven to be effective with a classification accuracy of 88,6%, but we wish to develop techniques to further improve this approach.

Linked open data (LOD) can be difficult to apply in domains that require a high level of data consistency. This is because, although, LOD sources contain large volumes of data, the quality varies a great deal. The research in this

© Springer International Publishing AG 2017
P. Różewski and C. Lange (Eds.): KESW 2017, CCIS 786, pp. 347–362, 2017.
https://doi.org/10.1007/978-3-319-69548-8_24

paper aims to investigate methods for exploiting LOD to aid in classification of business transactions and identify why this can be a problematic task. We examine two linked open resources; *Wikidata*, which is a collaboratively edited knowledge base operated by the Wikimedia Foundation and is intended to provide a common source of data, and *DBpedia*, which allows users to semantically query relationships and properties of Wikipedia resources, including links to other related datasets. Two main approaches to the problem are covered:

- Using extracted Linked Open Data to extend the baseline feature set
- Enhancing original/extracted data to better exploit the Linked Open Data

This paper gives a detailed description of the implementation of two approaches which do not lead to improvement of performance. We do, however, provide a thorough analysis of the results obtained from testing the system, giving valuable insight into why the use of LOD as a semantic feature enrichment tool is a difficult task. Due to the general nature of the techniques in this project, they can be transferred to other LOD sources and other applications within text classification.

The remainder of this paper is structured as follows. In Sect. 2 we present studies that are related to the conducted work in this paper. Section 3 describes the theoretical foundation upon which we have built our project as well as giving a detailed description of the data we have used and to what extent it is represented. Section 4 follows with providing a presentation of the experiments we have conducted and the results we obtained. The project is summarized in Sects. 5 and 6 by discussing our findings, providing recommendations for further work, and drawing our final conclusions.

2   Related Work

A project conducted by Skeppe [10] attempts to improve on an already automatic process of classification of transactions using machine learning. No significant improvements were made using a fusion of transaction information in either early or late fusion. The results do however show that bank transactions are well suited for machine learning, and that linear supervised approaches can yield acceptable scores.

Iftene et al. [13] present a system designed to perform diversification in an image retrieval system, using semantic resources like YAGO, Wikipedia, and WordNet, to increase hit rates and relevance when matching text searches to image tags. Their results show an improvement regarding relevance when there is more than one concept in the same query.

In the research conducted by Ye et al. [14] a novel feature space enriching (FSE) technique to address the problem of sparse and noisy feature space in email classification. The FSE technique employs two semantic knowledge bases to enrich the original sparse feature space. Experiments on an enterprise email dataset have shown that the FSE technique is effective for improving the email classification performance.

Poyraz et al. [15] perform an empirical analysis the effect of using Turkish Wikipedia (Vikipedi) as a semantic resource in the classification of Turkish documents. Their results demonstrate that the performance of classification algorithms can be improved by exploiting Vikipedi concepts. Additionally, they show that Vikipedi concepts have large coverage in their datasets which mostly consist of Turkish newspaper articles.

Xiong et al. [7] present a simple and effective method of using a knowledge base, Freebase, to improve query expansion, a classic and widely studied information retrieval task. By using a supervised model to combine information derived from Freebase descriptions and categories to select terms that are useful for query expansion, experiments done on the ClueWeb09 dataset with TREC Web Track queries demonstrate that these methods are almost 30% more successful than strong, state-of-the-art query expansion algorithms.

We have combined feature enrichment using external LOD resources with classification of real bank transactions. This is an important intersection that needs further research. We hope to have laid a foundation upon which others can continue research in the domain of classification of financial data.

3   Data and Methods

3.1   Original Data Set

The original data set used in this project consists of **220618** records of unstructured real Norwegian bank transaction texts from Sparebank1. The transaction texts have a corresponding category (C) and sub-category (SC) of which the transaction belong in. There is a total of 10 main categories and 63 subcategories. In this project, we will only do experiments with the main categories which make for a good indication of the performance of our selected approaches. The main categories are shown in Table 1 and an example of a transaction text can be seen in Table 2.

By conducting a Simple Random Sample of 400 transactions and manually labeling them by the main category, we achieved an accuracy of 93%. The mislabeling was mostly due to poor quality of the data where business names were

Table 1. Main categories and their IDs

ID	Main category name	Main category name in English
42	Bil og transport	Automobile and Transport
43	Bolig og eiendom	Housing and Real-Estate
44	Dagligvarer	Groceries
45	Opplevelse og fritid	Recreation and Leisure
47	Helse og velvære	Health and Well Being
48	Hobby og kunnskap	Hobby and Knowledge
49	Klær og utstyr	Clothes and Equipment
103	Annet	Other
104	Kontanter og kredittkort	Cash and Credit
181	Finansielle tjenester	Financial Services

Table 2. Data entry example

Description	SC	C
Rema 1000 Norge AG 05.01	61	44

not present in the text. Accuracy scores which are close to 93% is therefore considered to be great accuracy scores.

3.2   Bag-of-Words Model

The Bag-of-Words (BoW) model is used to convert text to a representation which is better suited for many machine learning algorithms. This technique is commonly used in natural language processing and information retrieval. In our application of the model, it is used as a tool for feature generation from the bank transaction texts. Each text is represented as a multiset (bag) of terms contained in the text when generating features for a corpus of texts. Given a corpus of the texts $X = x_1, x_2$ (see Table 3), the BoW representation produced is shown in Table 4. The resulting matrix has a row for each text and a column for each term in the corpus. The value of a cell is the frequency of the represented term in a given text. The features can then be used as input to a predictive model such as the one in this project.

Table 3. Corpus example

X	Sentence
x_1	Greg has a table
x_2	A table is a table

Table 4. Bag-of-words example

X	Greg	has	a	table	is
x_1	1	1	1	1	0
x_2	0	0	2	2	1

3.3   Logistic Regression

The Logistic Regression (LR) classification algorithm is linear and estimates a probability of a class Y given a feature-vector X. It does this by using a logistic function to find the relationship between the class and the feature-vector. It assumes that the distribution P(Y|X), where Y is the class and X is the feature-vector, is on a parametric form and then estimates it from the training data. The probability P(Y|X) of X belonging to class Y is given by the sigmoidal function which we can see in Eqs. 1 and 2.

P(Y|X) is estimated by linearly combining the features of X multiplied by some weight w_i and applying a function $f_i(Y, X)$ on the combinations. f_i is a function which returns a relationship value between a feature of a class and a feature in a feature-vector in the form of true or false based on the probability being over a certain threshold. Some features are more important than others, so the weight w_i denotes the "strength" of the feature.

$$z(Y, X) = \sum_{i=1}^{N} w_i f_i(Y, X) \tag{1}$$

$$P(Y|X) = \frac{1}{1 + exp(-z(Y, X))} \tag{2}$$

In this project, we need a classifier which can handle multiple classes. A Softmax Regression classifier, a variant of LR classifier, can handle more than two classes. In Softmax Regression, we replace the logistic function in Eq. 2 with the SoftMax function as we see in Eq. 3 which gives the probability of each class [8].

From the expression in Eq. 3 it can be shown that $\sum_{y \in Y} P(y|X) = 1$. This leads to the following classifier in Eq. 4 for a feature-vector X which outputs the class $\hat{y}$ if only the class of the feature-vector is needed and not the probability itself.

$$P(y|X)_{y \in Y} = \frac{exp(z(y, X))}{\sum_{y' \in Y} exp(z(y', X))} \tag{3}$$

$$\hat{y} = argmax_{y \in Y} P(y|X) \tag{4}$$

3.4   Baseline

A baseline refers to a set of configurations and techniques applied to our system used as a basis for defining change measuring improvement. In our system, the baseline approach is a standard machine learning approach to text classification which involves using a BoW representation and Softmax Regression for classification.

Preprocessing. Before the text can be used in a classification algorithm, some preprocessing steps are applied to the data. To remove noise from the data, we clean the description string to remove all punctuation, numbers, and words shorter than three characters (see Fig. 2a and b). The text is then transformed to vector representation using the BoW model (see Fig. 2c) and is ready for use in a classification algorithm.

Classification. The classification algorithm we apply in the baseline approach is Logistic Regression using a Softmax scheme which creates a true multinomial classifier of which can be used to classify data based the highest probability yielded of the likelihood of belonging to one of the multiple classes. The choice of a classifier is based on finding out that the Logistic Regression produces promising results where a more complex classifier like a Feed-Forward Neural Network does not improve the results [11]. From this, we can conclude that the data used in this project is linearly separable and a simpler classifier like a Softmax Regression classifier will be sufficient.

3.5   Wikidata and DBpedia

Wikidata and DBpedia are both LOD knowledge bases for extracting structured data from the web. Wikidata is a user-curated source of structured information which is included in Wikipedia and DBpedia provides structured data from the Wikipedia and Wikimedia Commons [1].

Both LOD sources are structured in a hierarchy consisting of objects where their hierarchical relationships are described with RDF-triples. A RDF-triple contains three components; subject, predicate and object [6]. An example of an RDF-triple in Wikidata or DBpedia related to the project can be seen in Fig. 5.

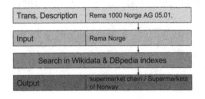

Fig. 1. Wikidata and DBpedia description extraction example

Fig. 2. Transaction representation example. (a) Transaction description. (b) Transaction description cleaned (c) Bag-of-Words representation.

We want to acquire the meaning of the words in the transaction texts with the use of Wikidata and DBpedia. A visible trend in the original data is that a company name usually is present in the transaction texts so an approach would be to find the company in Wikidata and DBpedia, and find information about what industry the company operates in.

One API-call per transaction text would be very time consuming since our system can only do few API calls per second and the original data set is of size 220618. Since both Wikidata and DBpedia support queries through a SPARQL-endpoint that is capable of returning thousands of results and we are looking for something specific, a less time-consuming approach would be to find all companies present in Wikidata and DBpedia and store the names along with company information to a local file. We also do not need all the information that Wikidata and DBpedia has to offer. An assumption of what would benefit training a prediction model the most would be a short description which specifically states something about what industry the company operates in. The closest predicate of which we could find that would fit our needs in Wikidata was *Description* and *Subjects* in DBpedia. The *Industry* predicate was considered and seemed more promising than *Subjects* in DBpedia but, unfortunately, relatively few companies were described with this predicate. The query results for Wikidata and DBpedia were stored in separate local files, and the two were indexed to separate indexes with the company name as key and the description as value by using *Whoosh* [4] for quick and reliant look-up.

From the companies in our index, we can find useful information about companies in the transaction texts as seen in Fig. 1. First, the transaction text is cleaned like it is in the baseline. Then we search our index for the first and/or the second word in the transaction text which represent the company name and if a result is returned, we extend our transaction text. After this the process of

Fig. 2 is applied, and the new transaction text is converted to a BoW representation. Depending on the information used to extend the original data is from the Wikidata index or the DBpedia index, the approaches are called the Wikidata approach or the DBpedia approach. If the information is extended from both Wikidata and DBpedia the approach is called the Wikidata & DBpedia approach.

3.6   WordNet

WordNet is a large lexical database of the English language and can be used for searching for definitions, synonyms and other information [2]. It can also be used for simple translation from a supported language to English before doing a search. The information about the word will be returned in English.

Natural Language Toolkit [3] provides a module that can be downloaded so that WordNet is available locally. This means that no calls to an API are needed. This will make the process of searching for information about words much faster.

By using the WordNet module that *Natural Language Toolkit* provides, a word can be sent in, and synonyms are returned if there are any (see Fig. 3).

We are trying to bring even more semantic meaning into the transaction texts by using synonyms. Often the same meaning is represented by using words that are synonyms. For instance, a café can be represented by the word coffee shop. By adding synonyms to a text, we can create similarities between two texts that are initially viewed as dissimilarities since the words are written differently.

With the help of WordNet we intend to extend the transaction texts, and also the descriptions we receive from Wikidata and DBpedia, with synonyms so that similarities between two or more records that originally are not represented will be more transparent as they now share more words. As seen in (see Fig. 4) the transaction text is cleaned like it is in the baseline and then split to get each word separate. Each word is sent to the WordNet module, and the synonyms are returned. The words are then concatenated to one text string again. Data which i.e. contain the word 'bar' would now share this word with a text which contains the word 'cafe'. The general similarities between two texts are now clearer after extending the data with synonyms. After extending the transaction text with synonyms the process in Fig. 2 is applied, and the new transaction text is converted to a BoW representation.

3.7   Yandex Translation

Yandex is a technology company that builds intelligent products and services powered by machine learning [5] and one of these services is a translation API that seems fit to translate the transaction texts in the original data set.

The transaction texts used in this project are in Norwegian, and this could create problems when using the selected LOD sources which returned descriptions are in English. By translating the original data to English, we hope to compensate for the possible problems created by extending data with data on

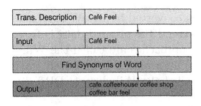

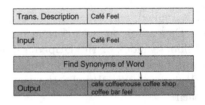

Fig. 3. WordNet - Extraction of synonyms for a word

Fig. 4. WordNet - Extraction of synonyms for a transaction

a different language. The translated transaction texts will hopefully share more words with the descriptions from the LOD sources.

The Yandex Translate module can translate the original data word for word instead of translating the whole texts. This is done since it can be unfavorable to change the idiomatic meaning of the text and rather replace the individual words with their translations.

As shown in Fig. 6 the translation is extracted by first cleaning the transaction text like it is in the baseline and then splitting the texts into separate words. We then translate each word respectively with the translation API. The returned translation of each word is then concatenated into one text string which constitutes the new transaction text. The process of Fig. 2 is then applied, and the translated transaction texts are then converted to a BoW representation.

4   Results

4.1   Experiments

All experiments have been conducted with the following parameters:

- Bag-of-Words size of 4000.
- Softmax Regression classifier.
- Classification classes are the 10 Main Categories.

There is a total of 87199 distinct terms in the transaction texts. Using a BoW size of 4000 means using the 4000 most frequently occurring terms. By iteratively incrementing the BoW size we found that the accuracy begins to stabilize at a size of 4000, thus making it a reasonable size to use.

In our experiments we use a number of subsets of the **Original Data Set (ODS)**, which is defined in Sect. 3.1. Since the sizes of these subsets are significantly smaller than ODS, we would also expect the results to be different. These subsets are defined as:

- **Wikidata Exclusive Subset (WES)** - This subset consists of only the bank transactions which yield a match in the Wikidata data source. The subset contains **113263** transactions and is used only in the Baseline, and Wikidata approaches.

- **DBpedia Exclusive Subset (DES)** - This subset consists of only the bank transactions which yield a match in the DBpedia data source. The subset contains **125765** transactions and is used only in the Baseline and DBpedia approaches.
- **Wikidata & DBpedia Exclusive Subset (WDES)** - This subset is the union of WES and DES. The subset contains **136474** transactions and is used only in the Baseline, and Wikidata & DBpedia approaches.

For every experiment the data set is divided into a training and test set, respectively 80% and 20% of the data set. The results given are averages of performance measures over ten iterations, shuffling the training and test set each time.

The evaluation metrics used are Accuracy (Micro-Averaged Recall), Macro-Averaged Recall, Macro-Averaged Precision and F-Score [9].

4.2   Baseline

Table 5 shows the Baseline model's evaluation scores on a BoW representation using various data sets. Table 6 shows the performance in each Main class.

Table 5. Baseline approach results on various data sets

Approach	Data Set	Accuracy	Recall	Precision	F-Score
Baseline	ODS	88.60%	92.80%	85.35%	88.53%
Baseline	WES	94.39%	94.13%	90.57%	92.17%
Baseline	DES	94.26%	94.65%	90.67%	92.46%
Baseline	WDES	94.08%	94.40%	90.70%	92.39%

Table 6. Per main class results for the Baseline approach

ID	Precision	Recall	F-Score
42	0.91	0.94	0.92
43	0.91	0.92	0.92
44	0.94	0.97	0.95
45	0.94	0.84	0.89
47	0.90	0.90	0.90
48	0.78	0.90	0.84
49	0.86	0.93	0.89
103	0.84	0.93	0.88
104	0.88	0.96	0.92
181	0.97	0.96	0.97

4.3   Use of Linked Open Data

Wikidata. This experiment shows the model's evaluation scores after the descriptions from Wikidata have been used to extend the data in ODS (see Table 7) and WES (see Table 8) before vectorizing it on a BoW representation.

DBpedia. This experiment shows the model's evaluation scores after the descriptions from DBpedia have been used to extend the data in ODS (see Table 9) and DES (see Table 10) before vectorizing it on a BoW representation.

Table 7. Wikidata approach results on ODS

Approach	Accuracy	Recall	Precision	F-Score
Wikidata	84.65%	89.85%	81.56%	84.92%
Wikidata + Translation	84.99%	88.97%	82.01%	84.90%
Wikidata + Translation + Synonyms	79.87%	85.27%	76.23%	79.79%

Table 8. Wikidata approach results on WES

Approach	Accuracy	Recall	Precision	F-Score
Wikidata	92.69%	92.19%	87.89%	89.77%
Wikidata + Translation	93.15%	92.11%	88.33%	89.98%
Wikidata + Translation + Synonyms	91.83%	90.96%	87.04%	88.72%

Table 9. DBpedia approach results on ODS

Approach	Accuracy	Recall	Precision	F-Score
DBpedia	86.48%	90.65%	83.74%	86.58%
DBpedia + Translation	86.65%	89.85%	84.20%	86.59%
DBpedia + Translation + Synonyms	81.74%	86.19%	78.33%	81.46%

Table 10. DBpedia approach results on DES

Approach	Accuracy	Recall	Precision	F-Score
DBpedia	93.48%	93.21%	89.23%	90.99%
DBpedia + Translation	93.53%	92.87%	89.24%	90.84%
DBpedia + Translation + Synonyms	92.41%	91.54%	87.46%	89.19%

Combination of Wikidata & DBpedia. This experiment shows the model's evaluation scores after the descriptions from Wikidata and DBpedia have been used to extend the data in ODS (see Table 11) and DES (see Table 12) before vectorizing it on a BoW representation.

Table 11. Wikidata & DBpedia approach results on the ODS

Approach	Accuracy	Recall	Precision	F-Score
Wikidata & DBpedia	82.97%	88.55%	79.48%	83.02%
Wikidata & DBpedia + Translation	83.48%	87.86%	79.73%	82.95%
Wikidata & DBpedia + Translation + Synonyms	79.71%	84.63%	75.27%	78.70%

Table 12. Wikidata & DBpedia approach results on WDES

Approach	Accuracy	Recall	Precision	F-Score
Wikidata & DBpedia	92.13%	92.01%	87.98%	89.75%
Wikidata & DBpedia + Translation	92.42%	91.53%	87.64%	89.33%
Wikidata & DBpedia + Translation + Synonyms	91.13%	90.29%	86.24%	87.94%

5 Discussion

5.1 Linked Open Data as Resources

As we see in Tables 5, 7, 9 and 11 the results produced using the Wikidata and DBpedia approaches show a performance decline compared to the Baseline approach. The observed results indicate that the Baseline approach itself was better suited for training a classification model than the proposed approaches experimented with was. The accuracy of the Baseline approach was **88,60%** and by using the Wikidata, DBpedia, and Wikidata & DBpedia approaches we can observe a decline in accuracy of **−3,95%**, **−2,12%** and **−5,63%**.

We also notice a corresponding drop in the other performance measures. As we have shown with the approaches in the previous section and the research presented in Sect. 2, it is indeed possible to improve accuracy using feature enrichment techniques. Expanding the feature set allows the classifier to find more distinct patterns on which to make decisions. Unfortunately, this was not the case with the data we collected from Wikidata and DBpedia. By further analysis of each LOD source, we discuss possible justifications for our results.

First and foremost, the hit-ratio, denoting how many transactions yielded a match in the LOD sources, was relatively small. By counting the number of transactions that produced a result in each data source we see that only a little over half of the original data yielded a hit in Wikidata, DBpedia their combination with **51,34%**, **57,01%** and **61,86%**.

Combining the Wikidata and DBpedia approaches was an attempt to increase this hit-rate, but still yielded a relatively low number. The reason for this was the great amount of overlap in which transactions yielded a match in the data sources. There were **102554** transactions in ODS which yielded a result in both Wikidata and DBpedia. Only **10709** of the transactions were found exclusively in Wikidata and **23211** transactions were found exclusively in DBpedia.

The low hit-rates of all three approaches indicate that the LOD sources are not extensive enough, separate or combined, for our use and are not likely to contribute positively when training our classification model.

Having observed these low hit-rates, we conducted experiments where we used the subsets of ODS which contained only transactions which yielded a match in the LOD sources. We did this to gain insight into how the LOD approaches could potentially perform compared to the Baseline approach given that all of the original data yielded a match in the LOD sources.

As we can see in Fig. 8, all LOD approaches increased substantially to the better. However, the performance was still worse in all of the LOD approaches than in the Baseline approach. This shows that much of the error from using Wikidata and DBpedia is introduced by the fact that a lot of the original data does not yield a result in the LOD sources. We had a theory that the increase in performance could be explained by the subsets having a label distribution which favored classes with a higher recall score. We can, however, see from the label distribution in Fig. 7 that the label distribution, with the exception of the classes **44** and **47**, is approximately the same for the original data set and its subsets. We could, therefore, conclude that the performance increase rather indicates that the removed transactions within each class were harder to classify.

The low hit-rate could be explained by the nature of the bank transaction texts. As stated in Sect. 3.1, our dataset consists of Norwegian transaction texts where many contain Norwegian company names. This makes it more difficult for us to get results from Wikidata and DBpedia since they contain relatively few Norwegian companies. Both LOD sources are focused on a more general level which makes deeper knowledge on a specific topic hard to obtain from them e.g. companies on a country basis. Smaller companies that operate in only one country are, understandably, not a priority when covering information on a global scale. On the other hand, larger companies and companies that are internationally known tend to give results even though they may be based in only one country. The information that can be extracted from Wikidata and DBpedia seems to be too general for the purpose of this project and does not give information to the extent that we require.

A side-effect of Wikidata and DBpedia covering information on a more general global basis is that the returned information might not represent the correct information. By this we mean that many results are *False Positives* which would make the information we extend the original data with incorrect and misleading. By conducting a Simple Random Sample test for each LOD source, we could see indications of this. Each Simple Random Sample test consisted of 100 transactions which yielded a result in each of the LOD sources. The evaluation was a subjective analysis since there is no actual correct answer to this test and therefore results may or may not represent the true results. The sample data of the tests revealed a low amount of true results which were interesting (see Fig. 9).

If we assume that this Simple Random Sample test is representative of the rest of the data that comes from Wikidata and DBpedia, then this is a clear indication that we are introducing many words which do not describe the transactions they are assigned to. We should then expect to observe a performance decline in both the results for Wikidata (see Table 7) and DBpedia (see Table 9) approaches compared to the results for the Baseline approach (see Table 5). This is also shown in performance decline in the experiments performed on the Wikidata and DBpedia (see Tables 8 and 10) subsets when compared to the Baseline on each respective data set (see Table 5).

From the Simple Random Sample, we see that Wikidata yields a higher percentage of correct and meaningful results than DBpedia since a lot of the results from DBpedia give little meaning. From this perspective, we would believe that Wikidata approach would perform better than the DBpedia approach. However, as we observe in the results, this theory does not hold up, and DBpedia performs a little better. This indicates that extending the transaction descriptions with data that does not contribute to distinguishing between classes, produces better performance results. This suggests that the data from the LOD sources do not help in this classification problem. These results could also explain why the combined approach performed worse than both the Wikidata and DBpedia approaches because even if one of the LOD sources return a correct result, the other one may return an incorrect result.

The data returned from both Wikidata and DBpedia was of variable length and content. Two companies that operate in the same industry would often have a description that was written differently, and no standard format was used. This could be another possible source of error. Conformity could have been an advantage for classification since a decision boundary would be more pronounced in the data. The description from Wikidata and DBpedia mainly consists of free-text which makes the description of many companies that operate in the same industry highly variable. This error could also be thought to make the performance of the combination of the two approaches to decline even further, which we believe is another reason for the poor result.

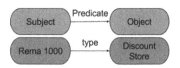

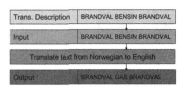

Fig. 5. RDF-Triple structure and example

Fig. 6. Yandex - Extraction of translation for a transaction

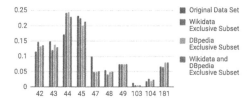

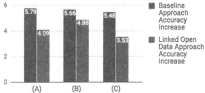

Fig. 7. Normalized label distribution over different data sets

Fig. 8. (**A**) Percent accuracy change of Baseline approach and the Wikidata approach from using ODS to WES. (**B**) Percent accuracy change of Baseline approach and the DBpedia approach from using ODS to DES. (**C**) Percent accuracy change of Baseline approach and the Wikidata & DBpedia approach from using the ODS to WDES.

5.2   Correction of the Proposed Approaches

In order to remedy the shortcomings of the LOD sources, we have used two methods in an attempt to correct this. First, we translate the original data to English and then extend both the transaction descriptions and the data from Wikidata and DBpedia with synonyms.

The translation of the original data showed improvement as seen from the result of all approaches (see Tables 7, 9 and 11). We believe that this increase in performance come from the reason that translated original data share more words with the extracted LOD than with the original data itself. The performance increase is true for all proposed approaches. However, even though there was an increase, the difference was not significant. The observed improvement obtained by using the translated original data instead of just the original data for

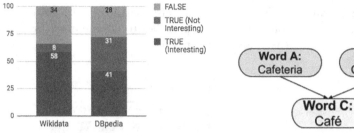

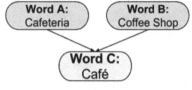

Fig. 9. Comparison of hit-value of wikidata and DBpedia

Fig. 10. An example of replacing synonyms with the same word

the proposed approaches on ODS was **0,34%**, **0,17%** and **0,51%** for Wikidata, DBpedia and the combined approaches respectively.

Extending the translated original data with synonyms in addition to Wikidata and DBpedia did however not result in a performance increase. As seen in Tables 7, 9 and 11, the performance in the experiments is clearly reduced. We believe that the way synonyms were used to extend the different approaches further contributes to making the problems observed even more significant by looking at the data returned by Wikidata and DBpedia. When we extend with synonyms to create similarities we also, as a side-effect, further reduce the conformity of the transaction texts. This is an effect created by adding many new words to the transaction texts. We also believe that since the data returned might be incorrect, the synonyms only enhance the observed error and therefore also create errors of greater significance. This side-effect was not taken into account when selecting approaches. For these reasons we can see that extending the translated original data and LOD with synonyms only contribute to a less clear decision boundary to perform classifications on.

A proposition for a better correction approach would be to replace words rather than just adding them to the text. If word **A** and word **B** are synonyms, replace them with a word **C**. See Fig. 10 for an example.

Using the translation approach may also have this effect since it is a possibility that synonyms could be translated to the same word. A more strict filtering method for choosing which words to find synonyms for could also be beneficial since many of the synonyms we extended contributed to confusion when finding a pattern of which we make classifications based on. However, the correction is likely to only result in a small performance increase since the data we extract from Wikidata and DBpedia still is insufficient for use in this project.

6   Conclusion

Firstly, we can conclude that Wikidata and DBpedia are not fit to be use as data sources in the classification of bank transactions. For a domain like this, consistency and conformity are critical. Due to the nature of LOD, the granularity of the searches in the LOD sources is important to get useful information of which we can use to extend the bank transactions with. We found that by using Wikidata and DBpedia, we get very few results on specific domains like businesses, primarily Norwegian, and too much information which either was too lacking or too descriptive to make improvements in our classification problem. A finding in this research, however, is that the data which is possible to extract from Wikidata and DBpedia is better suited for internationally known companies. This means that the results potentially could have been better if the experiments were conducted on a different data set where the companies mainly were internationally known companies and not country specific companies.

Despite our two attempts to correct the shortcomings of the data extracted from both LOD sources, the yielded results from the attempted approaches were still inferior to the Baseline approach. We believe that with a better approach of how to make use of synonyms we could have produced better results, although, still limited by the quality of the LOD.

The concept of a structured web is interesting, and using all of this available information shows potential. If the LOD sources continue to grow and conformity is introduced to the structured data then LOD may prove useful in projects like this in the future.

References

1. Wikidata DBpedia. http://wikidata.dbpedia.org/. Accessed 10 June 2017
2. Fellbaum, C., "What is WordNet?". In: Brown (2005). WordNet and wordnets. https://wordnet.princeton.edu/. Accessed 15 June 2017
3. Natural Language Toolkit. NLTK Project (2017). http://www.nltk.org/. Accessed 13 June 2017
4. Chaput, M.: About Whoosh (2012). http://whoosh.readthedocs.io/en/latest/intro.html#about-whoosh. Accessed 09 June 2017
5. Yandex (2017). https://yandex.com/company/general_info/yandex_today/. Accessed 30 May 2017
6. RDF Working Group: Resource Description Framework (RDF) (2004). https://www.w3.org/RDF/. Accessed 29 May 2017
7. Xiong, C., Callan J.: Query expansion with freebase. In: Proceedings of the 2015 International Conference on the Theory of Information Retrieval, 27–30 September, Northampton, Massachusetts, USA (2015)
8. Bishop, C.M.: Pattern Recognition and Machine Learning. Information Science and Statistics. Springer, New York (2006)
9. Van Asch, V.: Macro-and micro-averaged evaluation measures (2013). https://www.semanticscholar.org/
10. Skeppe, L.B.: Classifying Swedish Bank Transactions with Early and Late Fusion Techniques. Master thesis, KTH Royal Institute of Technology, Stockholm (2014)
11. Perlich, C.: Which is your favourite Machine Learning Algorithm? (2016). http://www.kdnuggets.com/2016/09/perlich-favorite-machine-learning-algorithm.html
12. Vollset, E., Folkestad, E.: Automatic Classification of Bank Transactions. Master thesis, Norwegian University of Science and Technology, Trondheim (2017)
13. Iftene, A., Baboi, A.M.: Using semantic resources in image retrieval. In: 20th International Conference on Knowledge Based and Intelligent Information and Engineering Systems, KES 2016, vol. 96, pp. 436–445. Elsevier (2016)

14. Ye, Y., Ma, F., Rong, H., Huang, J.Z.: Improved email classification through enriched feature space." In: Li, Q., Wang, G., Feng, L. (eds) Advances in Web-Age Information Management (WAIM) (2004)
15. Poyraz, M., Ganiz, M.C., Akyokus, S., Gorener, B., Kilimci, Z.H.: Exploiting Turkish Wikipedia as a semantic resource for text classification. In: International Symposium on Innovations in Intelligent Systems and Applications (INISTA), pp. 1–5 (2013)

Author Index

Printed in the United States
By Bookmasters